Game Development with Unreal Engine 5 Volume 1

Design Phase

Tiow Wee Tan

Apress®

Game Development with Unreal Engine 5 Volume 1: Design Phase

Tiow Wee Tan
Department of Game Development – The Game School
Inland Norway University of Applied Science
Hamar, Norway

ISBN-13 (pbk): 978-1-4842-9823-7   ISBN-13 (electronic): 978-1-4842-9824-4
https://doi.org/10.1007/978-1-4842-9824-4

Copyright © 2024 by Tiow Wee Tan

This work is subject to copyright. All rights are reserved by the Publisher, whether the whole or part of the material is concerned, specifically the rights of translation, reprinting, reuse of illustrations, recitation, broadcasting, reproduction on microfilms or in any other physical way, and transmission or information storage and retrieval, electronic adaptation, computer software, or by similar or dissimilar methodology now known or hereafter developed.

Trademarked names, logos, and images may appear in this book. Rather than use a trademark symbol with every occurrence of a trademarked name, logo, or image we use the names, logos, and images only in an editorial fashion and to the benefit of the trademark owner, with no intention of infringement of the trademark.

The use in this publication of trade names, trademarks, service marks, and similar terms, even if they are not identified as such, is not to be taken as an expression of opinion as to whether or not they are subject to proprietary rights.

While the advice and information in this book are believed to be true and accurate at the date of publication, neither the authors nor the editors nor the publisher can accept any legal responsibility for any errors or omissions that may be made. The publisher makes no warranty, express or implied, with respect to the material contained herein.

> Managing Director, Apress Media LLC: Welmoed Spahr
> Acquisitions Editor: Spandana Chatterjee
> Development Editor: James Markham
> Editorial Assistant: Kripa Joseph

Cover designed by eStudioCalamar

Cover image designed by Freepik (www.freepik.com)

Distributed to the book trade worldwide by Springer Science+Business Media New York, 1 New York Plaza, Suite 4600, New York, NY 10004-1562, USA. Phone 1-800-SPRINGER, fax (201) 348-4505, e-mail orders-ny@springer-sbm.com, or visit www.springeronline.com. Apress Media, LLC is a California LLC and the sole member (owner) is Springer Science + Business Media Finance Inc (SSBM Finance Inc). SSBM Finance Inc is a **Delaware** corporation.

For information on translations, please e-mail booktranslations@springernature.com; for reprint, paperback, or audio rights, please e-mail bookpermissions@springernature.com.

Apress titles may be purchased in bulk for academic, corporate, or promotional use. eBook versions and licenses are also available for most titles. For more information, reference our Print and eBook Bulk Sales web page at http://www.apress.com/bulk-sales.

Any source code or other supplementary material referenced by the author in this book is available to readers on GitHub. For more detailed information, please visit https://www.apress.com/gp/services/source-code.

If disposing of this product, please recycle the paper

Table of Contents

About the Author vii

About the Technical Reviewer ix

Acknowledgments xi

Introduction xiii

Chapter 1: Getting Started with Unreal Engine 5 1

What Is Unreal Engine 5? 2

The Benefits of Unreal Engine 5 2

    Expanding Horizons 2

    Realistic Visuals 3

    Reimagining Possibilities 3

    Procedural Foliage: Unreal's Solution 3

    Unleashing Visual Fidelity 4

    Unleashing Creativity 4

    Elevating Control 4

    Chaos Unleased 4

    Empowering Sound Designers 5

Setting Up the Development Environment 5

    Installing Unreal Engine 5 and Dependencies 5

    Installing Visual Studio Community 2022 and Dependencies 10

Other Preparations 13

Prepare New Project 14

    Run on Low-Quality Mode 19

    Show Engine Content 20

Summary 22

iii

TABLE OF CONTENTS

Chapter 2: From Heightmap to Large Open-World Landscape 23

Set Up a New Level 24

Prepare the Heightmap 26

    Convert EXR to PNG with GIMP 27

Create Landscape from Heightmap 31

Ensure Correct Scale from Perspective 38

Convert to World Partition 42

    Save the Level Before Conversion 42

    The Conversion 46

    World Partition Editor 50

    World Partition Runtime Settings 56

Summary 60

Chapter 3: Auto-Blend Landscape Materials 61

Material Setup 62

Attach Material to Landscape 64

Auto-Blending Height-Based Materials 67

    Important Nodes to Look Into 68

    The Design of Materials 79

    Improvement with Slope Attribute 95

Summary 99

Chapter 4: Revitalizing Visuals: Asset Import and Procedural Creation 101

Quixel Bridge and Megascans 101

    Asset ID 104

    Surface (Texture) Assets 106

    3D Assets and 3D Plants 108

Visual Improvement on AutoBlend_Height_MAT Landscape Material 113

    General_Template_MF 115

Procedural Content Generation (PCG) 137

    Set Up the PCG Volume 138

    Important PCG Nodes to Look Into 145

TABLE OF CONTENTS

    Design of Content Volume PCG Graph 154

    Final Visualization Result 176

  Summary 178

Chapter 5: Enhancing Visual Realism with Runtime Virtual Textures and Material Blending 181

  Runtime Virtual Texture 181

    Color Map in RVT 182

    World Height in RVT 183

  Set Up an RVT Asset in UE5 185

  RVT Volume 187

    The Setup of RVT Volume 188

    Set RVT Volumes Inside Landscape Partitions 195

  RVT Output in Material Editor 197

    Set Up RVT Output Node 198

    See the Result on the RVT Assets 205

  Material Blend with RVT Assets 206

    Additional Logic to This Duplicated Material (RVT) 211

  Summary 220

Chapter 6: Mastering Lumen Global Illumination in Unreal Engine 5 223

  How Lumen Works 224

    Enabling and Configuring Lumen 225

  Evaluating Lumen's Lighting Effects 227

    Preparation Tasks 229

    Lumen in Post-Processing Effects 265

  Summary 275

Chapter 7: Harnessing the Power of Niagara: Practical Examples in Unreal Engine 5 277

  Getting Started with Niagara 278

    Creating First Particle Effect 278

    Learning the Basics of Particle Systems 282

TABLE OF CONTENTS

    Harnessing the Power of Niagara #1: Smoke from Landing Spaceship 303

    Harnessing the Power of Niagara #2: Spiral Effects on Magical Ball 331

    Harnessing the Power of Niagara #3: Dust Storm 361

  Summary 407

    Bringing Worlds to Life: Concluding the Design Phase 407

Index **411**

About the Author

Dr. Tiow Wee Tan holds the position of Associate Professor at the Department of Game Development – The Game School, Inland Norway University of Applied Sciences (INN), Norway.

His interests revolve around the development of extended reality (XR) for cross-disciplinary enterprise applications. He has actively participated in projects involving machine learning and artificial intelligence (AI) for gaming. Some of his notable works encompass

- The utilization of XR technologies within game engines like Unity3D and Unreal Engine
- Creation of hybrid XR research applications for fields such as healthcare, art, and tourism
- Serious game development exploring the integration of educational content with engaging gameplay mechanics
- Investigation into cognitive processes within AI-driven human-computer interaction environments to enhance user experience and system performance

About the Technical Reviewer

Simon Jackson is a long-time software engineer and architect with many years of Unity game development experience as well as an author of several Unity game development titles. He loves to create Unity projects as well as lend a hand to help educate others, whether it's via a blog, vlog, user group, or major speaking event.

His primary focus at the moment is the Reality Toolkit project, which is aimed at building a cross-platform Mixed Reality framework to enable both VR and AR developers to build efficient solutions in Unity and then build/distribute them to as many platforms as possible. He is also a board member of the MonoGame Foundation, aiming to secure and promote open source game development for all developers.

Acknowledgments

This book would never have been possible without the support and encouragement I received from various individuals. I am profoundly grateful for the faith my publisher put in me, for their insightful suggestions, and their endless patience.

Introduction

A Journey Beyond Boundaries

Welcome to an immersive journey through the realms of Unreal Engine 5 (UE5), the latest evolution in the universe of game development. This book, crafted with both the beginner and the seasoned developer in mind, aims to unfold the vast potential that UE5 holds. Through its pages, we invite you to explore the cutting-edge features that make UE5 a powerhouse for creating visually stunning and deeply interactive gaming experiences.

What This Book Covers

Unreal Engine 5 has set new standards in the industry, offering unparalleled tools for rendering, physics, and overall game design. This book is structured to guide you through these advancements, starting from the foundational concepts to more complex implementations. You will learn about the revolutionary Nanite virtualized geometry system, Lumen for dynamic global illumination, and how to utilize the Quixel Megascans library to bring realism into your worlds. Furthermore, we delve into procedural generation, Niagara VFX, and more, ensuring you have a comprehensive toolkit at your disposal.

Whom This Book Is For

If you are looking to enhance your skills with the latest UE5 features, this book is designed for you. Programmers will appreciate the deep dive into the logic of UE5's Procedural Content Generations, while artists will find the chapters on visual effects and environmental design enlightening. Educators and students will also find this volume invaluable for understanding the practical design applications of game development principles.

INTRODUCTION

Structure of the Book

The book is segmented into detailed chapters, each focusing on different aspects of UE5.

The initial chapters introduce UE5's interface and core functionalities, easing you into the workflow and basic tools.

Subsequent chapters cover landscape and environment design, leveraging UE5's powerful tools for creating expansive and dynamic worlds.

Advanced topics such as procedural generation, lighting, and material design are explored in depth, offering insights into creating more lifelike scenes and characters.

What You Will Learn

By the end of this book, you will have a solid understanding of Unreal Engine 5 and be equipped with the skills to

> Craft vast, open worlds with rich, dynamic environments

> Utilize UE5's advanced features to enhance visual fidelity and interactivity

Before You Begin

As you embark on this journey, we encourage you to explore, experiment, and push the boundaries of what you believe is possible in game development. Unreal Engine 5 is a tool of unlimited potential, and with this book as your guide, you're well on your way to unlocking your creative visions.

Beyond This Volume

The journey doesn't end here. As you progress, you will be prepared to explore the programming aspect of game creation with C++, which will be covered in depth in the follow-on volume. This future volume will delve deeper into interactive gameplay mechanics, ensuring your transition from design to development is both comprehensive and enlightening.

Let's begin this adventure together and create something extraordinary.

CHAPTER 1

Getting Started with Unreal Engine 5

Welcome to the fascinating world of game development with UE5! In these initial seven chapters of our comprehensive series focused on the design phase of the project, we lay the groundwork for what is to be an extensive exploration of UE5's vast capabilities. We'll delve into its robust features and the myriad benefits they bring, setting the stage for both novices and seasoned developers alike. These first seven chapters serve as the first step in a series dedicated to the design aspects of game creation, offering a solid foundation on which to build up the knowledge. As we conclude this volume on design phase, anticipate a seamless transition in the next volume, where we will embark on the programming phase, specifically harnessing the power of C++ to bring interactive elements to life within UE5.

This book is based on UE version **5.3.2**, providing a comprehensive guide tailored to the features, workflows, and screenshots based on this specific release.

It's important to note that as Epic Games continues to innovate and enhance the UE, features and functionalities may evolve in future versions. While the core principles and techniques described herein are designed to be as evergreen as possible, the readers should be aware that some procedures and interfaces may be amended or updated in subsequent releases, reflecting the ongoing development and improvement of the engine.

Users of future versions of Unreal Engine may need to adapt the knowledge gained from this book to align with the latest advancements and changes implemented by Epic Games.

CHAPTER 1   GETTING STARTED WITH UNREAL ENGINE 5

What Is Unreal Engine 5?

UE5 is an advanced and feature-rich game development engine created by Epic Games. It is widely recognized as one of the industry's leading game engines, powering numerous successful and visually stunning games. UE5 offers a wide range of tools, technologies, and features that empower developers to bring their creative visions to life. With its powerful rendering capabilities, advanced physics systems, dynamic lighting, and robust multiplayer support, UE5 provides an extensive toolkit for creating high-quality and immersive gaming experiences.

UE5 takes advantage of the latest hardware capabilities to deliver breathtaking visuals, realistic physics simulations, and seamless gameplay experiences. It provides developers with a vast array of features, such as advanced real-time rendering, dynamic lighting and shadows, advanced particle systems, and robust audio tools. The engine also supports a wide range of platforms, including PC, consoles, mobile devices, and virtual reality, allowing developers to reach a broad audience and create engaging experiences across different platforms.

The Benefits of Unreal Engine 5

UE5 provides numerous advantages that make it a popular choice among developers, and its amazing features will be discussed and used in the chapters mentioned for each following point.

Expanding Horizons

In Chapter 2, we will discover the unleashed advanced features and capabilities of UE5, which stands as a formidable tool for crafting awe-inspiring open-world landscapes. Developers harness its power to construct vast, immersive environments teeming with intricate details and lifelike authenticity. Within this digital realm, one can traverse sprawling forests, scale towering mountains, roam bustling cities, or tread through desolate wastelands – the possibilities are boundless.

Realistic Visuals

In Chapters 4 and 6, we can discover that UE5 is renowned for its cutting-edge graphics capabilities, which allow developers to create visually stunning and highly realistic environments. The engine's advanced rendering features, such as Nanite and Lumen, enable the rendering of incredibly detailed and complex scenes with ease. Nanite (in Chapter 4), the virtualized micropolygon technology, allows for the rendering of millions of polygons in real time, resulting in unparalleled levels of detail and visual fidelity. Lumen (in Chapter 6) provides dynamic global illumination, enabling realistic lighting and reflections that enhance the overall visual quality of the game. These advanced rendering features make UE5 a powerful tool for creating visually impressive games.

Reimagining Possibilities

Within UE5, the trio of groundbreaking tools – Quixel Bridge, Quixel Megascans, and MetaHuman – reshaped our game development approach, redefining what was possible. Quixel Bridge (in Chapter 4) became our invaluable asset management tool, providing access to a vast library of photorealistic assets. Quixel Megascans (in Chapter 4) elevated our virtual worlds with dynamic weather systems and immersive soundscapes. Meanwhile, MetaHuman (in the next volume) revolutionized character creation, allowing us to sculpt lifelike human characters with incredible fidelity. Together, these tools opened up new realms of creativity and immersion, propelling us into a new era of interactive storytelling.

Procedural Foliage: Unreal's Solution

In Chapter 4, we can discover that UE5 offers a powerful Procedural Content Generation Framework (PCG) for creating realistic outdoor environments with vegetation. The PCG framework allows the creation of procedural content and tools, from asset utilities to entire worlds, with easy integration into existing pipelines. Furthermore, we can experience the foliage mode, which enables us to quickly paint or erase sets of static mesh geometry. This allows us to populate a large outdoor environment with foliage in a short amount of time.

CHAPTER 1   GETTING STARTED WITH UNREAL ENGINE 5

Unleashing Visual Fidelity

In Chapter 5, we explore Runtime Virtual Texture streaming for UE5. This groundbreaking technology allows developers to render high-resolution textures without consuming excessive memory. By dynamically loading and unloading textures based on the player's viewpoint, developers can create immersive and detailed environments while optimizing memory usage and performance.

Unleashing Creativity

Also, in Chapter 7, we discover Niagara and its representation of the cutting-edge visual effects (VFX) system of UE, offering advanced features and capabilities. This empowering tool enables technical artists to independently develop additional functionality without relying on programmers. Its adaptable and flexible nature makes it highly versatile for various creative needs. UE5 comes bundled with Niagara, ensuring that users have immediate access to its potential.

Elevating Control

In the next volume, the focus of the book will shift toward the programming realm, specifically the intricacies of C++ in UE5. A key feature we'll explore is UE5's Enhanced Input System, a sophisticated upgrade from UE4's default input system. This new system is tailored for projects requiring complex input handling, including dynamic control remapping and advanced action sequences. We will delve into its features such as radial dead zones, chorded actions, and contextual inputs, all while learning how to leverage this system to refine raw input data within an asset-based framework. This volume promises to deepen your programming expertise, enhancing the interactive dimension of your UE5 projects.

Chaos Unleased

In the next volume, our exploration will extend into the realm of C++ programming, where we will engage with UE5's Chaos system. This robust feature introduces a new era of physics and destruction simulation, capable of managing diverse entities like rigid and soft bodies, cloth, fluids, and particles. The Chaos system is engineered for high-performance, scalable simulations compatible across various platforms. Utilizing C++ programming, developers can harness the full potential of Chaos to craft vividly interactive scenes and enhance gameplay with spectacular visual effects.

Empowering Sound Designers

Further into the forthcoming volume, we will delve into C++ programming and how it intersects with UE5's MetaSounds, an advanced audio system that grants designers unparalleled control over digital signal processing (DSP) graph generation. This control enables the creation of rich, realistic soundscapes, pushing the boundaries of audio quality and immersion in sound design. Through C++ integration, we can exploit the full capabilities of MetaSounds to elevate the auditory experience within their projects.

Setting Up the Development Environment

As a developer using UE5, it is crucial to have a properly configured development environment for an efficient and smooth workflow. This section will guide us through the process of setting up the development environment for working with UE5 and Visual Studio Community 2022. We will cover the installation of both Unreal Engine and Visual Studio, as well as the necessary dependencies to ensure seamless integration between the two tools, which we can follow throughout the remainder of this book.

Installing Unreal Engine 5 and Dependencies

UE is a powerful game development engine widely used in the industry. Version 5 brings exciting new features and improvements. To begin, follow these steps to install UE:

1. Visit the official Unreal Engine website (www.unrealengine.com) and navigate to the "Get Unreal" section.

2. Click the "Get Started" button and create an Epic Games account if you don't have one already.

3. Once you've created an account and logged in, you will be directed to the Unreal Engine download page. Select the "Unreal Engine 5" version and follow the option to save time and disk space, before clicking the "Install" button as shown in Figure 1-1.

CHAPTER 1   GETTING STARTED WITH UNREAL ENGINE 5

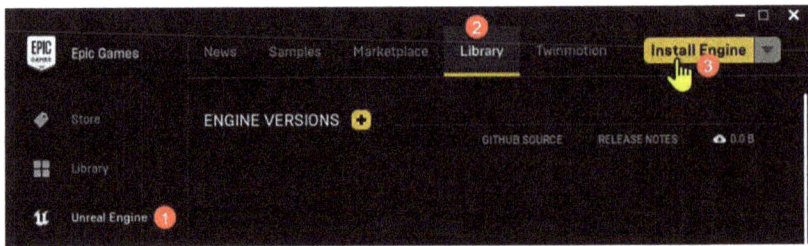

Figure 1-1. *Install Unreal Engine*

4. Note that the Unreal Engine is quite a large application to install. We can choose to install the engine on an alternate drive with a large capacity, as shown in Figure 1-2.

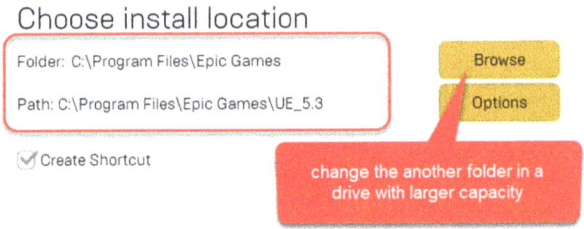

Figure 1-2. *Install to the appropriate folder*

5. To further reduce the engine installation size, there are some options that we can decide not to install from the Option button, as shown in Figure 1-3.

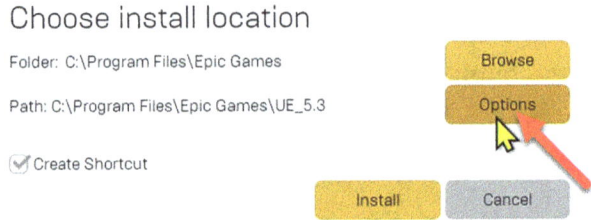

Figure 1-3. *Look into the installation option*

6. We are only interested in downloading and using the binary copy of the engine, so it is best to deselect the Engine Source, as shown in Figure 1-4.

CHAPTER 1   GETTING STARTED WITH UNREAL ENGINE 5

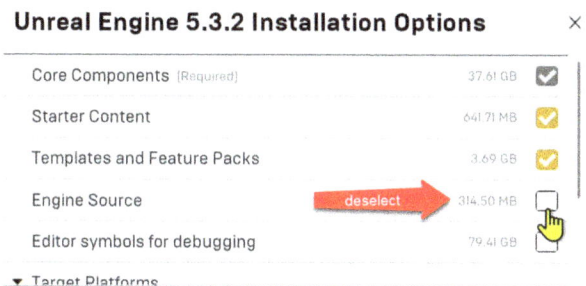

Figure 1-4. Deselect Engine Source to save hard disk space and installation duration

7. Since we are only making a game for the Windows desktop platform, we can deselect all other target platforms, as shown in Figure 1-5.

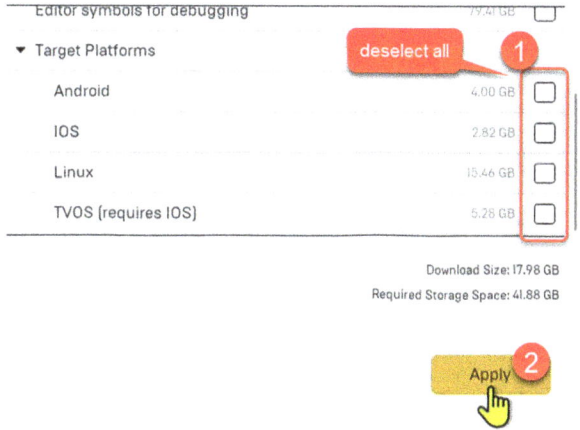

Figure 1-5. Deselect all other target platforms

8. After clicking the Install button (Figure 1-6), the Epic Games Launcher will begin downloading and installing the Unreal Engine. This process may take some time, as it involves downloading a significant amount of data and completing the installation and verification processes.

CHAPTER 1   GETTING STARTED WITH UNREAL ENGINE 5

Figure 1-6. *Start the installations*

9. After installing Unreal Engine, as we are going to work with Visual Studio for using C++ with Unreal Engine, let's proceed to install the necessary dependencies for the Visual Studio Integration Tool plugin by following the steps under the Marketplace section. It is easier if we search for this plugin using the search textbox as shown in Figure 1-7.

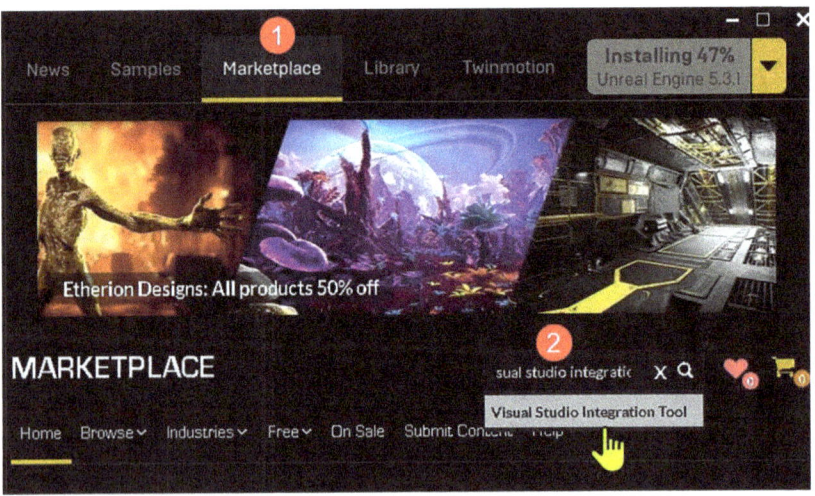

Figure 1-7. *Search for the Visual Studio Integration Tool plugin*

10. After Visual Studio Integration Tool plugin had been added to our library, we can now add this plugin to the UE that we have installed in our system (as shown in Figures 1-8 and 1-9).

CHAPTER 1   GETTING STARTED WITH UNREAL ENGINE 5

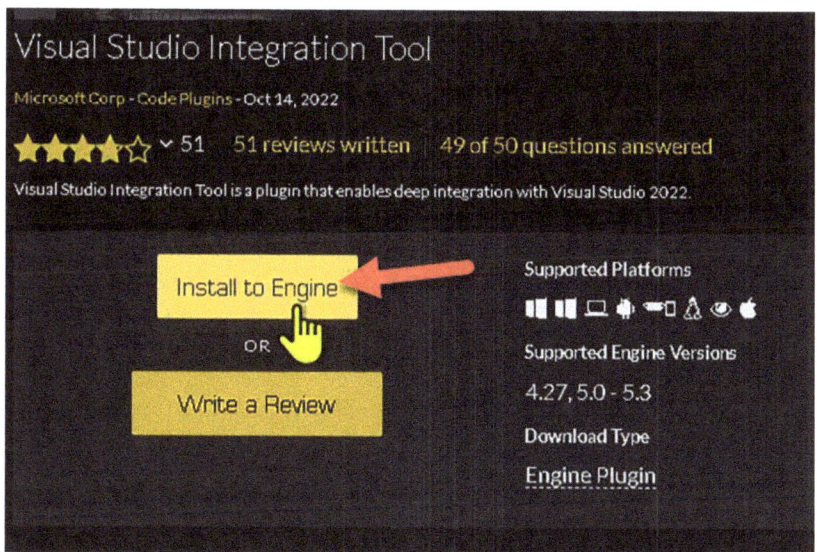

Figure 1-8. *Click the Install button after the plugin is downloaded*

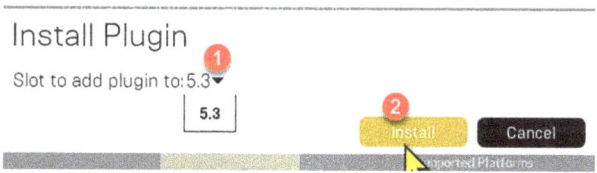

Figure 1-9. *Select the Engine version to install*

Congratulations! We have successfully installed UE5 and the required dependencies. Now let's move on to installing Visual Studio Community 2022.

If you're keen on diving into the nuts and bolts of game development with C++, then the following section is your gateway to deepening your understanding and skill.

We'll be navigating through the intricacies of C++ programming within Unreal Engine 5, providing a hands-on approach to bringing complex functionalities to life.

However, if C++ is not within your current scope of interest, feel free to skip the upcoming section. Rest assured, there's a wealth of knowledge in the remaining content that will continue to enrich your journey in game development.

Installing Visual Studio Community 2022 and Dependencies

Visual Studio is a highly regarded integrated development environment (IDE) widely utilized for programming in various languages, most notably C++. As we prepare to advance our project development into C++, detailed in the next volume of our book, installing Visual Studio will be an essential step. This powerful IDE will be our primary tool for writing, debugging, and compiling the C++ code that will bring interactivity and functionality to our UE5 creations. With Visual Studio, we'll have a robust suite of features at our disposal to develop complex game mechanics and systems within our project. Here's how we can install Visual Studio Community 2022:

1. Go to the official Visual Studio website (`https://visualstudio.microsoft.com`) and navigate to the "Downloads" section.

2. Click the "Community" edition to download Visual Studio Community 2022. This edition provides all the necessary features for Unreal Engine development.

3. Run the Visual Studio installer that we had downloaded.

4. In the installer, install the necessary modules and individual components (also, you might want to consider removing the unnecessary modules, as shown in Figure 1-10).

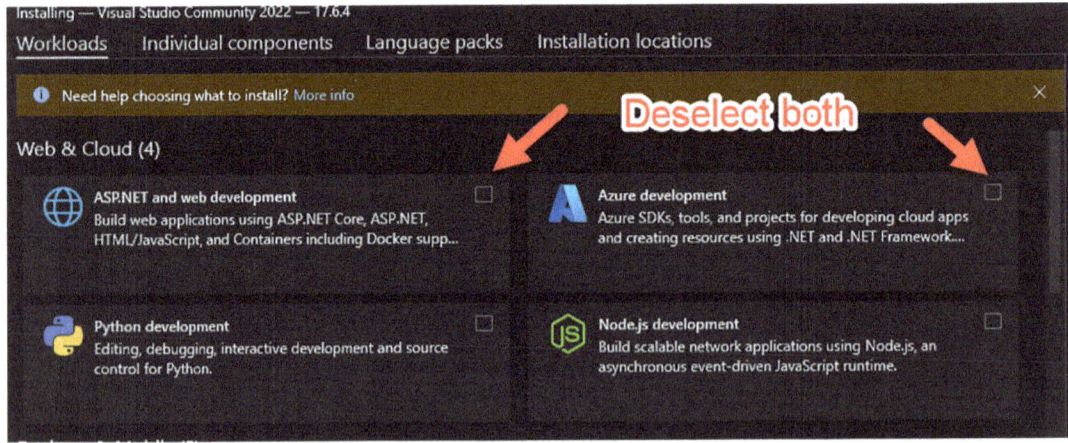

Figure 1-10. Deselect the default modules to save hard disk space and installation duration

5. The two important modules that we must install to allow us to use C++ with Unreal Engine are the .NET desktop environment and Desktop development with C++ (as shown in Figure 1-11).

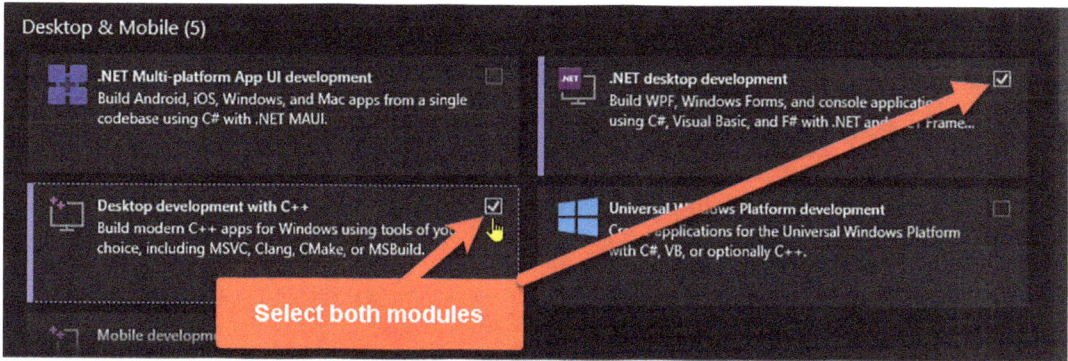

Figure 1-11. *Be sure to install these two modules*

6. Another module that we must also install is the Game Development with C++, as shown in Figure 1-12.

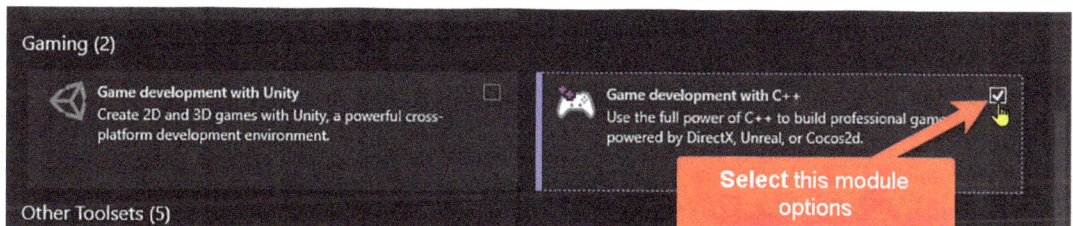

Figure 1-12. *Be sure the Game development with C++ is selected*

7. There are three important individual components (.NET 4.6.2 targeting pack, MSVC v143, and IDE support for UE) that we need to install to ensure the build with Unreal Engine, as shown in Figures 1-13 to 1-15.

CHAPTER 1   GETTING STARTED WITH UNREAL ENGINE 5

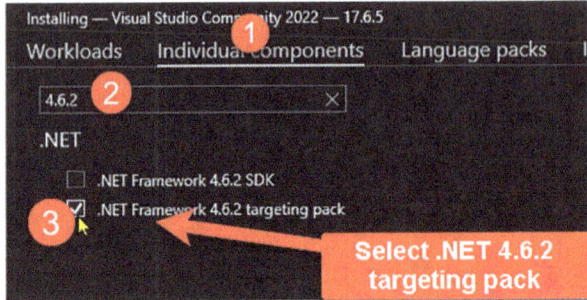

Figure 1-13.   *.NET Framework 4.6.2 targeting pack is selected*

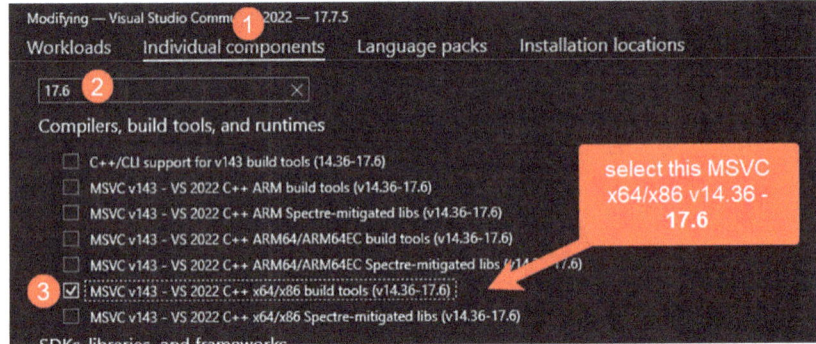

Figure 1-14.   *Be sure the MSVC v143 x64/x86 is selected*

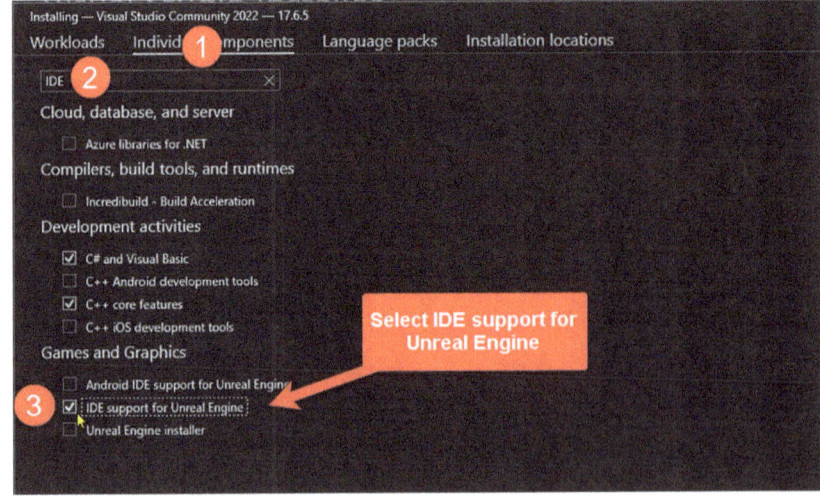

Figure 1-15.   *Lastly, be sure the IDE support for UE is selected*

8. Review the selected components and click the "Install" button to begin the installation process (as shown in Figure 1-16). This process might take some time as it installs various packages and dependencies.

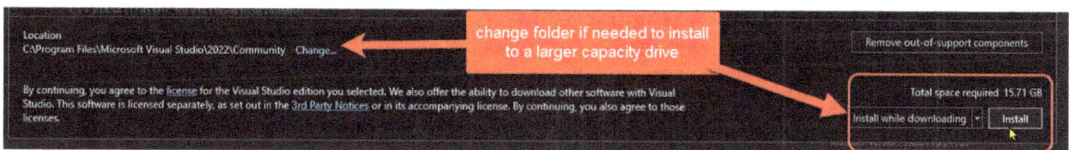

Figure 1-16. *All set to begin the installation*

9. Once the installation is complete, ensure to launch Visual Studio Community 2022 at least once to ensure the necessary initialize setup.

Other Preparations

In order to adhere to the requirements throughout the journey of this book, we will require additional resources and tools beyond what is available within UE5.

One of the additional software tools that we need is a 2D picture editor application, and you are welcome to use any desired tools. In this book, we will be utilizing the GNU Image Manipulation Program (GIMP), which is a free and open source raster graphics editor. The link to download GIMP can be found as follows:

GIMP editor: `www.gimp.org/downloads/`

Regarding 3D assets and animation files required for this book, we have the option to register for free on two websites. Furthermore, all the assets mentioned in this book are available for free download from these websites. The links to access each specific asset and animation will be provided in their respective chapters at a later stage.

These are the links to two websites that we will need to register before we follow on:

TurboSquid: `www.turbosquid.com/`

Mixamo (also works with Adobe login details): `www.mixamo.com/#/`

CHAPTER 1   GETTING STARTED WITH UNREAL ENGINE 5

Finally, we will utilize freely downloadable sound effect files from some audio library website where registration is not required. The link to access these audio files will be provided in the respective chapter as well.

Note If you prefer to do so, you are welcome to use your own 3D assets or animation files in any part of this book.

Prepare New Project

We are now going to launch UE (as shown in Figure 1-17) and prepare a new empty Unreal project (as shown in Figure 1-18) before the start of the next chapter. Since the engine is required to build this new C++ project, the initial start-up time for this project might take a little longer to complete before we can see the UE editor opened with the Visual Studio IDE.

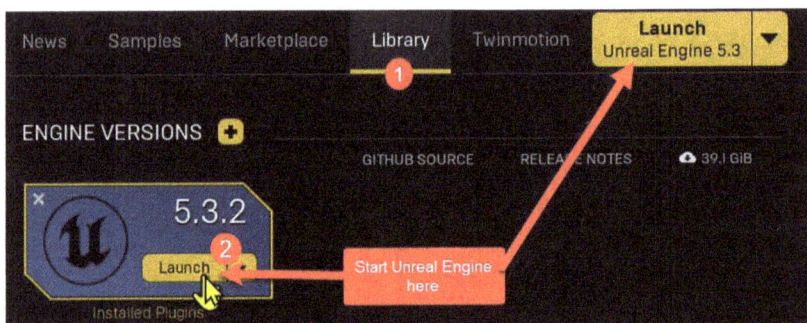

Figure 1-17. *Set up the plugin in the project*

If you're more inclined toward visual scripting and prefer to steer clear of C++, there's an alternative route for you.

When you reach Step 3, you can opt to work with UE5's Blueprint system instead. Blueprints provide a user-friendly, node-based interface that allows you to create complex game logic without delving into traditional code.

For those not interested in C++, Blueprints offer a powerful and accessible way to bring your game ideas to life. So, choose the path that best suits your skills and interests, and let's continue to build and innovate within UE5.

CHAPTER 1　GETTING STARTED WITH UNREAL ENGINE 5

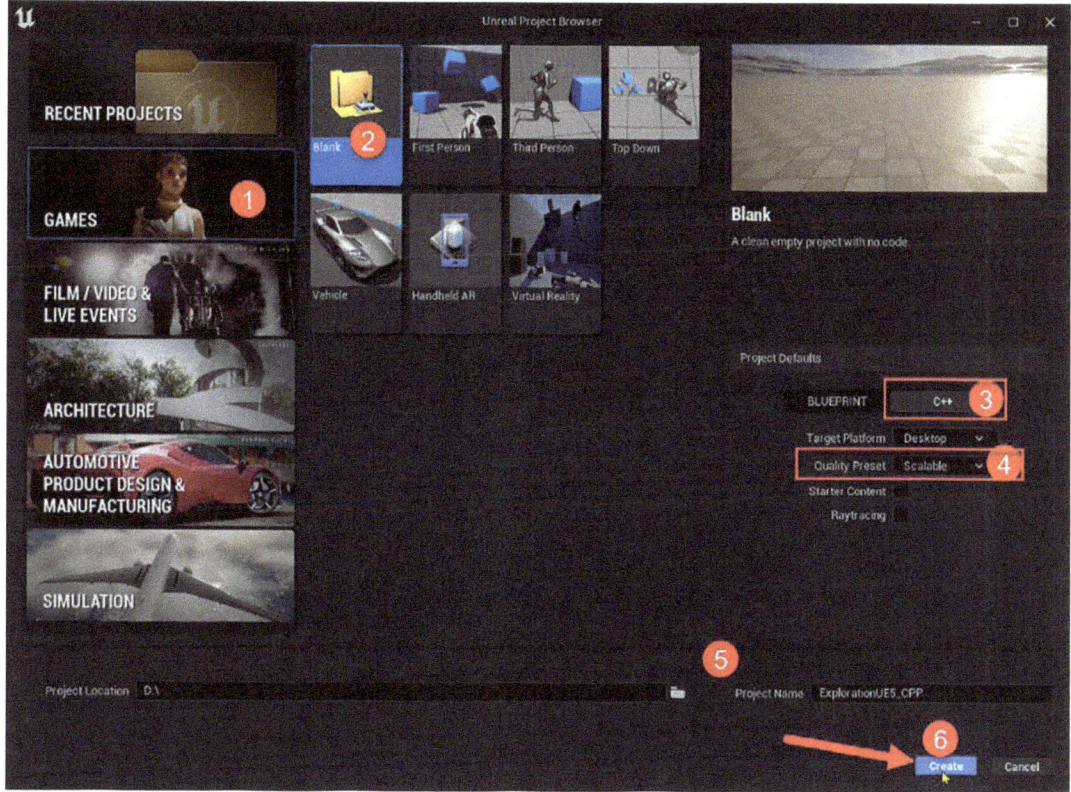

Figure 1-18. *Create an empty C++-based UE project*

After opening the project, it is necessary to perform the following steps before proceeding to the subsequent chapters in this book:

1. Get prepared for the later chapters by installing the necessary plugins (include the latest version of Visual Studio Integration Tool plugin) as shown in Figures 1-19 to 1-21 (only for C++ developers). We can restart the editor after selecting all the necessary plugin tools as shown in Step 2 in Figure 1-21.

CHAPTER 1   GETTING STARTED WITH UNREAL ENGINE 5

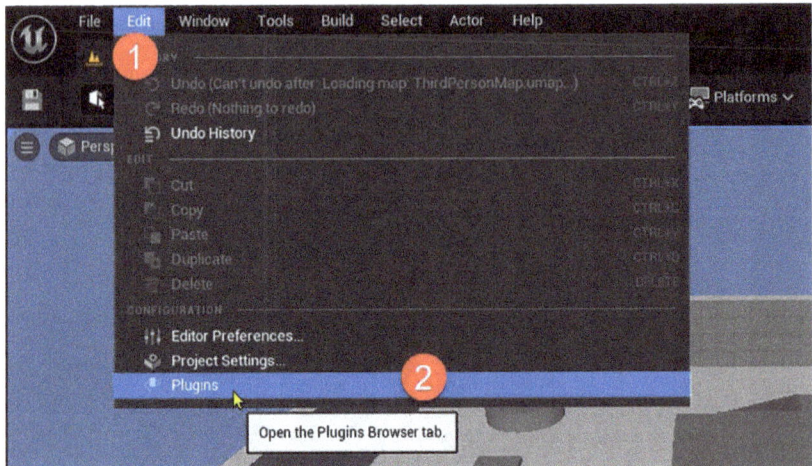

Figure 1-19. Open up the Plugin Browser inside the project

Figure 1-20. Select the Procedural Content Generation Framework

CHAPTER 1 GETTING STARTED WITH UNREAL ENGINE 5

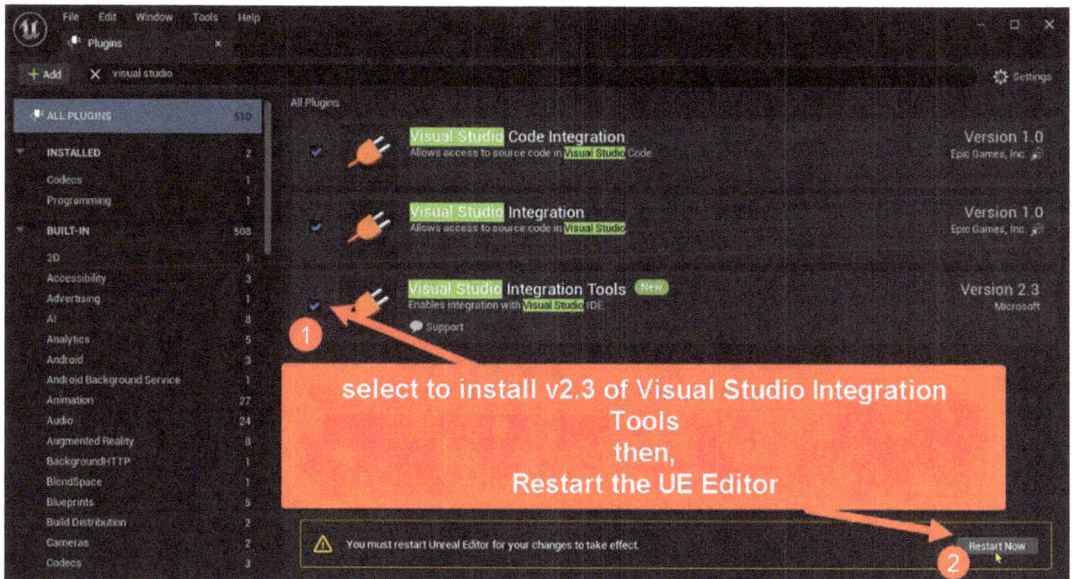

Figure 1-21. *Be sure Visual Studio Integration Tool is selected (only for C++ developers)*

2. Virtual Texturing Support will be discussed in this book, so we need to select to enable this feature within the project settings, as shown in Figures 1-22 and 1-23.

3. After completing the preceding steps, the entire project will be requested to restart by clicking the Restart button located at the bottom corner of the editor, as shown in Figure 1-24.

17

CHAPTER 1   GETTING STARTED WITH UNREAL ENGINE 5

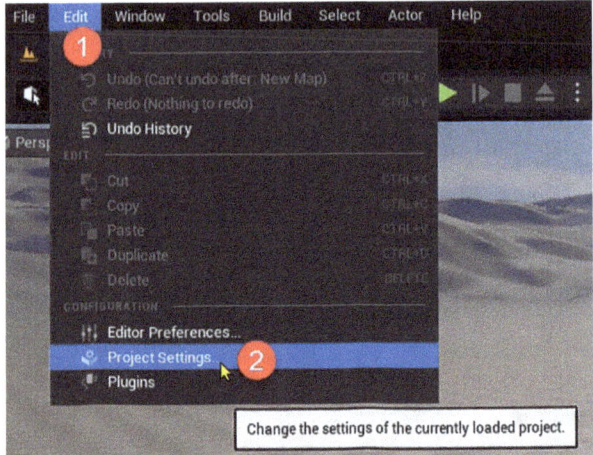

Figure 1-22. Open settings of this project

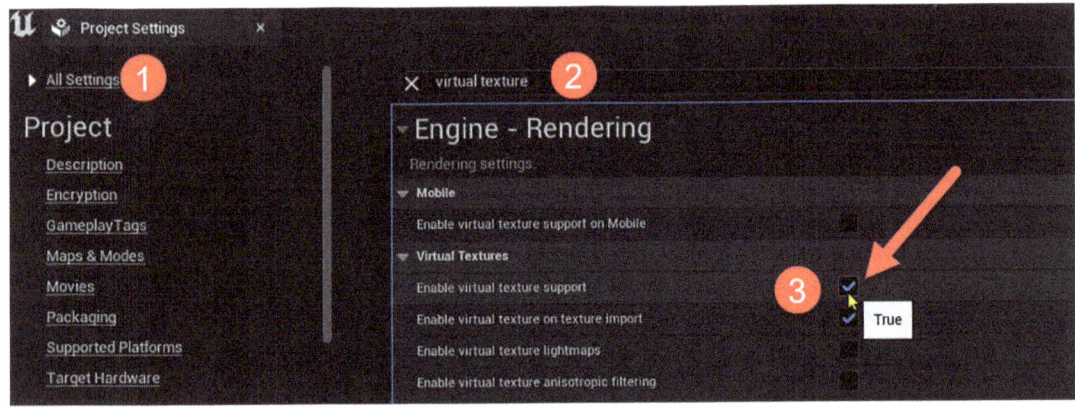

Figure 1-23. Enable virtual texture support

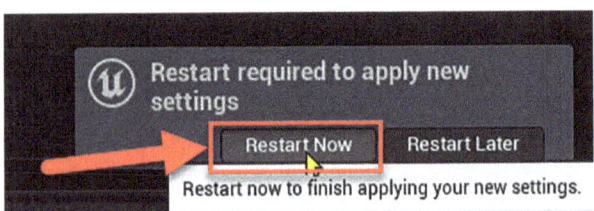

Figure 1-24. Restart the project

Run on Low-Quality Mode

Feel free to adjust the project settings to run on low-quality mode (as depicted in Figure 1-25) throughout the entire development process, if deemed necessary. This will ensure a smoother experience at the beginning of the development process.

The settings panel for graphical quality within the UE editor should have the scalability settings set to 'Low'. This configuration is typically used to reduce the visual fidelity in favor of improved editor performance, particularly useful during the development phase on machines with limited processing power or when working with complex scenes that can be taxing on the system.

Step 1 is to set the scalability settings to a low preset, which is confirmed by the detailed settings. This happened due to a manual change because the engine automatically scales down the settings in response to detected performance issues.

Step 2 indicates that all quality settings have been collectively adjusted to their lowest values. This includes options like Resolution Scale, View Distance, Anti-Aliasing, Post Processing, Shadows, Global Illumination, Reflections, Textures, Effects, Foliage, and Shading. Each of these settings has its corresponding quality level set to "Low," as indicated in the column under "Quality."

This adjustment is to ensure that the editor runs smoothly, allowing the developer to work without the hindrance of performance issues such as low frame rates or long loading times. It's a common approach taken to optimize the editing experience, especially when in the early stages of level design or when iterating rapidly on game mechanics, and visual fidelity is not the primary concern. Reducing the quality settings can also be helpful when trying to identify performance bottlenecks or when testing how the game would perform on lower-end hardware.

CHAPTER 1 GETTING STARTED WITH UNREAL ENGINE 5

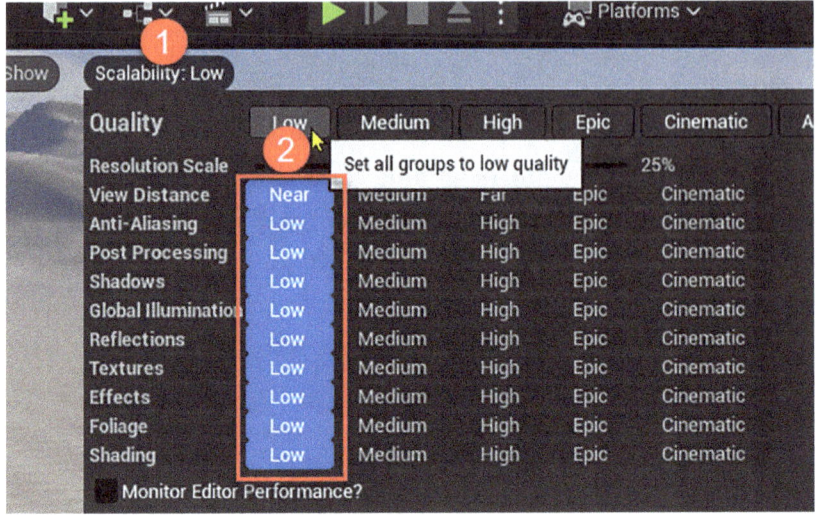

Figure 1-25. *Set low-quality settings*

Show Engine Content

Since we did not import any characters or assets at the beginning of the development process, we can leverage the existing assets already available within the engine itself (inside the editor).

To access these assets, we need to follow the steps outlined here. For instance, we have a 3D third-person character that we will utilize in the upcoming chapters. To enable the display of Engine Content, refer to the instructions provided in Figure 1-26.

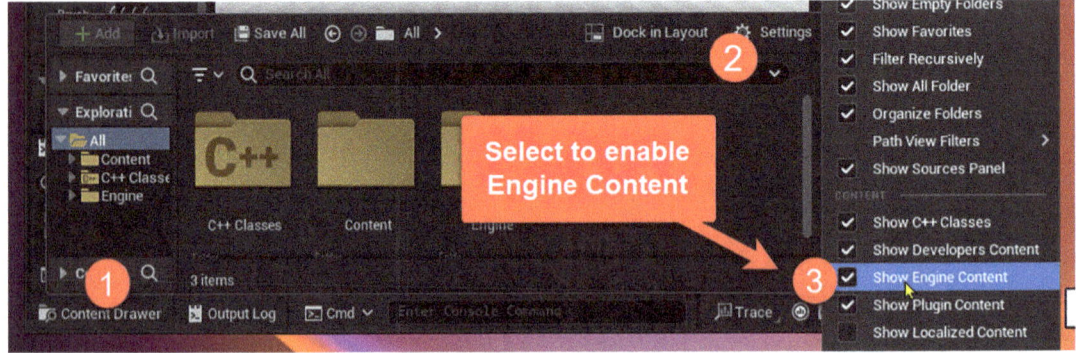

Figure 1-26. *Enable the Engine Content*

CHAPTER 1   GETTING STARTED WITH UNREAL ENGINE 5

Once the Engine Content is enabled, an additional folder named "Engine" will appear in the Content Drawer. For example, by entering "TPP" in the search box (under the Content Drawer in Step 1 and the Engine folder in Step 2), as shown in Figure 1-27, we can locate the available third-person character. The steps inside the figure are described as follows:

> Content Drawer: Step 1 shows the area typically where all the project's content is organized. We can find different assets, blueprints, and other resources necessary for game development here.
>
> Engine Content: Step 2 shows the area that specifically lists the content that is part of the engine itself, such as basic materials, textures, and engine-specific blueprints and classes.
>
> Search Bar: Step 3 shows the search bar, where we will type in 'TPP,' allowing all assets with 'TPP' in their names or related to a third-person perspective (which is what TPP commonly stands for in game development) to appear.

We will use this character in the later section after setting up the landscape in Chapter 2.

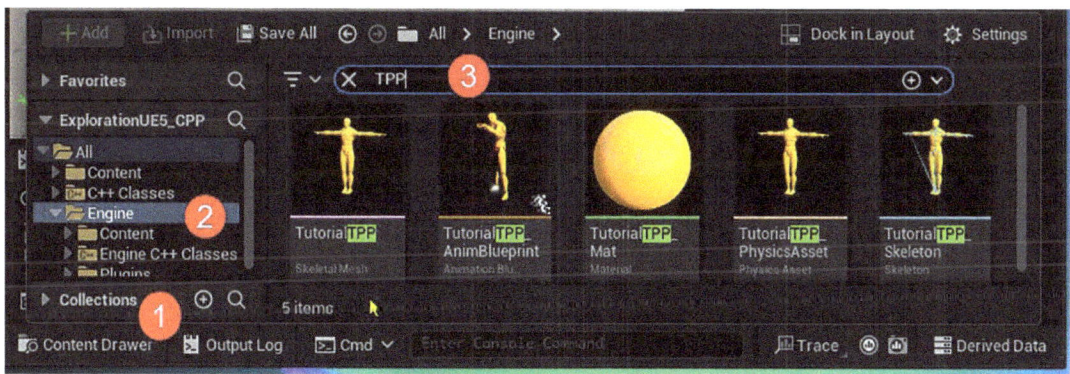

Figure 1-27. Look for TPP characters in the Engine folder

Summary

This first chapter introduced UE5, an advanced game development engine, and outlined the installation process. This process includes downloading and running the Epic Games Launcher to set up UE5 with all the required components. Additionally, we have gone through the installation of Visual Studio Community 2022 with the necessary modules to work alongside UE. Finally, we have also set up a new UE project with the necessary plugins, allowing us to start working on the game for this book.

With UE5 and Visual Studio Community 2022 successfully installed, we are now equipped with a powerful combination of tools to unleash their creativity and bring their game ideas to life. The next chapter will be about creating a terrain using a landscape in UE5, which is the first step in our journey of creating "The Making of Space Exploration Game."

CHAPTER 2

From Heightmap to Large Open-World Landscape

A landscape is one of the most important aspects of creating immersive and realistic worlds for games or simulations. However, creating a landscape can also be a complex and time-consuming task, especially when dealing with large and detailed worlds. That's why UE5 offers a powerful and flexible toolset for creating and managing a landscape with ease and efficiency.

In this chapter, instead of using the available landscape tool to sculpt the landscape ourselves, we will learn how to import heightmaps from external sources. This approach will save us a lot of tedious time when creating large landscape areas. Additionally, we will learn how to optimize and stream our landscape using one of the UE5 features, World Partition.

> A heightmap refers to a grayscale texture or image that encodes elevation data for a terrain or landscape. It is a commonly used technique for representing the height variation of the terrain surface in a simple and efficient manner.
>
> A heightmap is essentially a 2D grayscale image, where each pixel's color value represents the height or elevation of the terrain at that specific point. The height is usually measured in world units, such as meters or feet. In the heightmap, the color white corresponds to the highest elevation, while black corresponds to the lowest elevation.
>
> Shades of gray in between represent varying heights. The darker the pixel, the lower the terrain, and the lighter the pixel, the higher the terrain. The resolution of the heightmap determines the level of detail and precision of the terrain. Higher-resolution heightmaps capture more details but may also require more memory and processing power.

CHAPTER 2   FROM HEIGHTMAP TO LARGE OPEN-WORLD LANDSCAPE

By the end of this chapter, we will have a certain level of understanding of the landscape tool in UE5 and how to use it to create stunning outdoor environments.

Set Up a New Level

Continuing from Chapter 1, we will now create a new map (level) that will allow us to add our own landscape for our space exploration game as shown in the following steps alongside with the figures:

1. Opening a New Level: Figure 2-1 shows a drop-down menu with an option to "New Level…" highlighted. It suggests that clicking this will open a new interface where you can create a new level. The keyboard shortcut for this action is Ctrl+N.

2. Selecting a Level Type: In Figure 2-2, a dialog box titled "New Level" presents four options to choose from: "Open World," "Empty Open World," "Basic," and "Empty Level." We will select the Basic level as shown in Step 1 and use the "Create" button as shown in Step 2 to proceed the level creations.

3. Deleting an Object: Figure 2-3 shows a part of the interface where various objects in the level are listed. Since we are not interested to use the current floor object, we can remove the floor object from the level by pressing the "Delete" key from the keyboard.

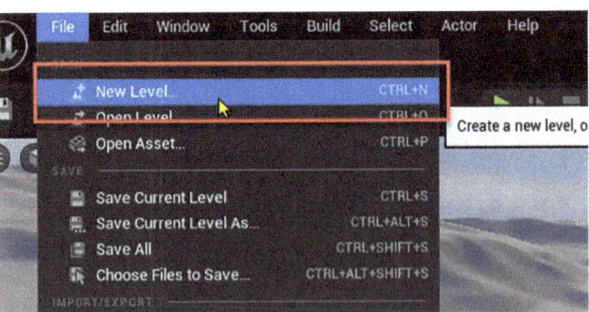

Figure 2-1. New level

CHAPTER 2   FROM HEIGHTMAP TO LARGE OPEN-WORLD LANDSCAPE

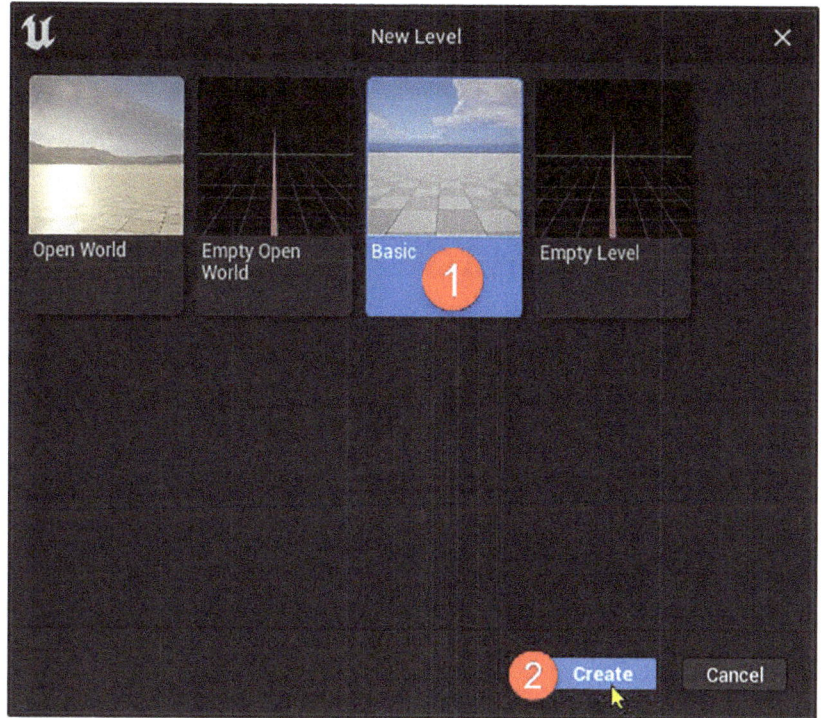

Figure 2-2. *Create a Basic level*

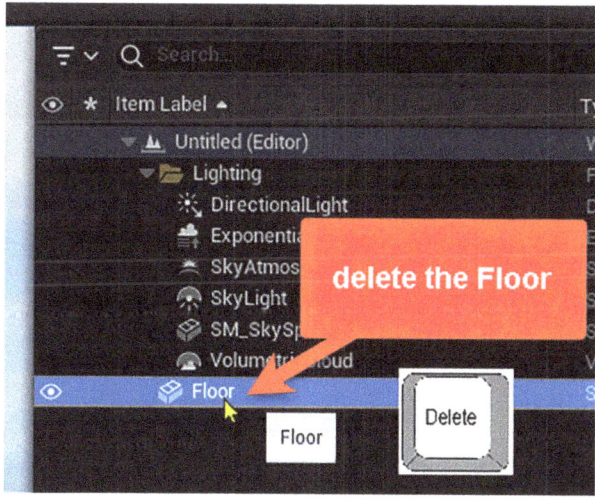

Figure 2-3. *Remove the Floor object*

Prepare the Heightmap

There are various methods to obtain a heightmap. One approach involves generating heightmaps based on real-world locations using websites like Tangram Heightmapper (https://tangrams.github.io/heightmapper/) or Terrain Party (https://terrain.party/). Alternatively, we can utilize tools such as World Machine (www.world-machine.com/) or Gaea from QuadSpinner (https://quadspinner.com/) to create heightmaps based on our imagination.

Moreover, there are specific websites that offer prebuilt libraries of heightmaps available for free download, which we can use in our development. In this book, we should download the Great Lake heightmap from the Motion Picture website through the following link: www.motionforgepictures.com/height-maps/ (as shown in Figure 2-4).

It is important to note that we need to adhere to certain guidelines to suit UE for the heightmap before importing it into the UE editor. The actual guidelines can be found on this UE documentation web link, specifically under the section "Heightmap Formats" (https://docs.unrealengine.com/5/en-US/creating-and-using-custom-heightmaps-and-layers-in-unreal-engine/).

In general, the accepted formats for creating/saving heightmaps from any application are either

- 16-bit grayscale PNG files
- 16-bit grayscale RAW files in little-endian byte order

CHAPTER 2　FROM HEIGHTMAP TO LARGE OPEN-WORLD LANDSCAPE

Figure 2-4. *Great Lake heightmap from www.motionforgepictures.com/ height-maps/*

Convert EXR to PNG with GIMP

In this example of the given heightmap, it is in a different picture format (EXR format), rather than PNG. So, it is necessary for us to convert it to a compatible format that suits UE, as mentioned earlier. While we are using the GIMP picture editor (Figure 2-5) to do this conversion, we can also use other picture editing software to perform the same task.

CHAPTER 2   FROM HEIGHTMAP TO LARGE OPEN-WORLD LANDSCAPE

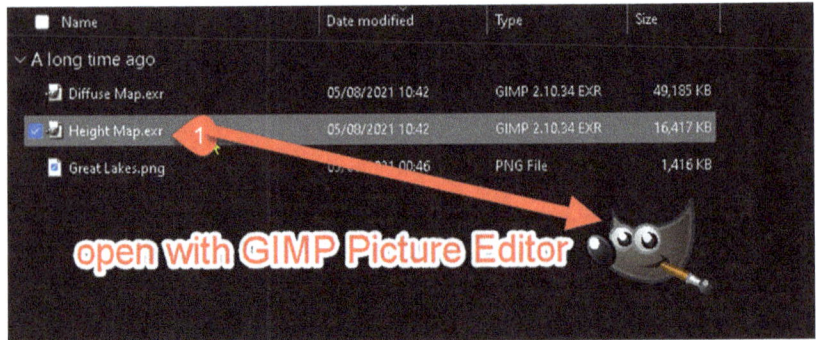

Figure 2-5. *Open the .exr picture with GIMP*

Inside the GIMP editor, we can save to 16-bit by changing the image mode to grayscale, which has only two channels: grayscale and alpha.

1. Exporting the Image

   As shown in Figure 2-6, we can export the image by navigating to the "File" menu and clicking "Export As…" (denoted by Step 1). This can also be achieved with a direct shortcut of Shift+Ctrl+E (as shown in Step 2).

2. Choosing the File Name and Type

   In the "Export Image" dialog box shown in Figure 2-7, type the desired file name; here, it is "Great_Lakes_Height_Map.png" (pointed out by Step 1).

   Make sure you're saving in the correct directory which in this case is a subfolder within "Downloads" named "Great Lakes."

   Select the file type you want to export the image as by clicking "Select File Type (By extension)," then choose "PNG image" (indicated by Step 3).

   Click the "Export" button (Step 4) to proceed to the next step.

CHAPTER 2   FROM HEIGHTMAP TO LARGE OPEN-WORLD LANDSCAPE

3. Setting PNG Options

   In the "Export Image as PNG" dialog box shown in Figure 2-8, we can choose the color values for the image. In this case, "16bpc GRAY" is selected from a drop-down menu (Step 1).

   Adjust any other necessary options such as the compression level or whether to save color profile information.

   Once all settings are configured, click "Export" (pointed out by Step 2) to finalize the export process.

Figure 2-6. *Export As... (to PNG)*

CHAPTER 2 FROM HEIGHTMAP TO LARGE OPEN-WORLD LANDSCAPE

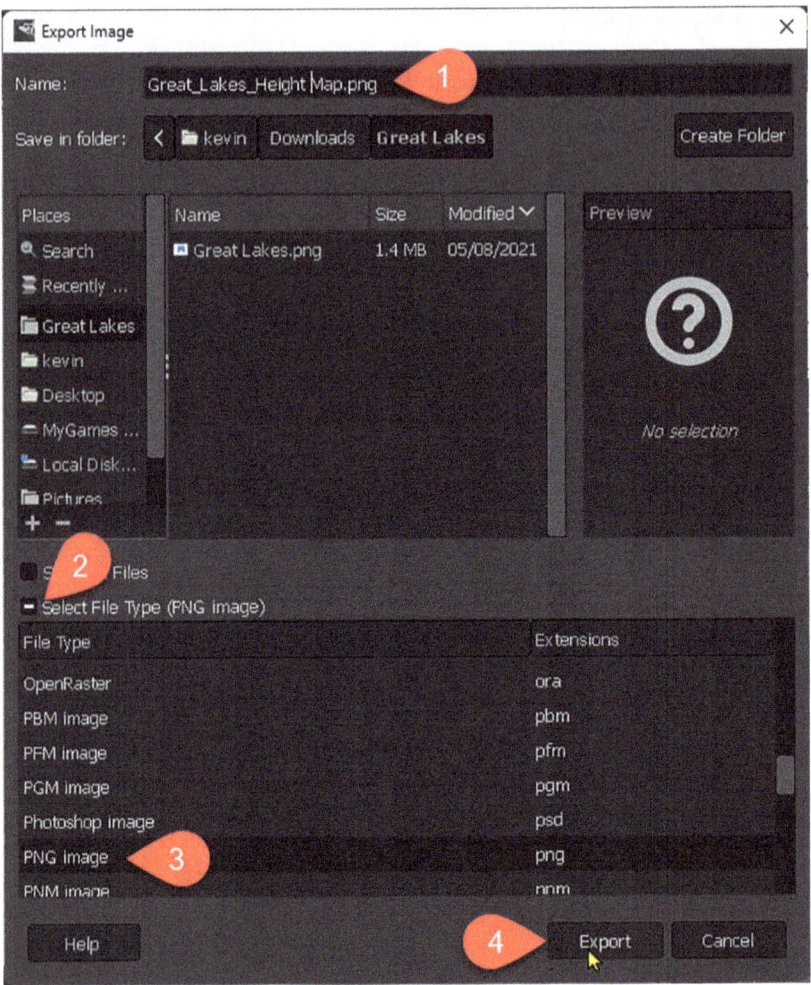

Figure 2-7. *Select PNG format to export*

CHAPTER 2 FROM HEIGHTMAP TO LARGE OPEN-WORLD LANDSCAPE

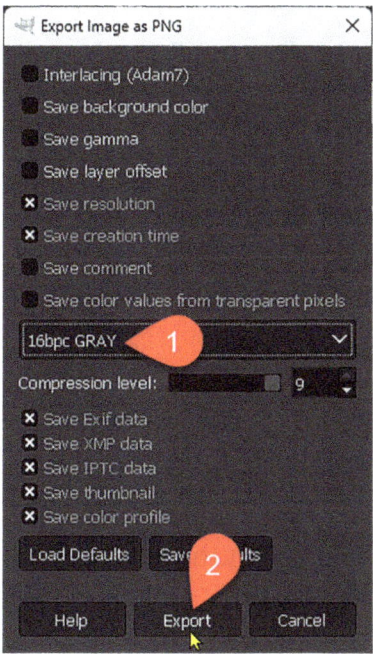

Figure 2-8. *Be sure it is a 16-bit grayscale*

Create Landscape from Heightmap

By default, we are working in Selection Mode inside the UE5 editor that allows us to select and manipulate objects within the level for editing purposes.

However, if we want to create a landscape in UE5, we need to switch to a different mode (the landscape mode) as shown in Figure 2-9.

1. We can follow Step 1 to activate the drop-down menu from "Selection Mode" into different modes.

2. As shown in Step 2, we will select "Landscape" mode (or with a direct keyboard shortcut of Shift+2), for editing terrain and landscape features.

This menu is typically used to switch between different editing modes, each specialized for different aspects of level and game development.

CHAPTER 2   FROM HEIGHTMAP TO LARGE OPEN-WORLD LANDSCAPE

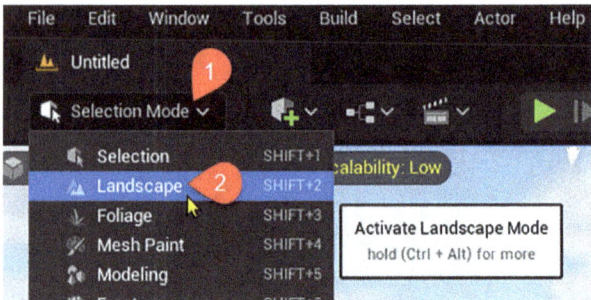

Figure 2-9. *Activate landscape mode*

Within the landscape mode, we can see a configuration tab that allows us to create a new landscape from scratch or import one from a file (heightmap), as we will follow in Figure 2-11.

As mentioned earlier, we are using the heightmap from the previous section (a grayscale texture that stores height data for an object) to help us set up our landscape. To successfully import a heightmap file into the UE editor, it's important to comprehend the various settings related to resolution, considering that heightmaps can be available in a multitude of resolution sizes. Different resolutions will require different settings in the UE5 landscape configurations.

We can look up the configuration using the following web link to the UE Landscape Technical Guide website: https://docs.unrealengine.com/5/en-US/landscape-technical-guide-in-unreal-engine/. In this case, we have a 2K resolution (2048x2048), and the recommended settings for the UE will be shown in Figure 2-10.

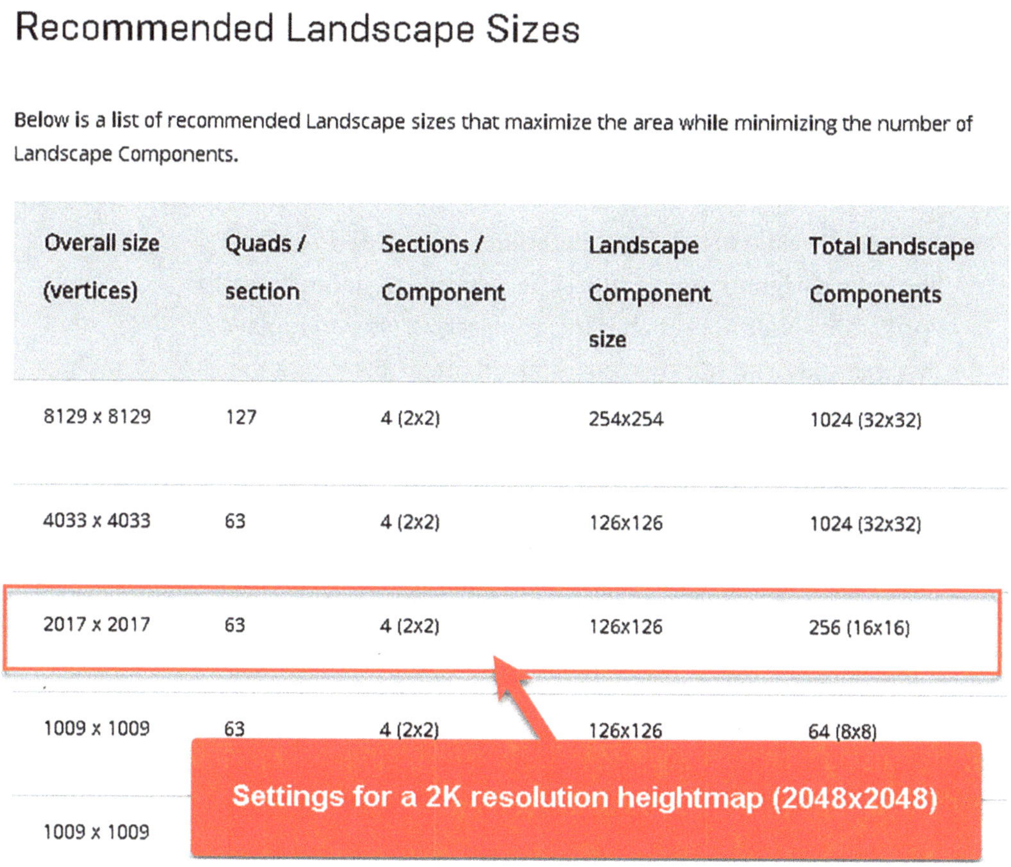

CHAPTER 2   FROM HEIGHTMAP TO LARGE OPEN-WORLD LANDSCAPE

Recommended Landscape Sizes

Below is a list of recommended Landscape sizes that maximize the area while minimizing the number of Landscape Components.

Overall size (vertices)	Quads / section	Sections / Component	Landscape Component size	Total Landscape Components
8129 x 8129	127	4 (2x2)	254x254	1024 (32x32)
4033 x 4033	63	4 (2x2)	126x126	1024 (32x32)
2017 x 2017	63	4 (2x2)	126x126	256 (16x16)
1009 x 1009	63	4 (2x2)	126x126	64 (8x8)
1009 x 1009				

Settings for a 2K resolution heightmap (2048x2048)

Figure 2-10. *The recommended settings for the heightmap with a specific resolution can be found at the UE Landscape Technical Guide website:* https://docs.unrealengine.com/5/en-US/landscape-technical-guide-in-unreal-engine/

The settings for our 2K Great Lakes heightmap can be seen in Figure 2-11 with explanations for each corresponding step.

1. Landscape Mode: This is the active mode, as indicated by the highlighted button at the top in Step 1.

2. Manage Tab: In Step 2, we select the Manage tab that provides tools and options for landscape management.

3. New Landscape Section: In Step 3, we select the "Create New" for creating or importing landscape data.

33

CHAPTER 2   FROM HEIGHTMAP TO LARGE OPEN-WORLD LANDSCAPE

4. Import from File: The selected option indicates that the user is importing a landscape from a file rather than creating one from scratch.

5. Select the File: In Step 5, we will select the heightmap file from the file browser button here.

6. Material: In Step 6, a default material named "MI_ProcGrid" has been assigned to the landscape, which will define its visual appearance. In the next chapter (Chapter 3), we will replace this material with our own customized materials to suit the landscape's desired style for our game.

7. Resetting the location can be done as in Step 7, by setting to 0, 0, 0.

8. Scale: Step 8 shows the X, Y, and Z scale values for the landscape, with X-axis and Y-axis (width and depth) set to 100.0 and Z-axis (height) set to 300.0, which will affect the overall dimension of the landscape.

Sometimes, using a smaller scale-Z value can result in a better-rendered shape of the landscape, rather than sharp pointy features.

In this Great Lakes example, we will use a larger scale-Z-axis value instead (compared to the value of scale-X-axis and scale-Y-axis) to highlight the height of the mountains in the landscape relative to its width.

The values for the following steps are based on the UE Landscape Technical Guide, as shown in the highlighted red box in Figure 2-10:

1. Section Size: Shown in Step 9, the selected section size is "63x63 Quads," which determines the number of quads per landscape section.

2. Sections Per Component: Shown in Step 10, we will use "2x2 Sections," which determines how many sections each component of the landscape will contain.

CHAPTER 2 FROM HEIGHTMAP TO LARGE OPEN-WORLD LANDSCAPE

3. Number of Components: Shown in Step 11, we will set to 16 horizontally and 16 vertically; this determines the number of components that will make up the landscape.

4. Overall Resolution: Shown in Step 12, we will set the resolution of the landscape once all components are combined, which is 2017x2017.

5. Import Button: Once we completed the preceding 12 steps, we complete the import process with this button.

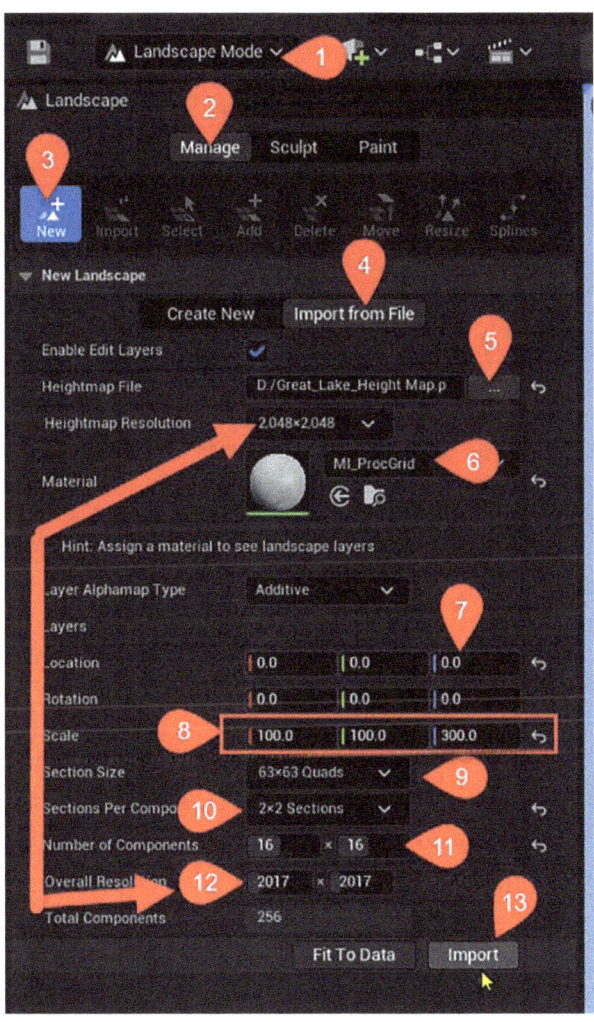

Figure 2-11. *Import heightmap*

35

After completing the import, we might find that the landscape does not appear in our scene (refer to Figure 2-12). This issue can arise from several factors, but a frequent cause is that the Exponential Height Fog may be obscuring the landscape (see Figure 2-13).

Figure 2-12. *Why is the landscape invisible?*

To resolve this, adjust the Z position of the fog object to a lower value. However, be sure to switch to Selection Mode as follows:

1. Switch to Selection Mode, as shown in Step 1 in Figure 2-13, or use the shortcut key (Shift 1) to ensure that the objects in the Outliners can be enabled and selectable for us to edit the desired values.

2. Activate the Outliner: As shown in Step 2, we can pull out the Outliner on the right-hand side to show various scene components.

3. The "ExponentialHeightFog" is selected, as shown in Step 3, which is a type of volumetric fog used to create atmospheric effects.

For instance, in this scenario, we've reduced it to a value that is ten times lower than the original (shown in Figure 2-14) as `Location Z = -68,500`, corresponding to the scale set in Step 8 (as outlined in Figure 2-11).

CHAPTER 2 FROM HEIGHTMAP TO LARGE OPEN-WORLD LANDSCAPE

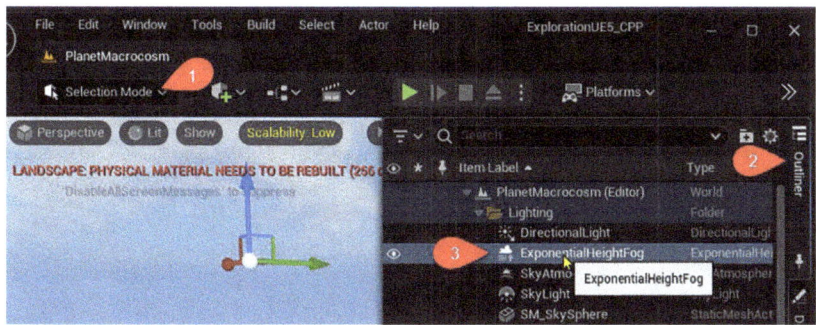

Figure 2-13. *Look for the Exponential Height Fog*

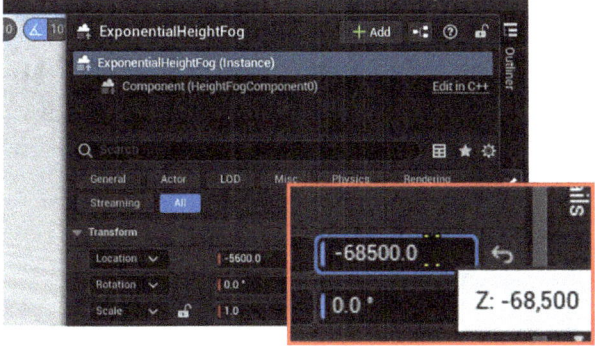

Figure 2-14. *Adjust the Z value (10x lower in this case)*

Rendering Issues with Artifacts: Should the landscape exhibit artifacts along its edges, as shown in Figure 2-15, this typically suggests an improper use of the Overall Resolution, as outlined in Step 12 of Figure 2-11. A properly rendered landscape will display clean and smooth edges, exemplified in Figure 2-16.

CHAPTER 2 FROM HEIGHTMAP TO LARGE OPEN-WORLD LANDSCAPE

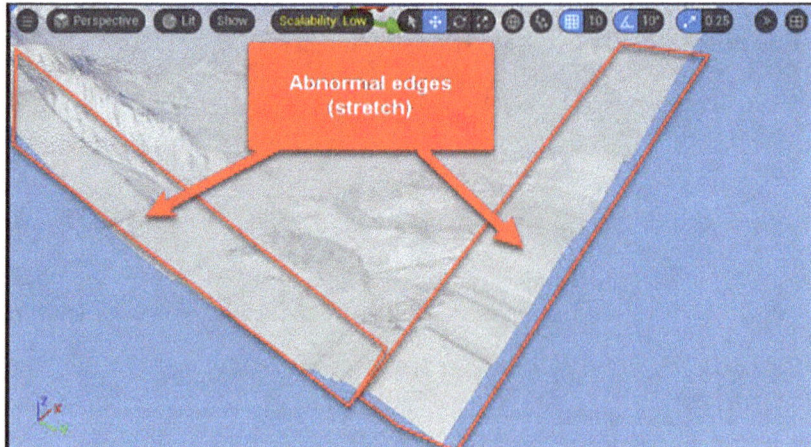

Figure 2-15. *Artifact issues by wrong import settings*

Figure 2-16. *Result from the correct import settings*

Ensure Correct Scale from Perspective

Before continuing to embark on this creative journey, the first step is to determine a standard size for a character within the landscape to ensure that the landscape is the size that we expect. By establishing this standard, we gain a valuable reference point, enabling us to better comprehend the vastness and proportions of the environment in relation to real-world scales.

Since we have not imported any characters into our project, we will take advantage of the existing third-person control that is available inside the Engine Content, just as we had prepared in Chapter 1, under the section "Show Engine Content."

We can mouse-drag the `TutorialTPP` character from the Engine folder to any part of the scene above the landscape (shown in Figure 2-17).

1. Selection Mode: Be sure that the editor is set to "Selection Mode," as indicated by Step 1.

2. Content Drawer: As shown in Step 2, we need to activate the Content Browser by selecting the Content Drawer from the bottom bar.

3. Engine Folder: As indicated in Step 3, we need to select the "Engine" folder. If you cannot locate this folder, please refer to the section titled "Show Engine Content" in the previous chapter for guidance.

4. Search Field: As stated in Step 4, you can simplify the process of locating assets by typing "tutorialtpp" into the Content Browser's search bar, which filters and displays only the assets containing that text.

5. TutorialTPP Assets: As shown in Step 5, various assets associated with the "TutorialTPP" appear in the Content Browser.

6. Asset Placement: As shown in Step 6, the placement of the asset is demonstrated by dragging the TutorialTPP character asset onto the terrain within the level.

If desired, one can press **the "END" from the keyboard** to place the asset on the landscape directly.

CHAPTER 2   FROM HEIGHTMAP TO LARGE OPEN-WORLD LANDSCAPE

Figure 2-17. *Add the third-person character from the Engine folder*

If necessary, we can adjust the scale of the landscape object (shown in Figure 2-18) to ensure the perspective from the character toward the landscape is suitable for our game. In the planet of our space exploration game, Figure 2-19 will make the landscape look a bit small compared to the standard character. Figure 2-20, on the other hand, shows the suitable scale compared to the character and its perspective placed on the landscape itself.

CHAPTER 2   FROM HEIGHTMAP TO LARGE OPEN-WORLD LANDSCAPE

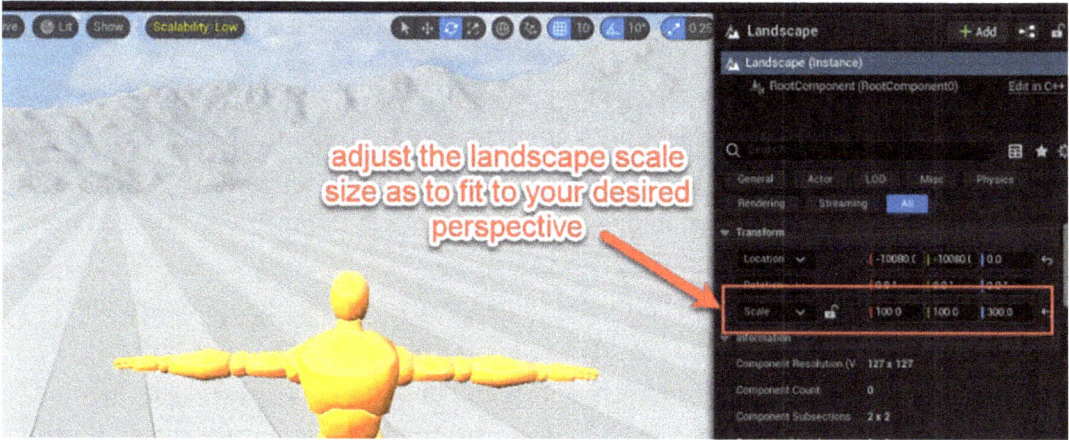

Figure 2-18. *Readjust the scale of the landscape to ensure your desired perspective view*

Figure 2-19. *The scale of this landscape looks a little too "small"*

Figure 2-20. *This looks about the correct perspective*

41

Convert to World Partition

UE5 aims to revolutionize the way developers create and handle vast open-world environments with its "World Partition" system. In traditional game development, large open worlds are often divided into smaller, manageable chunks to fit into memory and improve performance. However, managing these chunks and transitions between them can be complex and lead to performance issues.

With World Partition, UE5 seeks to eliminate those issues by providing a more efficient and seamless approach. World Partition is a feature that addresses the challenges of creating massive, detailed worlds and improves performance by loading and streaming only the necessary parts of the environment dynamically, depending on the player's location and view.

World Partition is a system that intelligently divides the game world into smaller cells using runtime data and the Hierarchical Level of Detail (HLOD). Each cell contains specific environment elements like geometry, textures, and foliage. As the player moves through the world, only relevant cells are loaded, reducing memory usage and improving rendering performance.

This dynamic loading enables developers to create detailed and expansive worlds without sacrificing performance or manual optimization. Moreover, World Partition supports advanced streaming and culling techniques to enhance visual fidelity and ensure a smooth player experience.

Save the Level Before Conversion

Since we have created one large piece of landscape, we need to perform a conversion in order to split it into partitions and enable World Partition to work with the dynamic loading process. Before the conversion, we need to save the current level into a physical file in our Content folder. This file will be loaded by the conversion tool, and then the landscape will be converted into a grid-based cell landscape.

Before we perform the conversion, it is wise to save the level as shown in Figure 2-21, where we can use the Save Current Level option or the shortcut key, Ctrl+S.

CHAPTER 2 FROM HEIGHTMAP TO LARGE OPEN-WORLD LANDSCAPE

1. As indicated in Figure 2-22, we are going to save the level in the folder named Level (Step 1), which we can right-click to create a new folder, and rename it.

2. Enter any desired name for the Level (Step 2).

3. Click the Save button (Step 3).

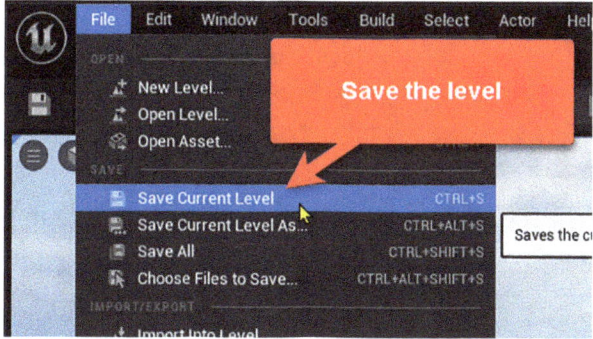

Figure 2-21. *To save the level map*

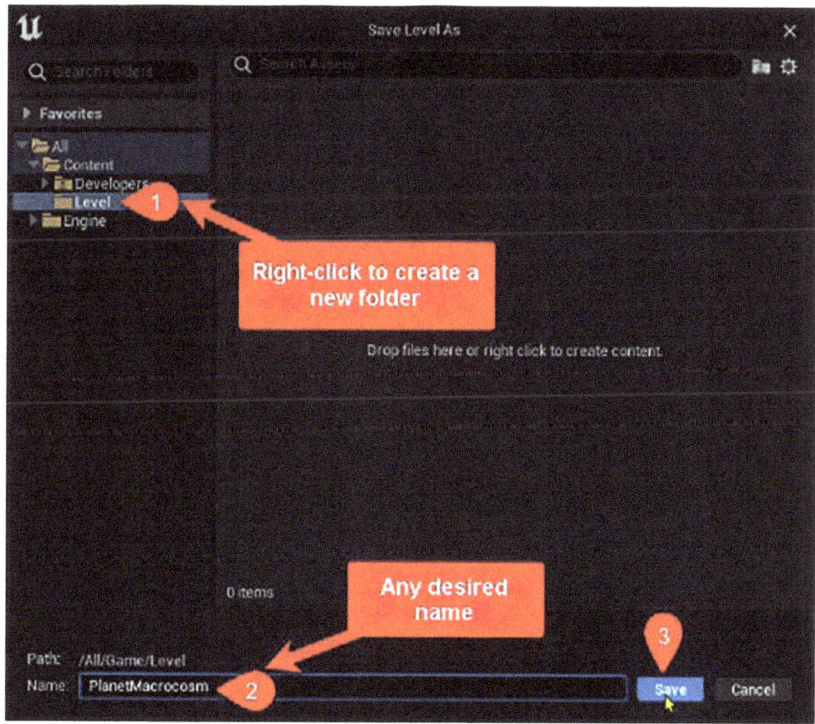

Figure 2-22. *Save the level named PlanetMacrocosm into a Level folder*

43

CHAPTER 2 FROM HEIGHTMAP TO LARGE OPEN-WORLD LANDSCAPE

For convenience, consider setting this level as the Start-Up Map, ensuring it loads automatically each time the project is opened. We can achieve this by following the steps outlined here:

1. Go into the Project Settings (Figure 2-23) from Step 1 to select Edit and Step 2 to look into Project Settings.

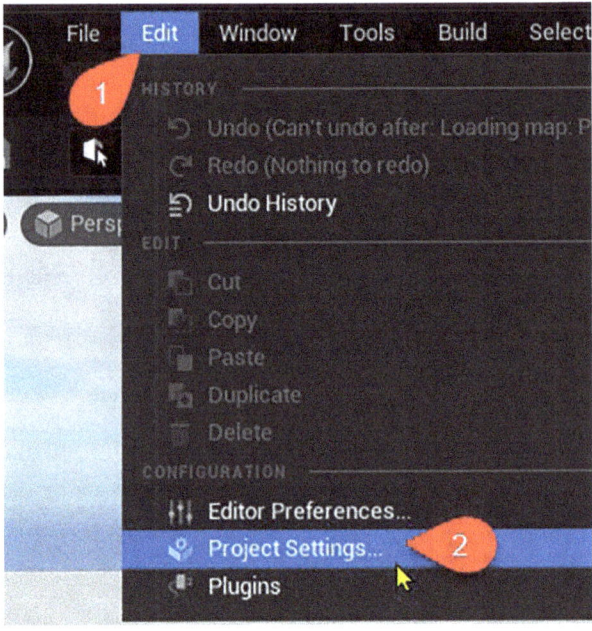

Figure 2-23. *Look into the Project Settings*

2. We can set our Editor Start-Up Map (Figure 2-24).

    a. Step 1: Select the Maps and Modes option from the Left selection options.

    b. Step 2: Select the Editor Start-Up Map to pull out the list of available levels in the project.

    c. Step 3: Select the recently saved level map.

44

CHAPTER 2 FROM HEIGHTMAP TO LARGE OPEN-WORLD LANDSCAPE

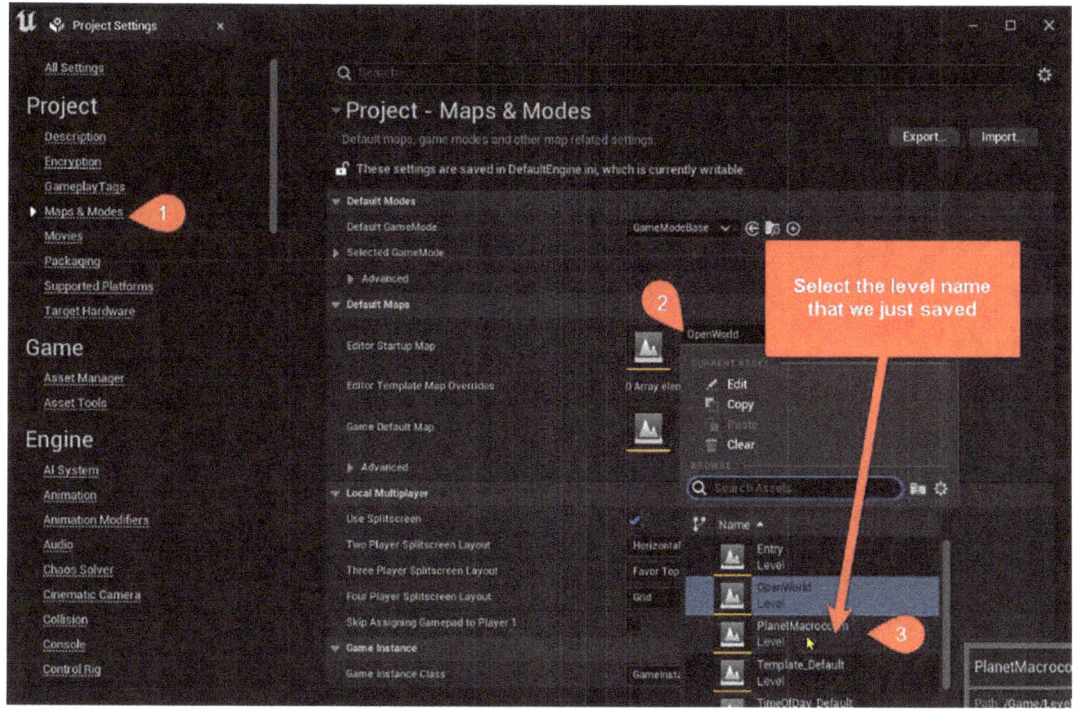

Figure 2-24. *Set the Editor Start-Up Map*

3. At the same time, we will also select Game Default Map (Figure 2-25) to our current saved level. This ensures that the map will be loaded as the default when we restart the project and when we deploy to play the game.

CHAPTER 2 FROM HEIGHTMAP TO LARGE OPEN-WORLD LANDSCAPE

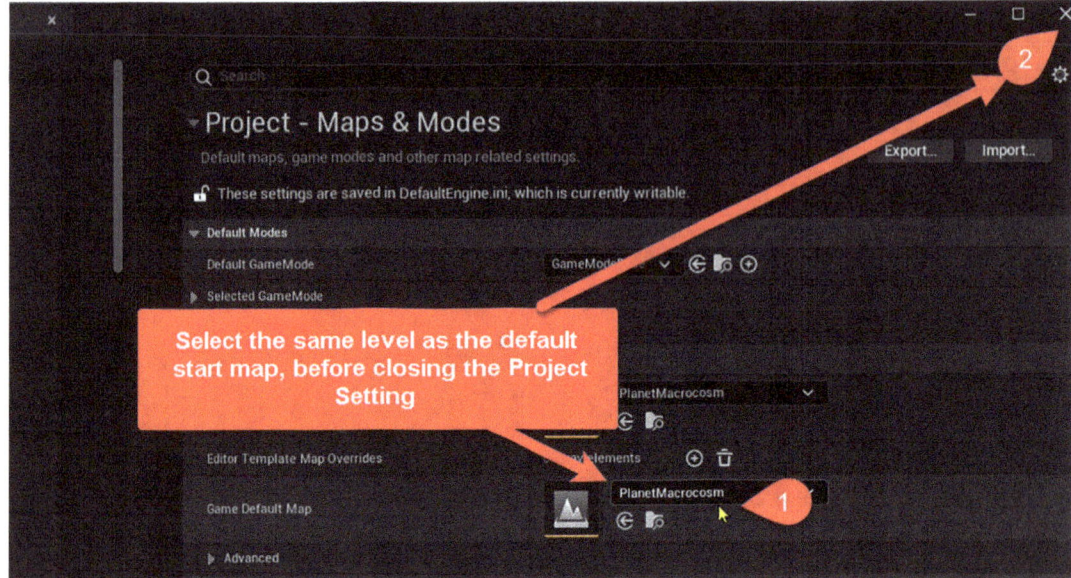

Figure 2-25. *Set the Game Default Map to start*

The Conversion

The conversion to World Partition can be found under the Tool menu, with the "Convert Level" option (Figure 2-26). We will then select the currently saved level that contains the landscape for the conversion (Figure 2-27).

CHAPTER 2   FROM HEIGHTMAP TO LARGE OPEN-WORLD LANDSCAPE

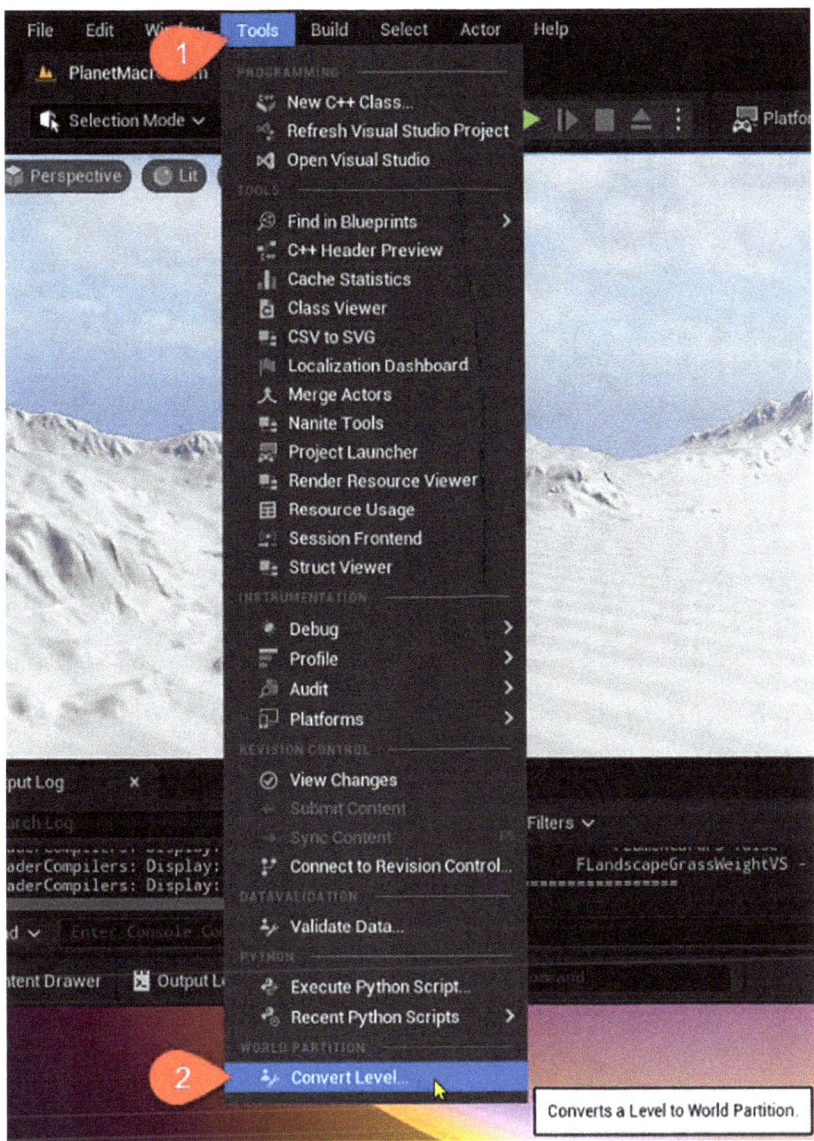

Figure 2-26. *Tool to convert the landscape to World Partitions*

CHAPTER 2 FROM HEIGHTMAP TO LARGE OPEN-WORLD LANDSCAPE

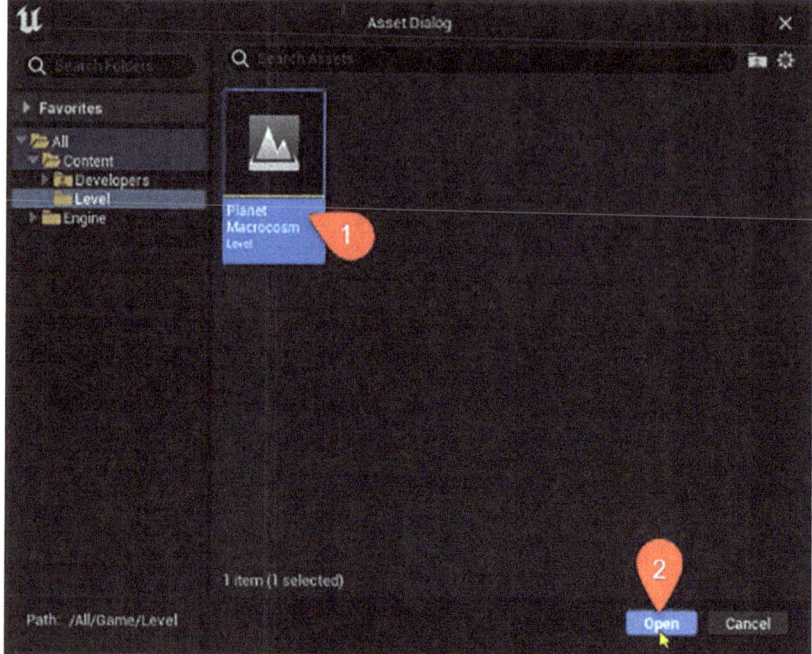

Figure 2-27. *Select the map that contains the conventional landscape*

We also have the option to convert "In Place" or not (Figure 2-28). Enabling the "In Place" option will convert the landscape to World Partition directly in the level file itself, while disabling the "In Place" option will mean that the World Partition will be saved to another level file with the postfix of _WP.

For this example, we will enable the "In Place" option for the conversion. After the conversion, we can see the converted partitions listed under the Landscape object (Figure 2-29).

If we prefer to have more options for the conversion, we can utilize the "Advanced" options within the "Convert Settings" dialog box in Figure 2-28 to provide granular control over the conversion process from traditional levels to the World Partition system in UE.

Options include setting the command to be responsible for the conversion, deciding whether to delete the original levels post conversion, generating initialization files, enabling a report-only mode for a dry run, opting for verbose output for detailed process logs, skipping the validation of stable GUIDs to potentially speed up the process, merging only sublevels instead of the entire level structure, and specifying whether to save foliage types to the content folder, likely for reuse after conversion.

Chapter 2   From Heightmap to Large Open-World Landscape

These settings offer users the ability to customize the conversion to fit their project's specific needs and to manage the transition to World Partition with greater precision.

We will notice that the entire landscape is not rendered in our scene after the conversion. This is due to the default behavior of World Partition, which disables all of the partition lists (when the project is reloaded).

In the next section, we will continue to look into the details of how to enable/disable and configure different settings of the World Partition.

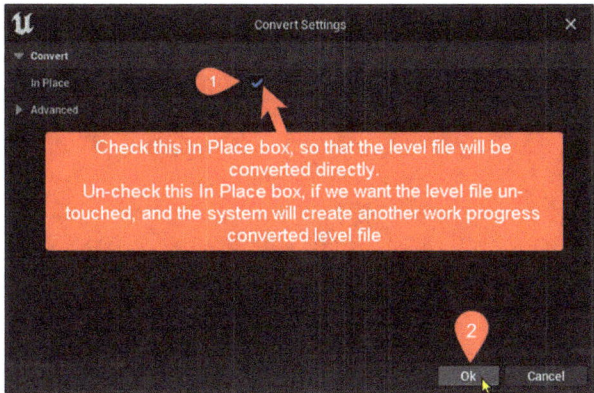

Figure 2-28. *In Place conversion or not?*

CHAPTER 2   FROM HEIGHTMAP TO LARGE OPEN-WORLD LANDSCAPE

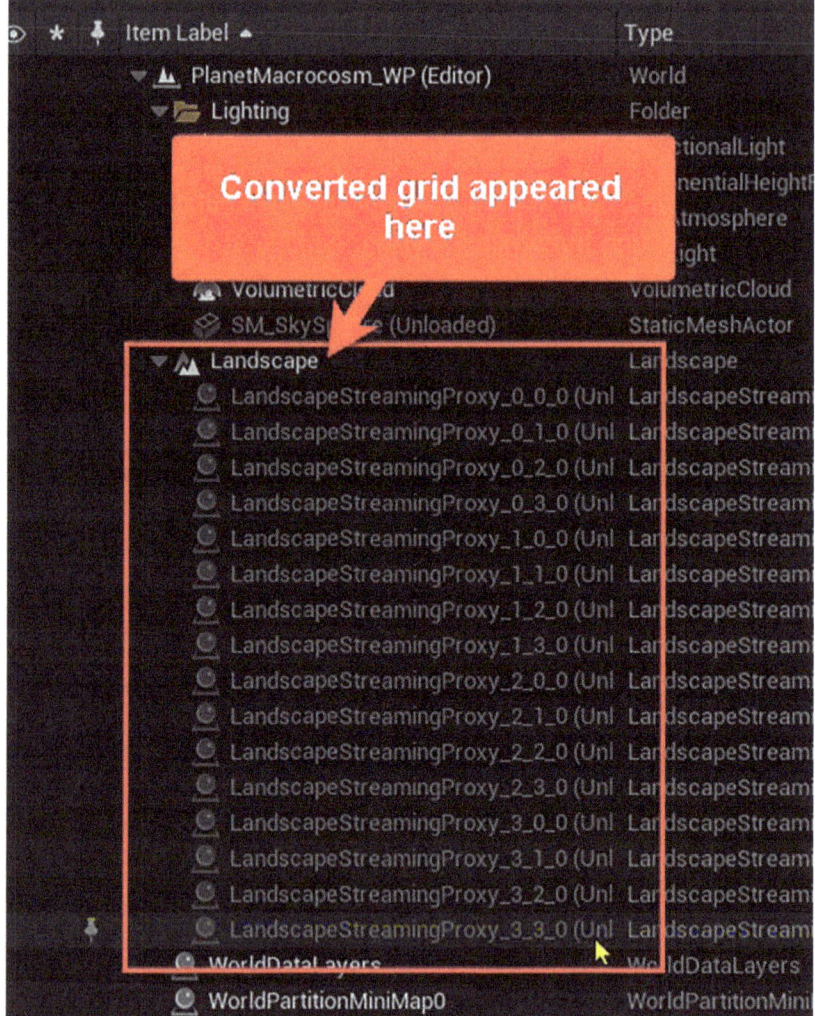

Figure 2-29. The landscape's World Partition is set

World Partition Editor

Note that the level is currently looking empty without any appearance of the landscape. This is the default behavior when the level is first loaded (as indicated at the corner of the level scene shown in Figure 2-30).

CHAPTER 2   FROM HEIGHTMAP TO LARGE OPEN-WORLD LANDSCAPE

We can use the World Partition Editor (accessible through Figure 2-31) to view the desired partition for this world landscape by following these steps:

1. Step 1: Activate the Window selection menu.

2. Step 2: Activate the World Partition submenu.

3. Step 3: Activate the World Partition Editor.

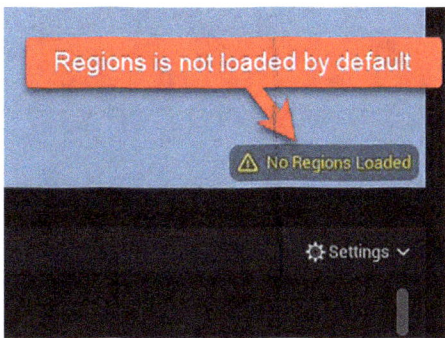

Figure 2-30. Notice the "No Regions Loaded" message at the corner of the level scene

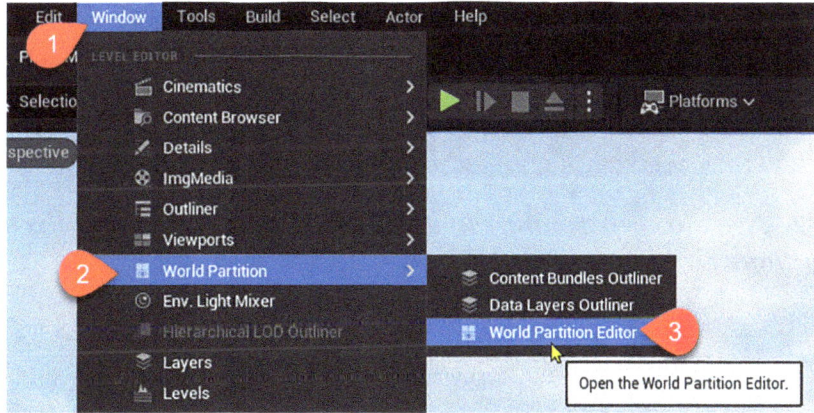

Figure 2-31. To look at the World Partition

Although we can see our camera located in the World Partition Editor (the little white triangle shape shown in Figure 2-32), we won't be able to see our partition grid for the landscape.

51

CHAPTER 2 FROM HEIGHTMAP TO LARGE OPEN-WORLD LANDSCAPE

It is necessary for us to use the mouse wheel to keep zooming in on the grid-mapped area. The actual partitions will be shown as gray-colored cells inside the black grid cells as shown in Figure 2-33.

Figure 2-32. *Need to keep zooming in using the mouse wheel to see the actual World Partition of the landscape*

CHAPTER 2   FROM HEIGHTMAP TO LARGE OPEN-WORLD LANDSCAPE

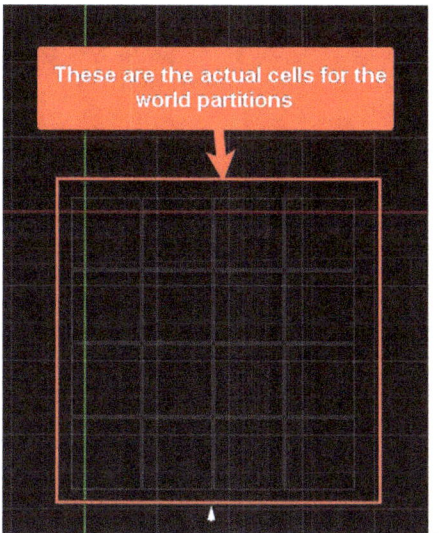

Figure 2-33. *The actual partition cells*

To enable us to view a specific part of the landscape, we can select a portion of the cell before right-clicking and choosing "Load Region From Selection" (shown in Figures 2-34 and 2-35). This concept applies similarly when we wish to view the entire landscape by selecting and loading all of the gray-colored cells (Figures 2-36 and 2-37). In summary, the instruction is as follows:

1. Open the World Partition Editor.

2. Zoom in to the view until you have squares.

3. Using the left mouse button, draw a box (as shown by the highlighted white box).

4. Right-click inside the drawn area (inside the white lines) and select the Load Region From Selection option.

53

CHAPTER 2 FROM HEIGHTMAP TO LARGE OPEN-WORLD LANDSCAPE

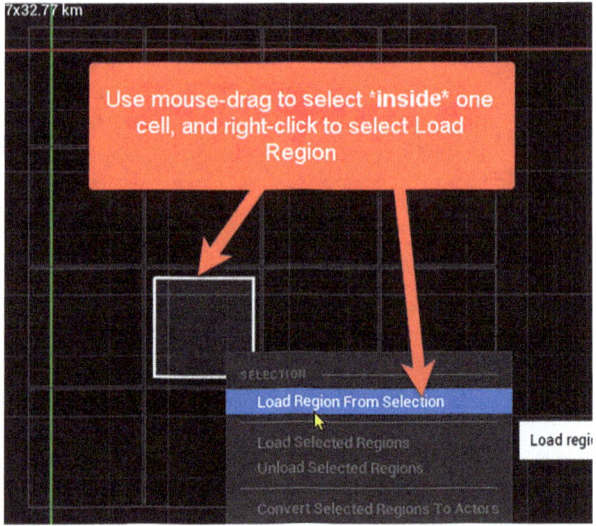

Figure 2-34. *To load the region*

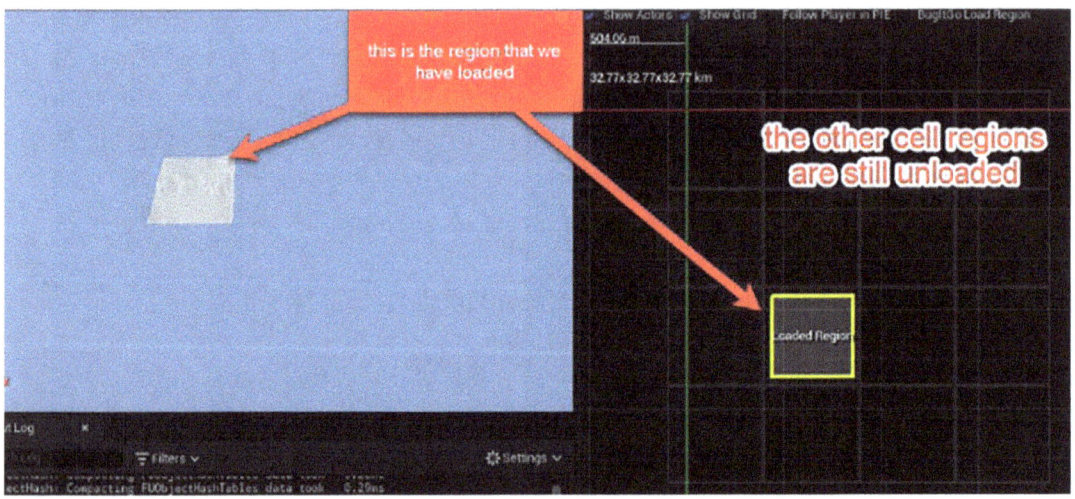

Figure 2-35. *The region is now loaded to render on the scene*

CHAPTER 2   FROM HEIGHTMAP TO LARGE OPEN-WORLD LANDSCAPE

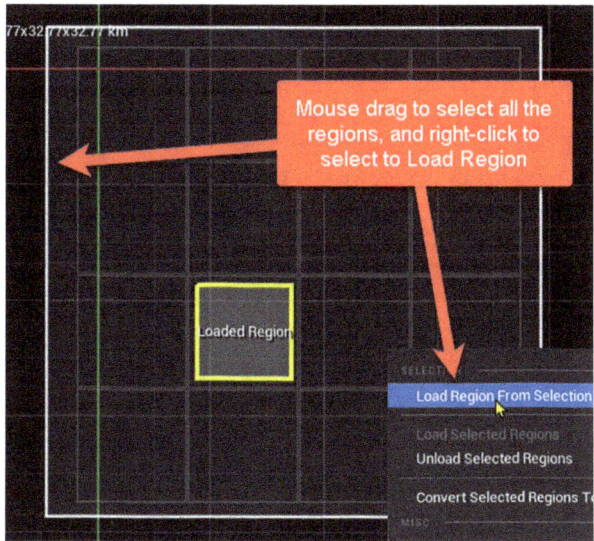

Figure 2-36. *To view the whole region (whole landscape)*

Figure 2-37. *All the regions are now loaded*

55

CHAPTER 2   FROM HEIGHTMAP TO LARGE OPEN-WORLD LANDSCAPE

World Partition Runtime Settings

In addition to the default values and settings for the conversions, we can customize the values for the cell partition settings, which can be found under World Settings at the top-right corner of the editor (shown in Figure 2-38).

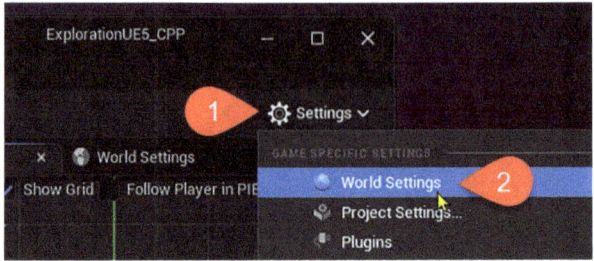

Figure 2-38. *To change some of the settings of the World Partition*

These settings will allow streaming to load or unload during the rendering process in gameplay. We can adjust the partition cell size, as shown in Figure 2-39, as well as the size of the "Loading Range," which indicates how big the range should be before the cell is loaded for rendering or unloaded from rendering (Figure 2-40).

It is entirely up to individual preference to set the value of the cell size and the loading range. It is important to strike a balance where we want the player to experience seamless transitions between different parts of the landscape (along with all the attached assets on that specific cell). At the same time, we need to consider loading a large area of landscape with all the assets on that area.

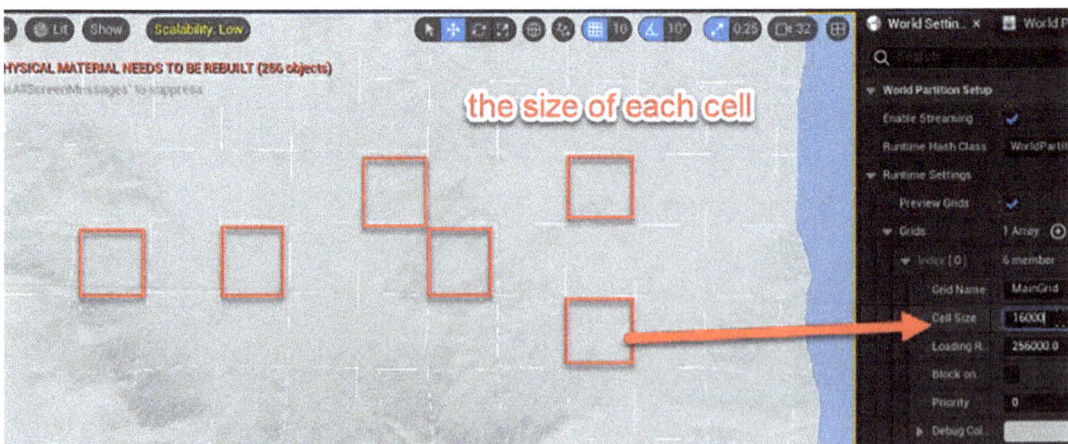

Figure 2-39. *Set the size of the cell in each partition*

56

CHAPTER 2 FROM HEIGHTMAP TO LARGE OPEN-WORLD LANDSCAPE

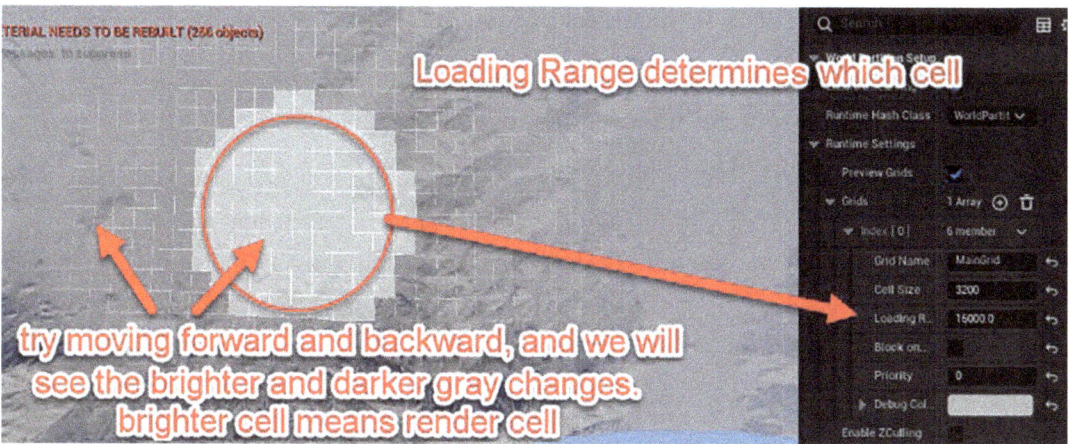

Figure 2-40. *Loading Range size*

Figure 2-41 shows different steps for us to try out when navigating the map, including setting the camera speed and using keys to move the camera around the map. Note that the camera moves efficiently if we lower down the camera to the minimum on the ground.

If needed, we can start navigating the camera viewpoint within the level after clicking the Play button, including the use of the WASD keys for movement, Q and E to raise and lower the camera view, and the mouse to turn the camera. It is also worth mentioning that pressing "Escape" will stop the navigation.

We will notice that certain partitions load when we get closer to the range and unload when we move away from the partition. All of these streaming processes to load and unload rely on the cell size and loading range.

We can also consider enabling Z-Culling, as shown in Figure 2-41, to improve rendering performance for large-scale scenes. This is to optimize rendering performance by avoiding the rendering of objects that are hidden or obscured by other objects in the scene.

CHAPTER 2   FROM HEIGHTMAP TO LARGE OPEN-WORLD LANDSCAPE

The summary of Figure 2-41 includes the following steps:

1. Camera Movement Speed: We can set the camera settings in the top toolbar to change the camera movement speed for better control during navigation.

2. World Setting Adjustment: This is to adjust the value under "World Settings" to ensure that the terrain is loaded and rendered effectively.

In UE, particularly when using the World Partition system, the cell size and loading range are essential settings that determine how the engine loads and manages game world content.

Cell Size: This setting determines the size of the grid cells into which the game world is divided. Each cell acts as a container for streaming in and out pieces of the level. A smaller cell size can result in more granular streaming, which can be beneficial for highly detailed environments where memory management is crucial. However, too small a cell size may cause overhead due to the increased number of cells being managed. Conversely, a larger cell size reduces the number of cells but can lead to less efficient streaming because larger chunks of the level are loaded at once. The optimal cell size is a balance between memory efficiency and streaming performance and largely depends on the complexity of the level and the target hardware's capabilities.

Loading Range: This setting specifies the distance from the camera at which cells are loaded. It essentially sets the radius around the player within which the game world will be fully rendered. A larger loading range can provide a more seamless experience with fewer pop-ins, as more of the world is loaded and ready to be displayed. However, a larger loading range will also require more memory and could impact performance if too much data is being loaded at once. A smaller loading range can improve performance and reduce memory usage but may result in more noticeable loading of assets in the player's field of view.

To determine the appropriate values for cell size and loading range, one must consider the level's design and the desired player experience. For open-world games where players can see far distances, a larger loading range may be

CHAPTER 2   FROM HEIGHTMAP TO LARGE OPEN-WORLD LANDSCAPE

necessary to maintain immersion. For more segmented or indoor levels, a smaller range might suffice.

Performance testing is crucial in this process. Developers should adjust these settings and then monitor the game's performance and memory usage, looking for the best balance that maintains smooth gameplay without exceeding hardware limitations. This often involves iterative testing and tweaking to find settings that work well with the game's specific content and the expected hardware specifications of the target audience.

3. Z-Culling Optimization: Another suggestion is to enable the "Z-Culling" within the World Settings to optimize rendering performance.

4. Navigation Instructions: During the design time, we can start navigating the camera viewpoint within the level after clicking the Play button, including the use of the WASD keys for movement, Q and E to raise and lower the camera view, and the mouse to turn the camera. It is also worth mentioning that pressing "Escape" will stop the navigation.

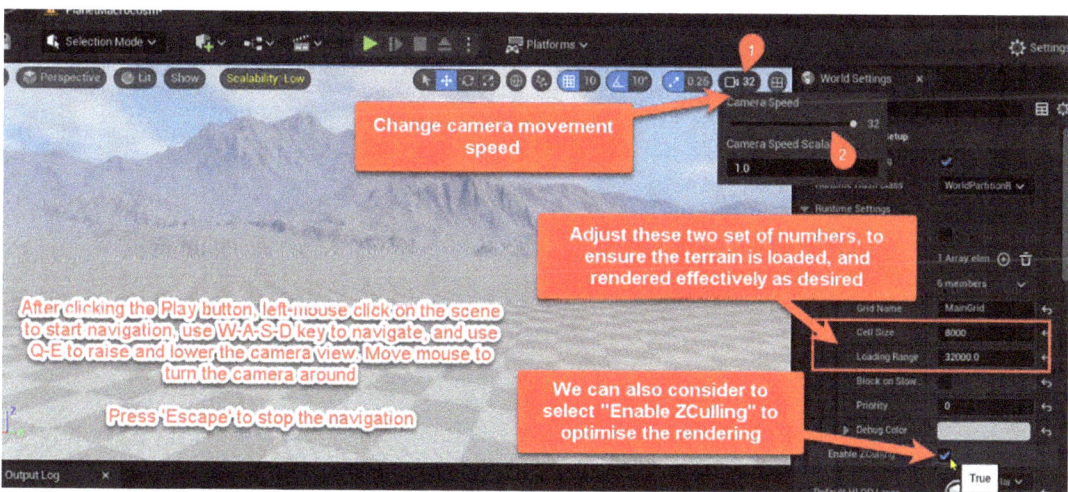

Figure 2-41. Set your own desired cell size and loading range size

59

CHAPTER 2   FROM HEIGHTMAP TO LARGE OPEN-WORLD LANDSCAPE

Summary

The chapter delved into the process of developing expansive open-world landscapes using UE5 and heightmaps. Despite the potential time-consuming nature of landscape creation, UE5 provides a robust toolset for efficient landscape management. The focus lies in importing heightmaps from external sources to save time when generating large areas while still allowing the utilization of existing landscape mode tools for sculpting or flattening the imported landscape.

The chapter guided us through the creation of a landscape from a heightmap, with adjustments to settings and the implementation of World Partition for optimal handling of vast open-world environments. World Partition intelligently divides the game world into smaller cells, dynamically loading necessary parts based on the player's location and view to enhance performance. The chapter also covered the World Partition Editor and offered insights into customizing streaming and rendering behavior during gameplay.

In the following chapter, the focus will shift toward enhancing the landscape's visual appeal by incorporating automated blended materials that can be painted onto the landscape.

CHAPTER 3

Auto-Blend Landscape Materials

In this chapter, we will explore a method for painting landscapes using auto-blend materials that rely on height and slope data. UE introduces groundbreaking advancements in landscape creation and material blending, making it a revolutionary tool for game developers and virtual environment creators alike.

Creating vast and detailed terrains manually through hand-painting can be a time-consuming and labor-intensive process. Artists are required to meticulously blend materials by hand, especially for large and complex landscapes. This method can lead to inconsistency in texture application, potentially creating visible seams or harsh transitions between different materials.

In contrast, the auto-blend material technique offers a more efficient solution. By taking advantage of height and slope data, it allows for seamless blending of different textures, such as grass, rocks, snow, and sand, across the landscape. This not only simplifies the workflow but also enhances the visual appeal of the virtual worlds, providing players with an immersive and breathtaking experience.

In conclusion, UE's auto-blend materials based on height and slope are a game-changer for landscape painting, elevating the art of virtual world creation to new heights. As we delve deeper into this chapter, we will unravel the secrets of leveraging these powerful features to craft captivating and realistic landscapes, opening up boundless opportunities for creativity and storytelling in the world of game development and beyond.

CHAPTER 3   AUTO-BLEND LANDSCAPE MATERIALS

Material Setup

To set up new materials in UE, follow the steps shown in Figure 3-1:

1. Open the Content Browser within our project.

2. Right-click in the empty space within the Content Browser to open the context menu.

3. Hover over "Create Basic Asset" to expand the submenu.

4. Click "Material" from the expanded menu options.

5. A new material asset will be created. We can rename it as Autoblend_Height_MAT.

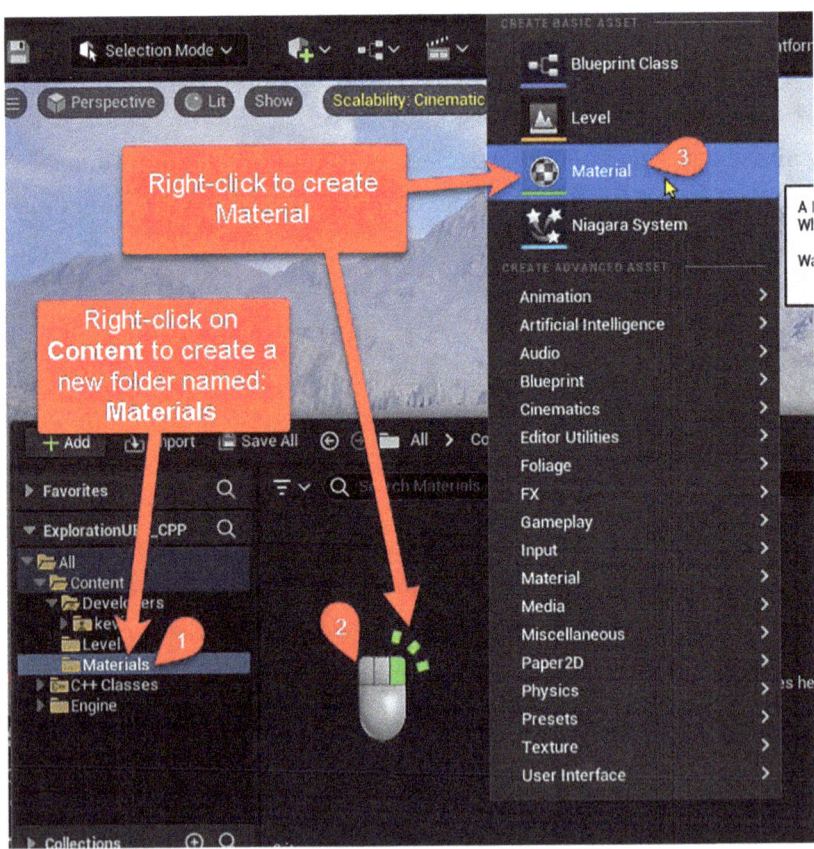

Figure 3-1. Create a new material and store it in a new folder named "Materials"

CHAPTER 3   AUTO-BLEND LANDSCAPE MATERIALS

Once we have created our base material, we will generate a Material Instance from it by following the steps demonstrated in Figure 3-2:

1. Once we have our base material, locate it within the Content Browser.

2. Right-click the material asset we want to create an instance of.

3. From the context menu that appears, select "Create Material Instance."

4. A new material instance will be created as a child of the base material. We can rename this instance (in this case, we named it `Autoblend_Height_MAT_Inst`).

5. Double-click the material instance to open it and adjust its parameters as needed.

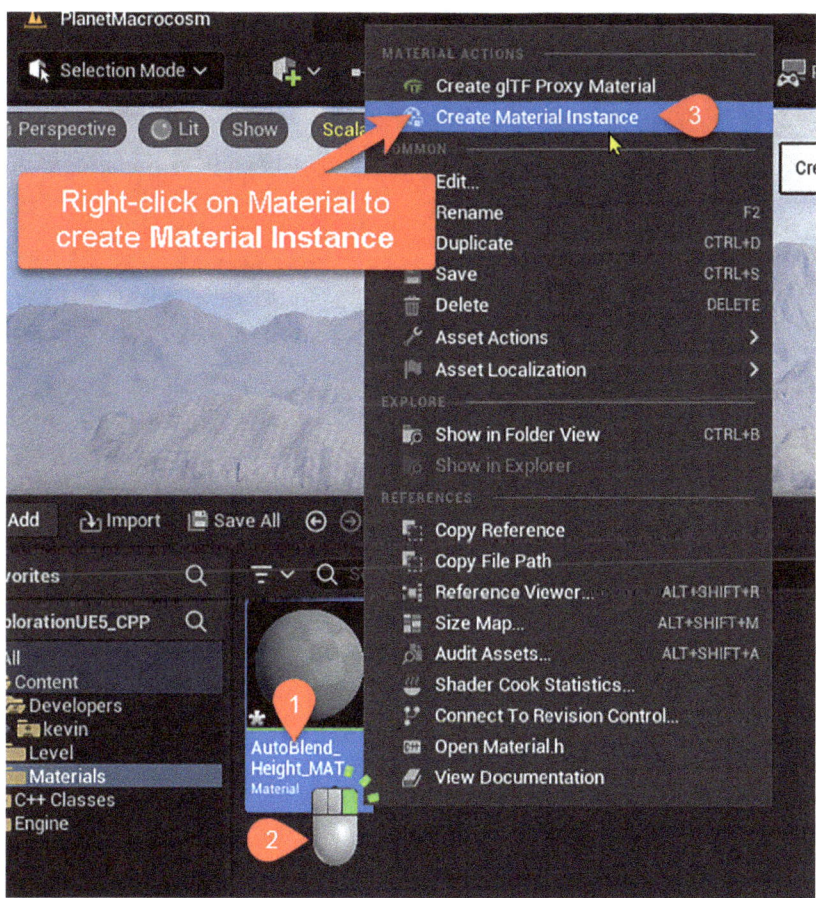

Figure 3-2. *Create a material instance based on the newly created material*

CHAPTER 3   AUTO-BLEND LANDSCAPE MATERIALS

The material instance inherits all the properties of the parent material, but allows us to change certain parameters (such as colors, textures, and other parameters) without affecting the parent material. This is useful for creating variations of the material for different objects in our game while keeping the underlying shader instructions consistent and optimized. This allows for an efficient workflow when managing multiple material variations across different game assets.

Attach Material to Landscape

We are now going to replace the material of the landscape with our newly created material instance with default material parent settings (Figure 3-3).

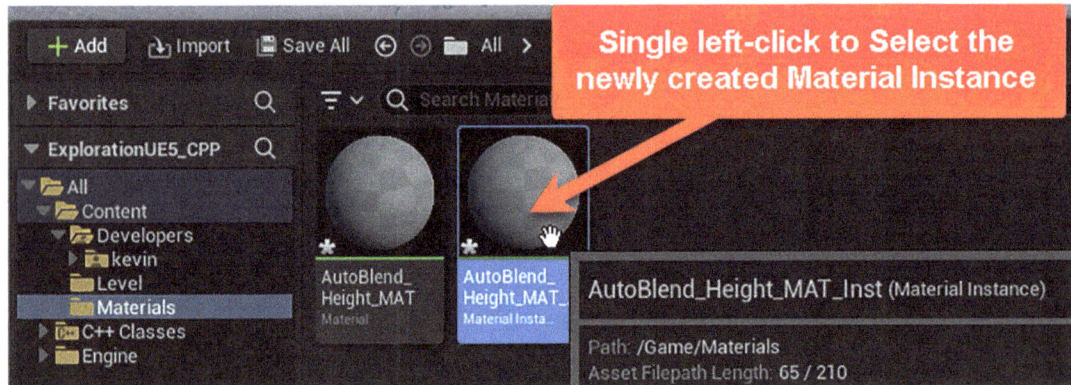

Figure 3-3. *Single left-click to select the material instance asset*

Since the landscape is now being converted into an Open World map with partitions, we need to ensure that we attach this new material instance to the landscape and its children partitions by selecting the landscape and all the children (Figure 3-4) as the following steps indicate:

1. In the World Outliner within the UE editor, locate the section where our landscape partitions are listed. These are typically named with a prefix like `LandscapeStreamingProxy`.

2. To select all the landscape partitions, click the first one in the list.

3. Hold down the `Shift` key on your keyboard.

CHAPTER 3   AUTO-BLEND LANDSCAPE MATERIALS

4. While holding Shift, click the last partition in the list to select all partitions between the first and the last.

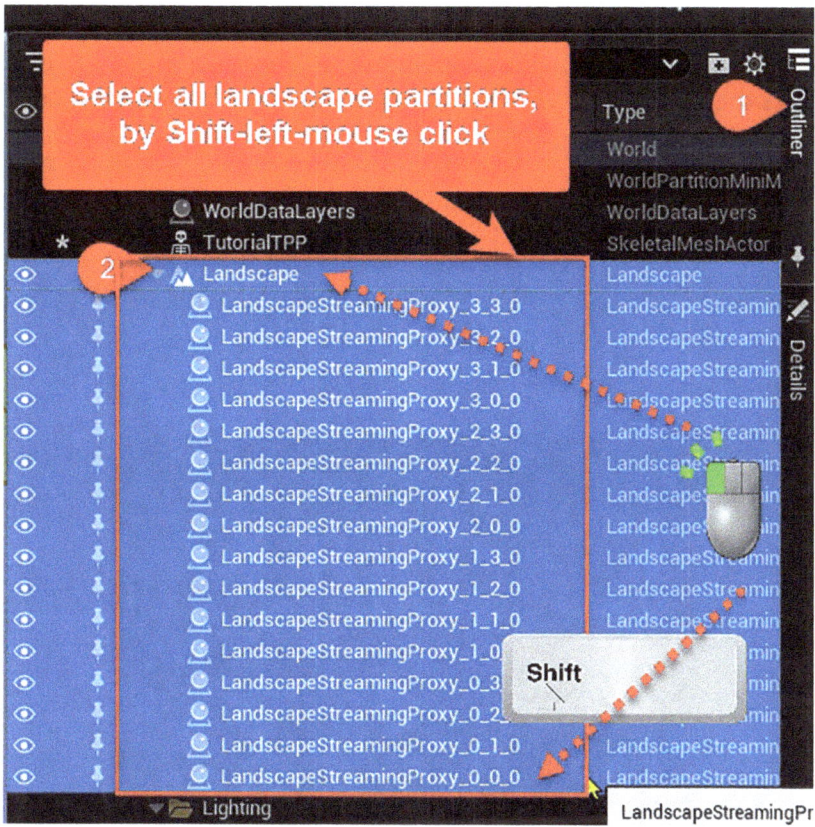

Figure 3-4. *Be sure to select the landscape and all of its children partitions*

And we can attach the materials to all of them (Figure 3-5) as indicated in the following steps:

1. With all the desired landscape partitions selected, go to the Details panel on the right-hand side of the editor.

2. Verify that all partitions are selected. We should see the number of objects selected indicated at the top of the Details panel.

3. In the Landscape section of the Details panel, find the setting for Landscape Material. It might display "None" if no material is currently assigned.

CHAPTER 3   AUTO-BLEND LANDSCAPE MATERIALS

Note The "Details" tab can be located differently due to the layout of your settings.

4. Click the circle with a dot inside it next to the Landscape Material field to open the Content Browser.

5. In the Content Browser, navigate to and select the material we wish to apply to the landscape partitions.

6. Once selected, click the "Use Selected Asset from Content Browser" to assign the material to all selected landscape partitions.

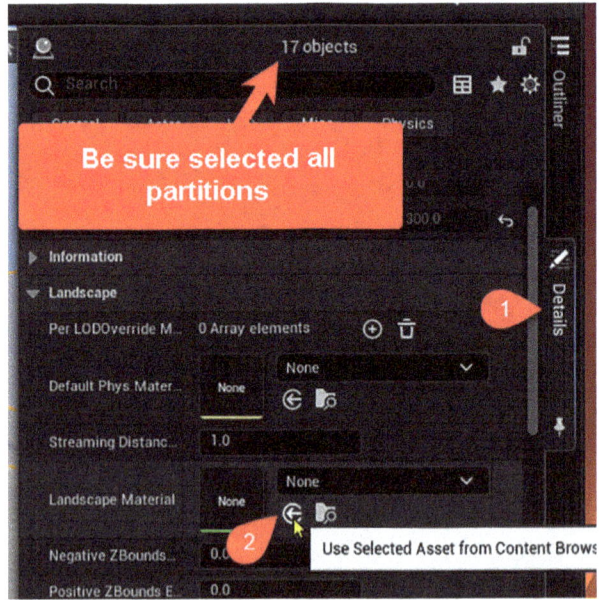

Figure 3-5. *To attach from the material instance we had selected earlier*

The landscape should now be updated with the new material we have selected, and we should see the changes reflected in the editor's viewport as shown in Figure 3-6.

CHAPTER 3 AUTO-BLEND LANDSCAPE MATERIALS

Figure 3-6. *All the partitions (including the parent landscape) should now be attached with the material instance*

Auto-Blending Height-Based Materials

After setting up the material instance, we will now proceed to actually build the material itself. To do this, we need to open up the material editor for this specific material, as shown in Figure 3-7. In the editor, we can see an empty scene example of the final material node, which is the same as the material itself.

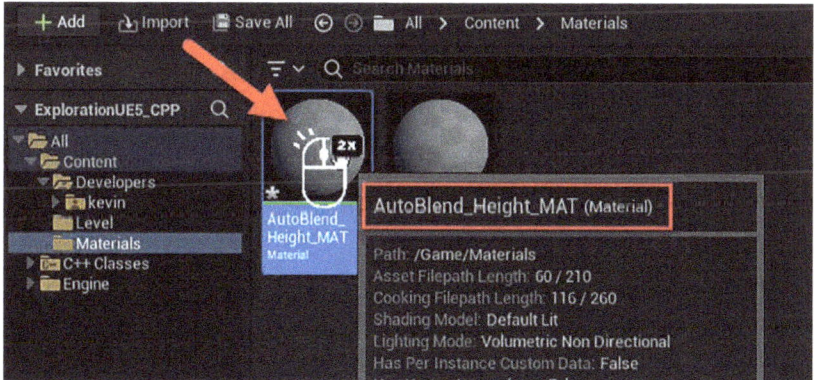

Figure 3-7. *Double-click to open up the material editor for this AutoBlend_Height_MAT*

67

CHAPTER 3   AUTO-BLEND LANDSCAPE MATERIALS

Important Nodes to Look Into

Before delving into the complete design of the auto-blend materials, it is important for us to examine some of the nodes that we are going to use in this design. This examination will enable us to gain a deeper understanding, and it will also provide us with the necessary skills to further manipulate the showcased design, based on what we have learned.

This session aims to nurture knowledge and understanding, focusing on the exploration and utilization of nodes in the material editor.

While the theory and concepts presented offer valuable insights, the readers are encouraged to use this as a practical exercise rather than merely an instructional guide.

By actively experimenting with the nodes in the material editor, you can not only fortify your comprehension of the subject matter but also gain hands-on experience, thus leading to a more in-depth understanding.

Remember, the goal is not just to learn, but to apply, experiment, and discover.

Named Reroute Declaration Node

In UE5, a valuable addition to the Material Editor is the ability to create named reroute nodes. These nodes serve an important purpose in complex material graphs by acting as identifiable reference points. They are akin to "variables" in traditional programming, allowing us to store and retrieve values without directly connecting nodes. This is particularly useful for organizing the material graph and simplifying the visual flow, making it more readable and easier to manage.

As shown in Figures 3-8 and 3-9, a named reroute node can be labeled and used to hold a value, which is determined by the type of input it receives. This stored value can then be accessed by other nodes across the material graph.

CHAPTER 3   AUTO-BLEND LANDSCAPE MATERIALS

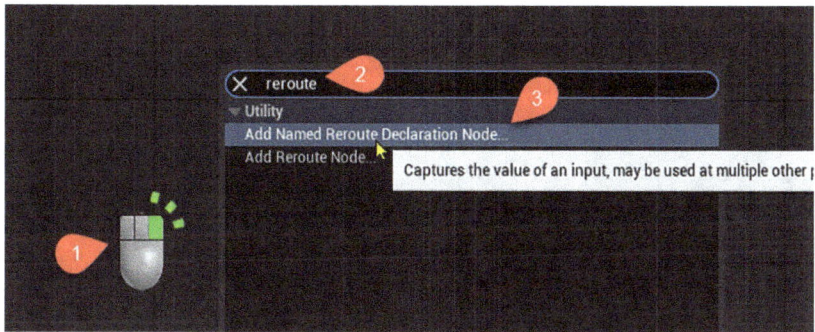

Figure 3-8.   *Create a named reroute declaration node*

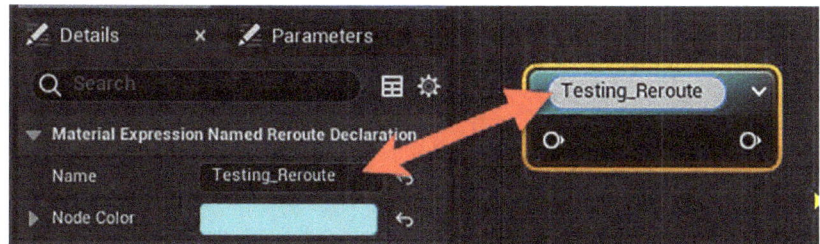

Figure 3-9.   *Name the reroute node*

This feature facilitates a more modular approach to graph construction, as shown in Figures 3-10 and 3-11, where the named reroute node "Testing_Reroute" stores a value that is later retrieved and utilized in different parts of the graph.

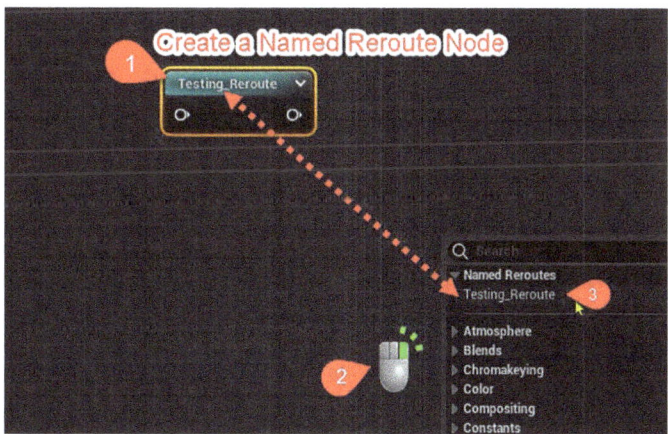

Figure 3-10.   *Set up a "continuous" named reroute node*

69

CHAPTER 3 AUTO-BLEND LANDSCAPE MATERIALS

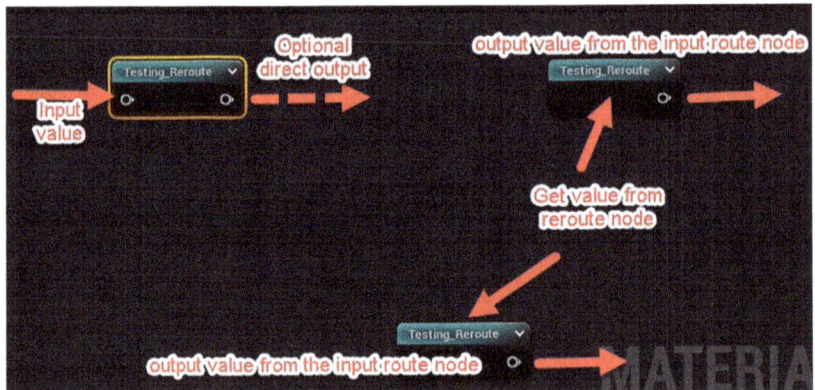

Figure 3-11. *Input and output to the named reroute node*

By employing named reroute nodes, you can create a more efficient and cleaner node setup, enhancing both the functionality and the readability of your material designs.

When working with reroute nodes in the Material Editor, it's crucial to understand the type of data they are intended to carry. A reroute node designated for a specific data type, such as a Vector3, should consistently handle the same type of data throughout our material graph.

For instance, if we establish a named reroute node labeled "ColourData" and assign it a Vector3 value representing color information, all nodes with the "ColourData" label must be utilized to route Vector3 color data only. This ensures data type consistency across our material's node network, preventing errors and maintaining the logical integrity of our material's functionality.

By adhering to this rule, we maintain clarity within our material setup, allowing us to confidently trace and modify data flows without concern for type mismatches, which could lead to unexpected results or compile errors in our material.

CHAPTER 3 AUTO-BLEND LANDSCAPE MATERIALS

Constant Node

The Constant node in UE's Material Editor is a simple yet powerful tool. It allows us to set up fixed values directly into the material network. We can easily set up a Constant node (shown in Figure 3-12) and assign its represented value, as shown in Figure 3-13.

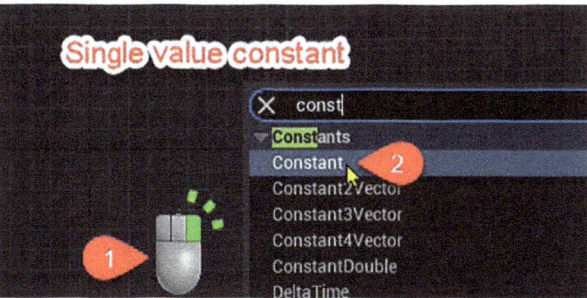

Figure 3-12. Constant node

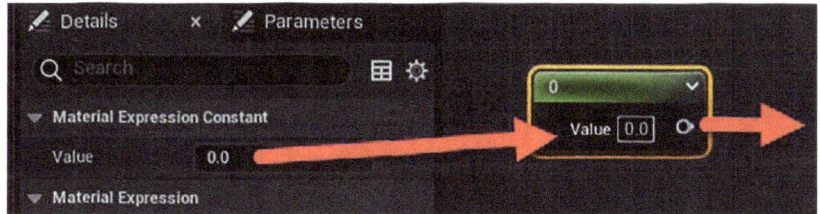

Figure 3-13. A single value set to the Constant node

Constant3Vector Node As RGB Color

To properly represent RGB as color in the material editor in UE, we need to use the Constant3Vector node (as shown in Figure 3-14). The first, second, and third values in this node represent the red (R), green (G), and blue (B) values, respectively (Figure 3-15).

It is crucial to understand that each of the RGB values is represented as a decimal number and must fall within the range of 0 to 1.

71

CHAPTER 3 AUTO-BLEND LANDSCAPE MATERIALS

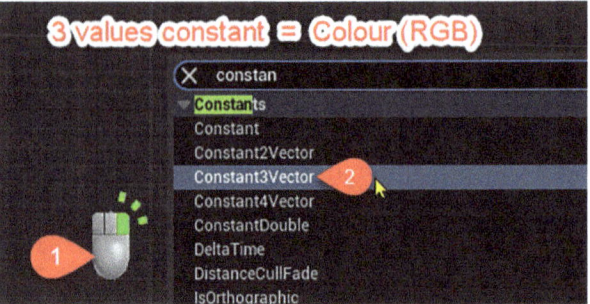

Figure 3-14. *Constant3Vector*

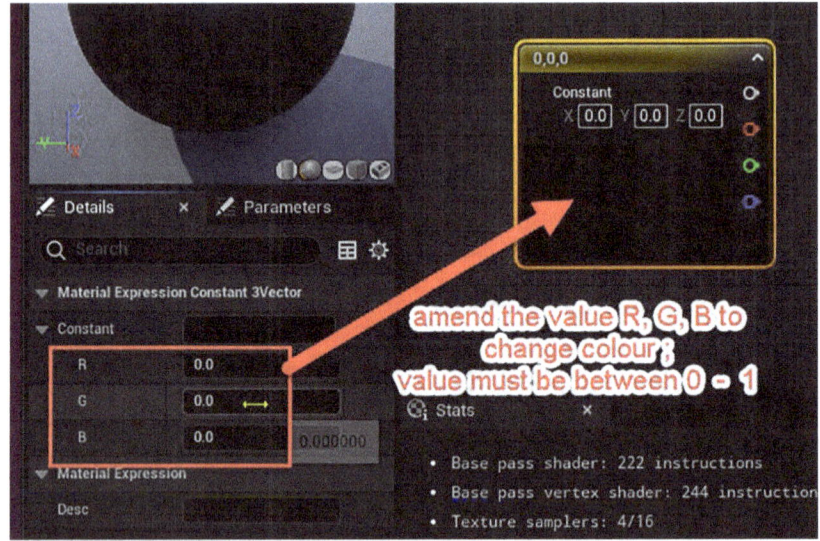

Figure 3-15. *Amend the value of RGB to see the color result*

Create a Material Parameter from a Node

We can parameterize a suitable node inside the material design in UE5. A material parameter refers to the process of converting a constant value within a material into a dynamic parameter that can be controlled and manipulated from outside the material instance.

CHAPTER 3   AUTO-BLEND LANDSCAPE MATERIALS

The integration of dynamic parameters in materials opens up a new level of flexibility. By setting up materials with this feature, we enable the assignment of values externally, right from the level editor's Details panel. This approach eliminates the need for frequent edits within the Material Editor itself, as it allows for on-the-fly adjustments within the context of the level's environment.

With this capability, the material's appearance can be tailored dynamically to fit the specific needs of different scenarios or objects, all from within the convenience of the level editor's interface. To create a parameter, we can right-click any of the nodes to which we can assign values (e.g., Constant node, Texture node) and then promote the node as a Parameter node, as shown in Figure 3-16 with the following steps to convert a constant value to a parameter within the Material Editor:

1. Click the constant node in the Material Editor to select it, and right-click the selected constant node to open the context menu.

2. In the context menu, look for the option "Convert to Parameter."

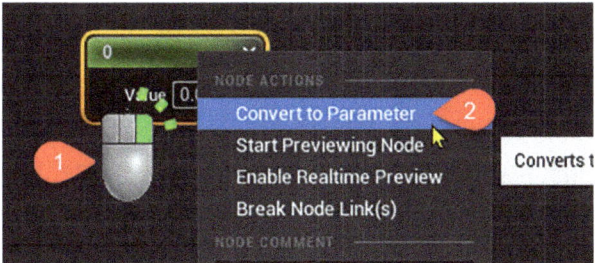

Figure 3-16. *Promote to a Parameterized node*

In the case of the Constant node, once it has been promoted to a Parameter, we can assign a default value and also set minimum and maximum values (if we want to restrict it from the outside) for the material's design (Figure 3-17).

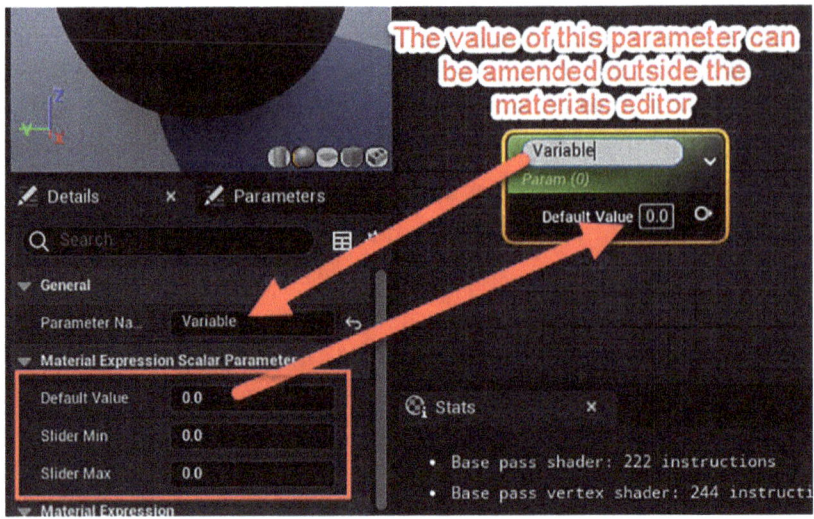

Figure 3-17. *Parameterized Constant node can be set with a default name, value, and min and max values*

The Parameterized Constant node is a powerful tool that enhances material flexibility by permitting the input of values from outside the Material Editor. This means you can adjust the parameters of your material in real time within the level editor, which is particularly useful for testing different looks without having to go back into the Material Editor for each change.

On the other hand, the named reroute node is utilized within the Material Editor to streamline and organize your material graph. By creating a named reroute node, you can store a specific value and then reference and reuse this value across multiple points in your graph. It acts as a fixed reference that simplifies the connections between nodes, especially in complex materials, but unlike the Parameterized Constant, it does not allow for external adjustments outside the Material Editor.

In summary, while both nodes contribute to the material's adaptability, the Parameterized Constant node allows for external control over values, and the named reroute node serves to internally manage and distribute values within the Material Editor.

CHAPTER 3 AUTO-BLEND LANDSCAPE MATERIALS

Material Layer Blend (Standard) Node

One of the important nodes that we are going to use in this auto-blend material is `MatLayerBlend_Standard` (shown in Figure 3-18 with the following steps):

1. In the Material Editor, right-click the node creation area to bring up the node search dialog, and start typing "`matlayerblend_stand`" into the search bar to filter the available nodes.

2. From the filtered list, find "`MatLayerBlend_Standard`". Click on it to create this node in the Material Editor workspace.

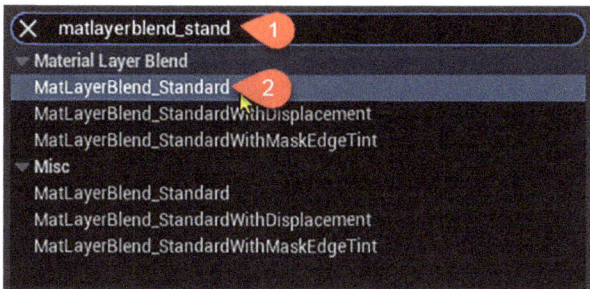

Figure 3-18. *MatLayerBlend_Standard*

The `MatLayerBlend_Standard` node is a versatile and essential tool for creating complex material blends. It enables artists and developers to combine multiple material layers and textures seamlessly, allowing for sophisticated surface effects.

This node allows us to set a material layer to represent the top part of the materials and another material layer to represent the base (bottom) part of the materials (as described in Figure 3-19). Then, we will input an Alpha value of each pixel (in our case, the Height value) to determine how much of the top layer and base layer materials should be combined as the result for each pixel in the level.

In summary, here we have the explanation of the node and its inputs:

- `Base Material`: This input is for the material that forms the bottom layer. It is the starting point of the blend and will be visible where the alpha (mask) is black or zero.

- `Top Material`: The input for the material that forms the top layer. This material layer will appear where the alpha (mask) is white or one.

CHAPTER 3   AUTO-BLEND LANDSCAPE MATERIALS

- Alpha: The alpha input determines the blend between the top and base materials. This is typically a grayscale value where black (0) will show the base material and white (1) will show the top material. Intermediate gray values will mix the two materials accordingly.

- The output of the "MatLayerBlend_Standard" node is a blended material that combines the properties of the base and top materials according to the height value specified by the alpha input.

This blended material can then be used for various effects, such as transitioning between two different terrains based on height or creating a wear-and-tear effect on surfaces.

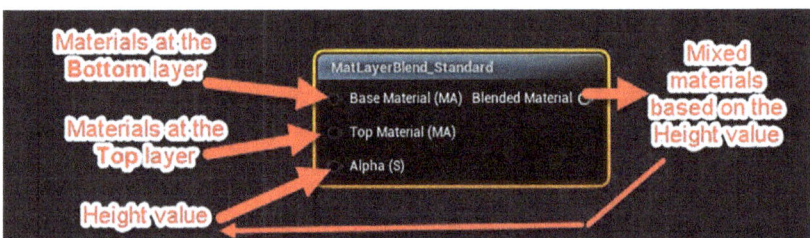

Figure 3-19.   *The setup for Material Layer Blend (Standard)*

Set Material Attributes Node

The Set Material Attributes node allows us to dynamically manipulate various material properties (Figure 3-20). With this node, we can change attributes such as color, roughness, metallic, emissive, and more during runtime or based on specific conditions in our material design.

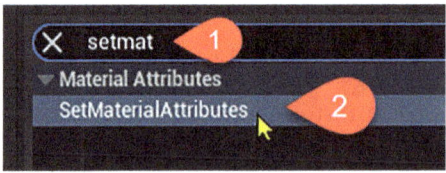

Figure 3-20.   *To create Set Material Attributes node*

By default, the node contains no attributes (properties) as shown in Figure 3-21. This flexibility enables us to set up the necessary attributes to create a single material, as shown in Figure 3-22.

CHAPTER 3   AUTO-BLEND LANDSCAPE MATERIALS

The `SetMaterialAttributes` panel is located at the bottom of the Material Editor window, just below the material preview area.

In the Material Editor, the `SetMaterialAttributes` node is initially presented without predefined attributes, allowing for a clean slate when beginning our material setup as shown in Figure 3-21.

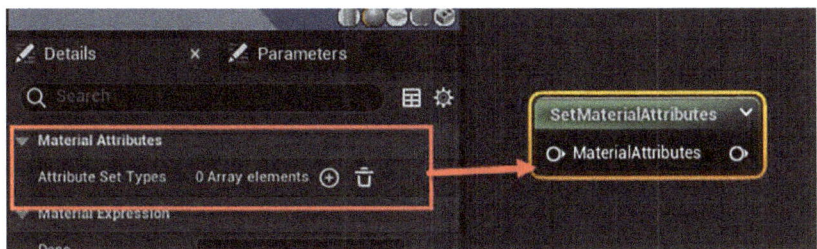

Figure 3-21. *Default Material Attributes node (with empty attribute)*

To define a material's properties using the `SetMaterialAttributes` node, we can add attributes as needed. The process of adding and configuring these attributes is detailed in Figure 3-22, which directly follows the initial description. This way, the progression from an empty attribute list to a populated one is clearly illustrated in consecutive figures.

Each attribute you add specifies a property of the material, such as `BaseColour` or `Metallic`. We can add these by clicking the plus icon (+) within the `Attribute Set Types` array and then selecting the desired attribute from the drop-down menu that appears. The step-by-step instructions described below allow us to build up the material's attributes, providing a more intuitive and cohesive reading and learning experience.

1. Array Element Indicator: This shows that there are currently zero attributes set in the "Attribute Set Types" array. You can add more attributes by clicking the plus icon (+).

2. Attribute Index: This part of the array shows the first attribute in the list, which is currently set to `BaseColour`. We can select different attributes from a drop-down menu for each index.

3. Add New Attribute: Clicking the plus icon (+) will add a new element to the "Attribute Set Types" array, allowing us to set additional material properties.

77

CHAPTER 3   AUTO-BLEND LANDSCAPE MATERIALS

4. Second Attribute Index: The second attribute in the array can be set to different attributes by selecting the drop-down icon.

5. Attribute Drop-Down Menu: Clicking the drop-down arrow next to an attribute allows you to change the attribute type. We change `Metallic` to `Specular` from the list. This indicates that the `SetMaterialAttributes` node is currently defining both the base color and the specular properties of the material.

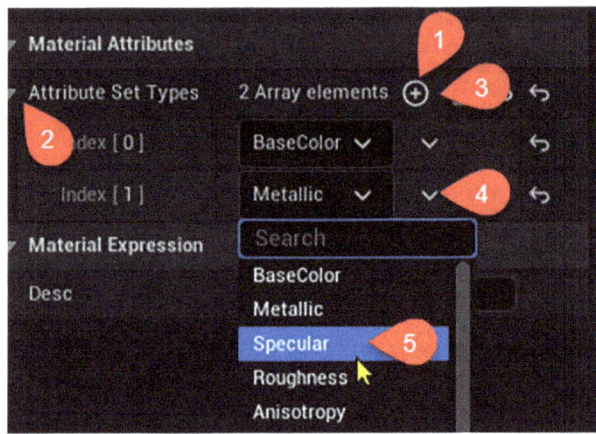

Figure 3-22. *Set up the attributes for this material node*

As shown in Figure 3-23, we have now set up a material node that contains only two attributes, `BaseColour` and `Specular`, but we can add further material attributes if desired.

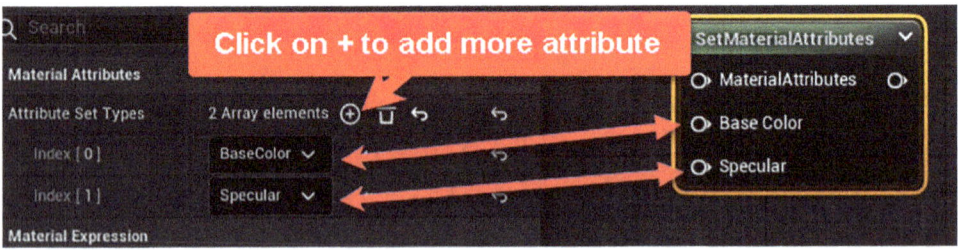

Figure 3-23. *Two attributes set up in this one material*

CHAPTER 3 AUTO-BLEND LANDSCAPE MATERIALS

The Design of Materials

In this section of our book, we delve into the design of materials within UE5, specifically focusing on constructing an auto-blend landscape material. To ensure clarity and continuity in our material design narrative, we start by building the foundational elements of our materials, explicating the reasons behind each step to foster a deeper understanding of the process.

We commence by using vibrant colors to temporarily represent the various layers in our material. These placeholders facilitate a visual understanding of how materials will interact and blend within the landscape. Currently, we will limit our design to three primary layers, although the methods we explore here can be expanded to accommodate more layers as needed.

In our design, we leverage a color gradient system to represent different elevational tiers of the terrain. Green signifies the lowest elevation, red marks the midpoint, and blue denotes the highest elevation. The transition between these points will be visually represented by a blend of the associated colors as illustrated in Figure 3-24. This figure displays our color gradient concept, illustrating the gradient transition between elevational points.

Figure 3-24. *Color gradient representation between positions*

As we progress into the more intricate aspects of material design, we now embark on the editing process of the `AutoBlend_Height_MAT`. This material, pivotal to our landscape, will dictate the visual narrative of our terrain by dynamically blending textures based on elevation.

In the upcoming sections, we will meticulously adjust the properties and parameters of this material, ensuring that each elevation level on our terrain is represented with the appropriate texture. The editing we perform here is the cornerstone of our auto-blendtechnique, and it will define the base upon which our landscape's realism is built.

79

In the event that we accidentally create a connection between nodes, we can perform the following steps to remove the connection link between nodes:

- Select the Node Connection: Hover over the wire that represents the connection you want to break. The wire will highlight when your mouse is directly over it.
- Break the Connection.
- Right-click the highlighted wire to open a context menu.
- From the context menu, select "Break Link" or "Disconnect" (the exact wording may vary based on the version of UE).

Alternatively, you can simply click and drag the wire away from either the input or output slot to break the connection without using the context menu.

Top Material

In the Material Editor, specifically within the workspace designated for constructing our layered materials, we'll assign blue (RGB: 0, 0, 255) as the Base Color within the SetMaterialAttributes node for the highest elevational areas, such as mountain peaks. This action designates blue as our top material representation.

It's important to note that this configuration is managed in a distinct area of the Material Editor, separate from the default "AutoblendHeight_MAT" material properties window that appears upon opening. This setup, routed through a node named Top_MAT, facilitates effortless adjustments and seamless blending with other material layers, addressing any potential confusion regarding the connection process in the initial stages.

Figure 3-25 outlines the setup for our Top Material using the color blue, where we can follow the steps to create and connect a constant vector to a SetMaterialAttributes node and then route it to a named material variable with the following steps:

1. Begin by navigating to the Material Editor, where we'll be working with the "Autoblend_Height_MAT" material, as mentioned previously.

CHAPTER 3   AUTO-BLEND LANDSCAPE MATERIALS

2. Create a Constant3Vector node by right-clicking an empty area in the Material Editor and selecting it from the list or by dragging out from an existing node and typing Constant into the search bar.

3. Set the Constant3Vector values to represent the color blue by setting the R (red) and G (green) values to 0 and the B (blue) value to 1.

4. Next, we need to add a "SetMaterialAttributes" node, crucial for defining the material properties. If you are following from the section where we described "SetMaterialAttributes," remember that this node initially appears empty.

   To create it, right-click in an empty area of the Material Editor associated with the "Autoblend_Height_MAT" material. Search for "SetMaterialAttributes" if it doesn't automatically show up in the menu that appears when you drag from the Constant3Vector node. Once you have the "SetMaterialAttributes" node in place, you'll need to manually add the Base Color property to the Material Attributes, as outlined previously.

5. Connect the first output pin (RGB represented as white color pin) of the Constant3Vector node to the Base Color input of the SetMaterialAttributes node.

6. Create a parameterized name by right-clicking and selecting "Create a new Material Parameter Collection."

7. Connect the output pin of the SetMaterialAttributes node to a newly created reroute named variable node, which in this case is Top_MAT (note that the color of this named reroute node can vary from what's in the screenshot).

8. For materials managed by the Top_MAT reroute named node, set it from the Base Color attribute within the SetMaterialAttributes node. This allows the Top_MAT node to be referenced throughout the material network.

During the process of constructing the material graph, we may find that certain nodes, such as `SetMaterialAttributes`, do not appear as options when dragging out from the output pin of another node, like a `Constant3Vector`. This can happen because the context-sensitive menu that appears when dragging out a connection is designed to show only the most relevant nodes based on the type of data being outputted.

If `SetMaterialAttributes` is not appearing in this context-sensitive list, do not be alarmed. This is an expected behavior in the UE Material Editor due to the specific compatibility checks performed by the editor.

To incorporate a `SetMaterialAttributes` node into our graph, we can follow these steps:

- Create the `SetMaterialAttributes` node separately by right-clicking an empty space in the Material Editor, searching for it in the list, and selecting it to place it in your graph.

- Once we have our `SetMaterialAttributes` node in the graph, we can then click and drag from the output pin of your `Constant3Vector` node (or any other appropriate node) and connect it to the `Base Color` input of the `SetMaterialAttributes` node.

- After the connection is made, ensure that the `SetMaterialAttributes` node's settings are configured as desired for our material to function correctly.

This method ensures us to have full control over the material graph and can access all the nodes you need, even if they don't automatically appear in the context menu.

This can be especially useful when working with complex materials that require precise control over attributes and parameters.

These actions will sever the connection between two nodes, allowing you to reroute or reorganize your material node graph as needed.

CHAPTER 3   AUTO-BLEND LANDSCAPE MATERIALS

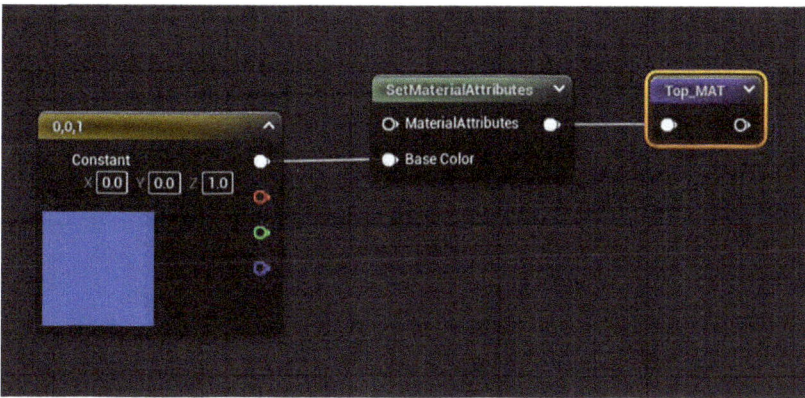

Figure 3-25. *Set up the Top Material (as blue color)*

To effectively manage our material graph, we employ the comment box feature in the Material Editor. This organizational tool helps in navigating complex networks by grouping related nodes under descriptive labels. In line with this, we've renamed the Top_MAT node to Blue_MAT, enhancing clarity and specificity within our material setup.

Figures 3-26 and 3-27 demonstrate how to select and group nodes within a comment box, labeled "Top layer materials." Figure 3-28 presents the completed group of nodes for the top material layer.

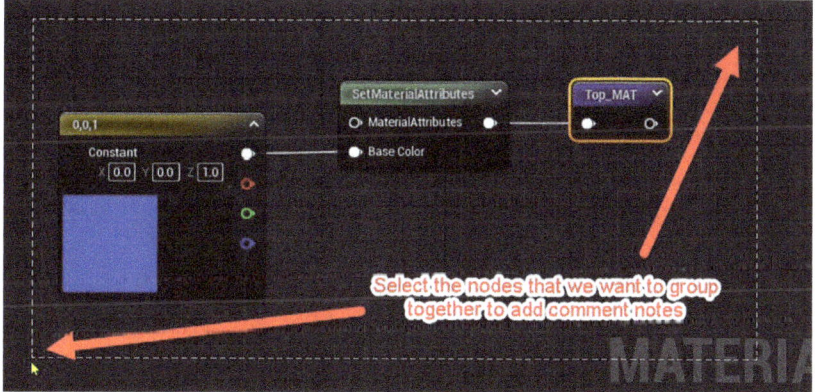

Figure 3-26. *Select the nodes that we want to comment out together*

CHAPTER 3   AUTO-BLEND LANDSCAPE MATERIALS

Figure 3-27. *Use appropriate comment for this section of the material design*

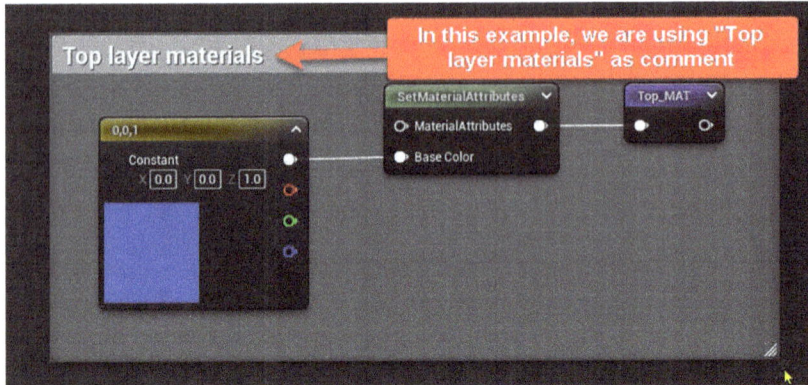

Figure 3-28. *Top layer material*

Middle Layer Material

Moving to the middle elevational zones, we select red (RGB: 255, 0, 0) as our `Base Color` and similarly route this through a node labeled `Middle_MAT`. Figure 3-29 showcases our Middle Layer Materials setup (note that the color of this named reroute node, `Middle_MAT`, can vary from what's in the screenshot).

CHAPTER 3   AUTO-BLEND LANDSCAPE MATERIALS

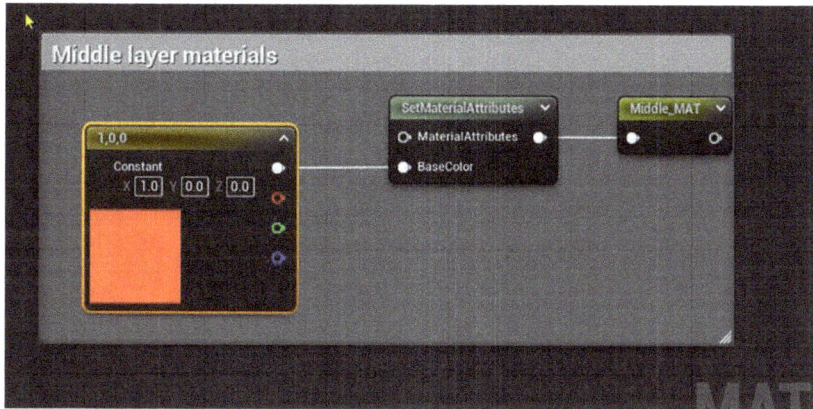

Figure 3-29. *Middle Layer Materials*

Bottom Layer Materials

For the lower elevation zones, like foothills or plains, green (RGB: 0, 255, 0) is chosen for the Base Color, directed through a node called Ground_MAT. Figure 3-30 displays the Bottom Layer Materials configuration (note that the color of this named reroute node, Ground_MAT, can vary from what's in the screenshot).

Figure 3-30. *Bottom Layer Materials*

Height Position for Individual Pixel

By employing the "World Position" node, which is identified as the node named "Absolute World Position" once placed in the node graph, and focusing on the Z-axis value, we precisely control how materials blend across the landscape. This value is routed to a node named Height_Position_Z. Figure 3-31 reveals the extraction of the Height Position value (note that the color of this named reroute node, Height_Position_Z, can vary from what's in the screenshot).

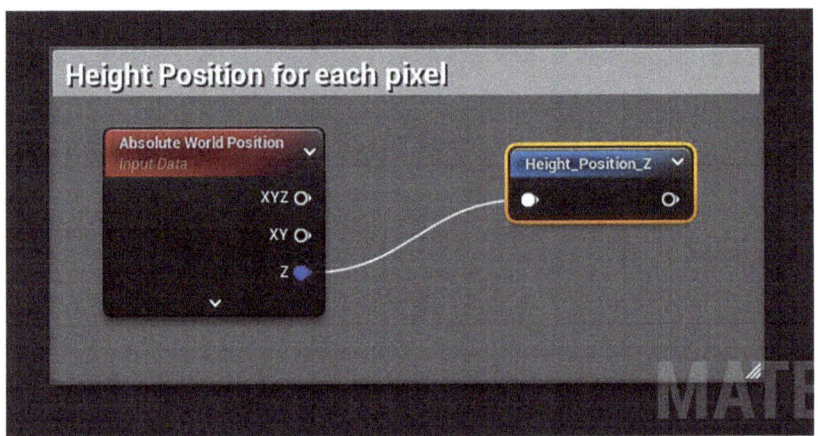

Figure 3-31. Height Position value

Bottom Half Layer Material

The MatLayerBlend_Standard node is then employed to blend the Ground_MAT and Middle_MAT based on the landscape's height information, using the "SmoothStep" node (from the **Misc** category) to define the transition between the defined elevational thresholds. Figure 3-32 illustrates the setup of the Bottom_Half_Layer_MAT using the MatLayerBlend_Standard node (note that the color of this named reroute node, Bottom_Half_Layer_MAT, can vary from what's in the screenshot).

CHAPTER 3 AUTO-BLEND LANDSCAPE MATERIALS

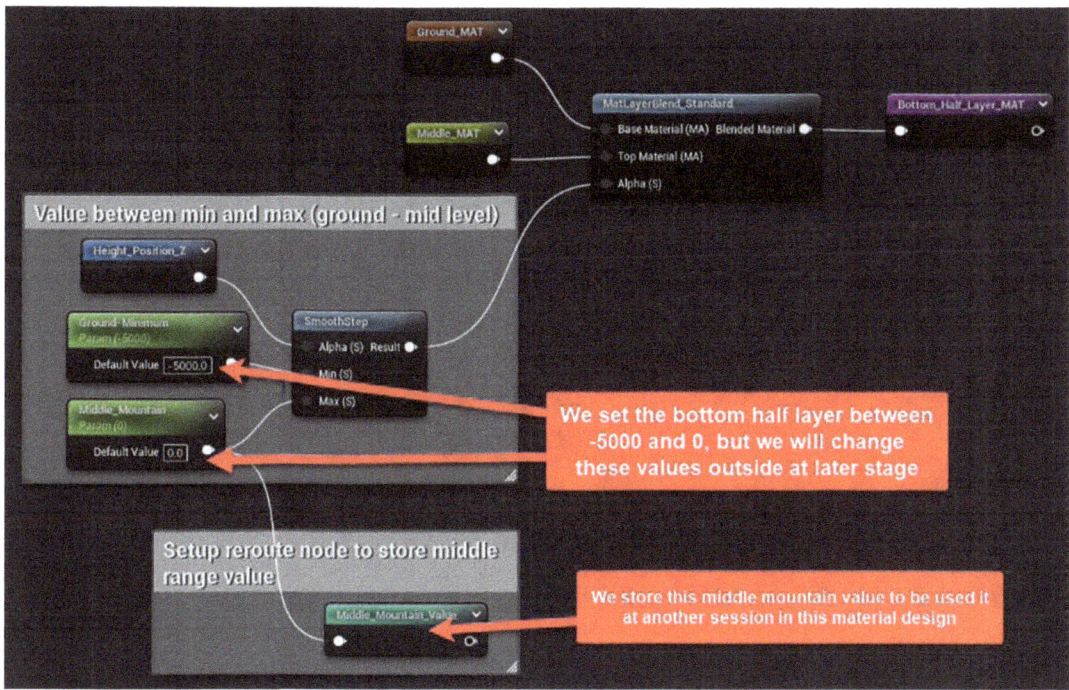

Figure 3-32. *Set up the Bottom_Half_Layer_MAT*

Note When searching for the "SmoothStep" node in the node selector, be aware that it appears in multiple categories. You'll find the correct version of this node under the "Misc" category.

Top Half Layer Material

For the upper half of our landscape material, we conduct a similar blending process, now incorporating the Bottom_Half_Layer_MAT with the Top_MAT to achieve a seamless transition across the entire elevational spectrum. Figure 3-33 shows the details of the design process for creating the Top_Half_Layer_Material.

CHAPTER 3 AUTO-BLEND LANDSCAPE MATERIALS

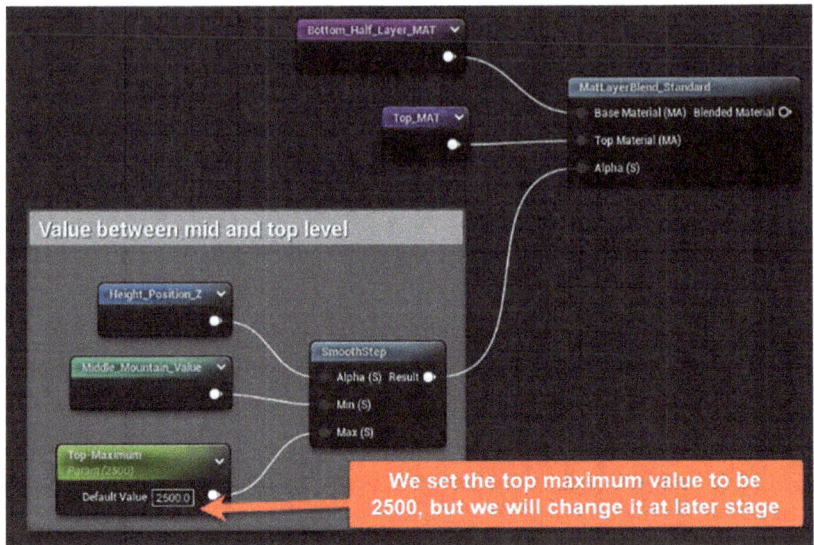

Figure 3-33. *Design of creating Top_Half_Layer_Material*

Note When searching for the "SmoothStep" node in the node selector, be aware that it appears in multiple categories. You'll find the correct version of this node under the "Misc" category.

The First Material Result

After blending our materials based on elevation, we can observe the results as a material attribute output. This output, visualized in Figure 3-34, shows the preliminary material result, ready to be applied to the landscape to achieve a terrain material that transitions between two different materials based on height. Here are the detailed steps:

1. In the Material Editor workspace, we connected a "`MatLayerBlend_Standard`" node that blends between a base material (connected to the "Base Material (MA)" input) and a top material (connected to the "Top Material (MA)" input). This blended result is then routed to the "Material Attributes" input of the final material node named "`AutoBlend_Height_MAT`".

88

CHAPTER 3   AUTO-BLEND LANDSCAPE MATERIALS

2. As Step 1 indicated, the detail property can be seen by clicking the background editor or on the `AutoBlend_Height_MAT` node itself, and the "`Details`" panel is often found on the left side of the Material Editor.

3. As indicated in Step 2, click on the "Details" tab and select the final material node ("`AutoBlend_Height_MAT`"). This instructs the material to use the blended attributes rather than a set of individual material parameters.

4. In Step 3, the "`Use Material Attributes`" checkbox is ticked as True in the "`Details`" panel under "`Material`" from the "`AutoBlend_Height_MAT`" node. This is to ensure the material attributes from the blend are utilized in the final material.

5. In Step 4, we set the output of the "`MatLayerBlend_Standard`" node to the Material Attributes of the "`AutoBlend_Height_MAT`" node.

These steps form the crux of the height-based terrain material blending, creating a material that transitions smoothly between the bottom and top layers at a certain point, which in many cases correlates with the height within the game environment. The result is a more realistic and visually appealing terrain.

CHAPTER 3   AUTO-BLEND LANDSCAPE MATERIALS

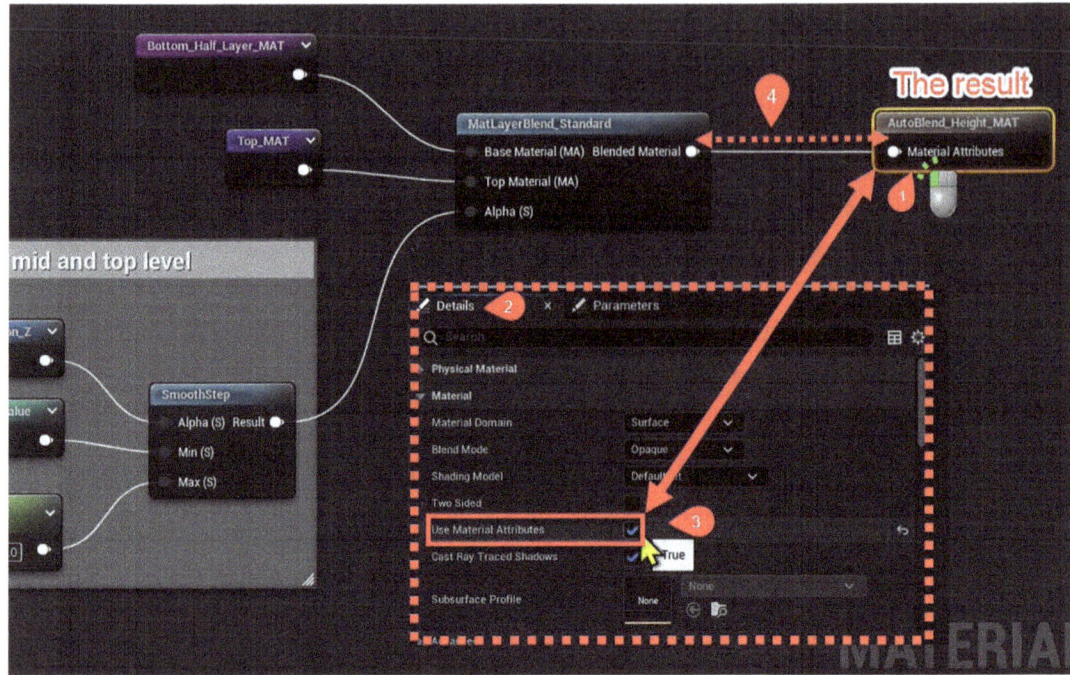

Figure 3-34. *The first final material output*

Visualize the Result

To see the fruits of our labor, we must apply the changes to the landscape. This is where we assign and adjust values for Ground_Minimum, Middle_Mountain, and Top_Maximum within the material instance editor. Figures 3-35 to 3-38 guide you through this application and parameter adjustment process.

Figure 3-35 shows the steps for applying changes to the material:

1. Ensure that you have selected the AutoBlend_Height_MAT material.

2. After making the desired modifications to the material, locate the "**Apply**" button in the toolbar at the top of the Material Editor, and click the "**Apply**" button to save the changes we have made to the material.

CHAPTER 3   AUTO-BLEND LANDSCAPE MATERIALS

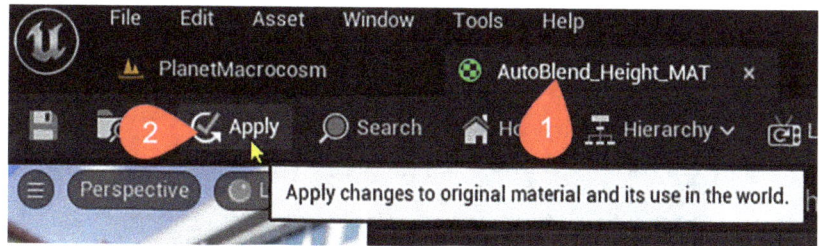

Figure 3-35. *Apply to enable the design material*

As shown in Figure 3-36, we will double-click to open up the instance of the `AutoBlend_Height_MAT`.

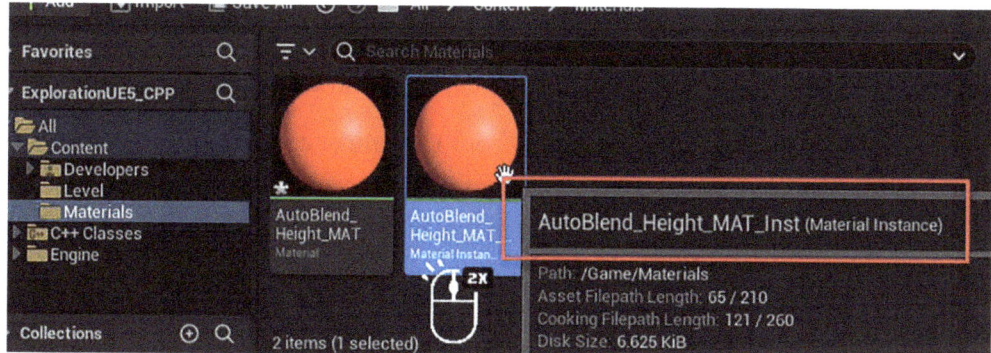

Figure 3-36. *Open up the material instance editor*

Figure 3-37 shows the details panel of the selected material instance, `AutoBlend_Height_MAT_Inst`. We will need to enable the checkboxes for the parameters to enable the corresponding parameters within the material instance. In this case, the parameters are `Ground-Minimum`, `Middle_Mountain`, and `Top-Maximum`. By checking these boxes, you can adjust these parameters directly in the Details panel without returning to the Material Editor.

91

CHAPTER 3   AUTO-BLEND LANDSCAPE MATERIALS

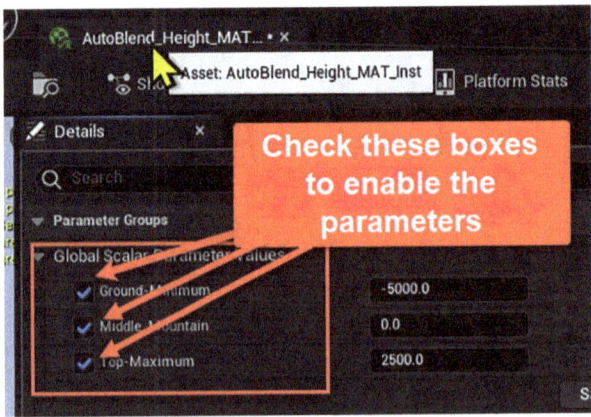

Figure 3-37. *Enable the parameters*

Figure 3-38 displays the application of these parameters in the landscape:

- The Color Gradient on the Landscape: The landscape is shown with a color gradient that visually represents different elevation levels. This gradient is the result of the material instance applied to the terrain.

- Parameter Value Indication: We can see the parameters Ground-Minimum, Middle_Mountain, and Top-Maximum, with their respective values. These are the min, mid, and max values that control the blending of the material based on the elevation of the landscape. By adjusting these values, you can fine-tune where each color appears on the terrain, effectively changing the elevation at which different materials blend into each other.

These figures show material instances can be used to dynamically alter material properties in UE and visually represent how changes to these parameters affect the material applied to a landscape. The gradient colors on the landscape serve as a visual guide to understand the elevation levels and the corresponding material blend that occurs at each level.

CHAPTER 3   AUTO-BLEND LANDSCAPE MATERIALS

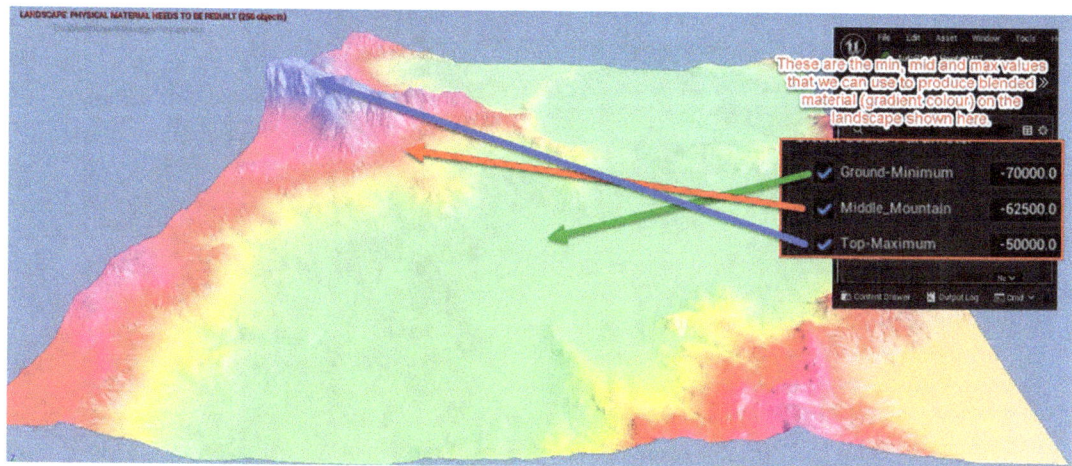

Figure 3-38. *Example of the parameter values used for this terrain*

Lastly, to accurately match our blended material to the terrain, we take into account an external heightmap. We place our character model at the lowest (as shown in Figure 3-39) and highest points (as shown in Figure 3-40) of the terrain to determine the range of Z values needed for our material parameters.

Figures 3-39 and 3-40 help in determining the Z value range for the terrain.

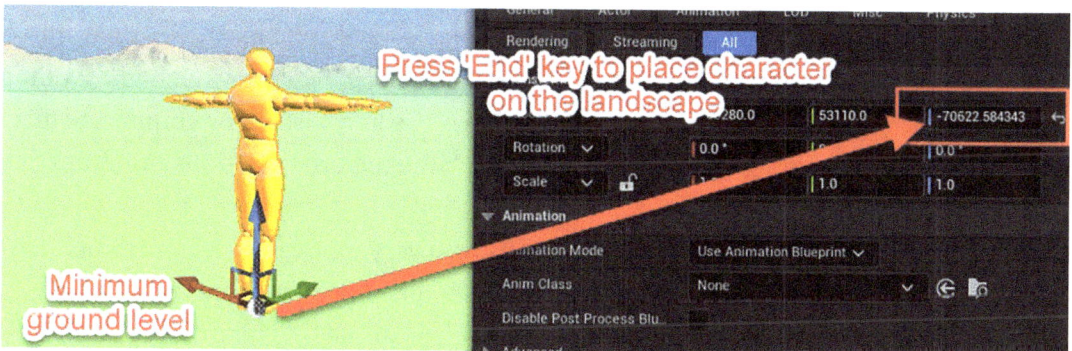

Figure 3-39. *To find out the Z value of the minimum ground value*

93

CHAPTER 3   AUTO-BLEND LANDSCAPE MATERIALS

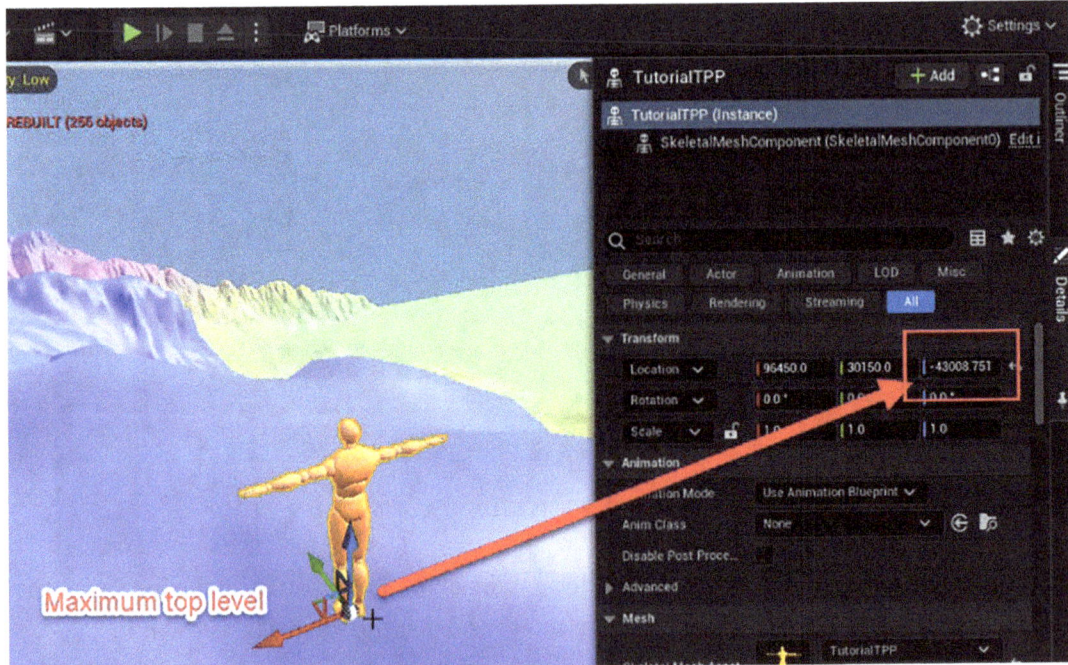

Figure 3-40. *To find out maximum Z of the terrain*

In the context of material blending, each elevation value – represented here by the Z-coordinate – determines where specific material layers will appear. For instance, the "Maximum top level" value shown in this screenshot is used to identify the highest point of our terrain at which the topmost material layer will be fully visible. By comparing this with other points, we establish a range within which the material can transition from one layer to another.

The reason for presenting different locations with varying Z values is to demonstrate how we can sample different elevation points on our terrain. This process is essential to calibrate the material's blending parameters, such as `Ground-Minimum`, `Middle_Mountain`, and `Top-Maximum`, to match the actual topography of our landscape.

By following these steps, we construct a coherent material graph that not only serves its intended function but also tells the story of its creation, allowing readers to understand the logic behind each connection and the role it plays in the larger design.

CHAPTER 3 AUTO-BLEND LANDSCAPE MATERIALS

Improvement with Slope Attribute

We have now implemented auto-blend materials for our landscape, but there are still important elements that we can improve. One such criterion is blending the slope of our landscape (as shown in Figure 3-41). Slope painting is a particularly important aspect when dealing with uneven, mountainous terrain.

By applying materials to the landscape based on the angle of the slope, game developers can create a more immersive environment. This technique adds depth and variation to the landscape, making it feel more organic and lifelike. Furthermore, it enhances gameplay mechanics, as different types of slopes can influence player movement and interaction with the environment.

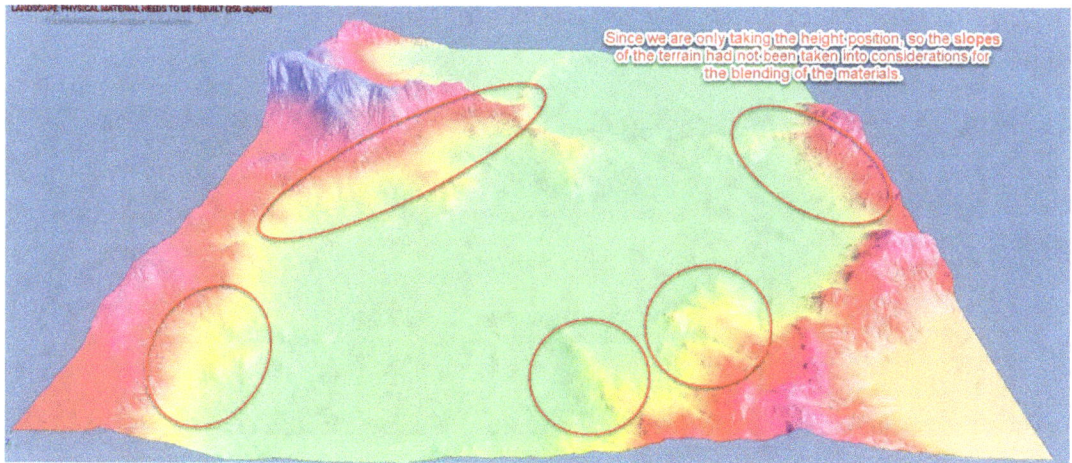

Figure 3-41. *Missing out the blending material on the slope areas*

Adding Slope Attribute in Material Design

The concept of blending materials on the slope is very similar to height-based blending, as we have seen earlier. However, to achieve this effect in UE, we need to add an additional blend layer that generates materials between the two existing materials, using the slope value as the alpha attribute.

To implement this, we should disconnect the link between the two nodes shown in Figure 3-42. Disconnecting the link can be achieved by following these steps:

1. Step 1: Right-click the pin itself.

2. Step 2: Select Break Node Link(s).

CHAPTER 3 AUTO-BLEND LANDSCAPE MATERIALS

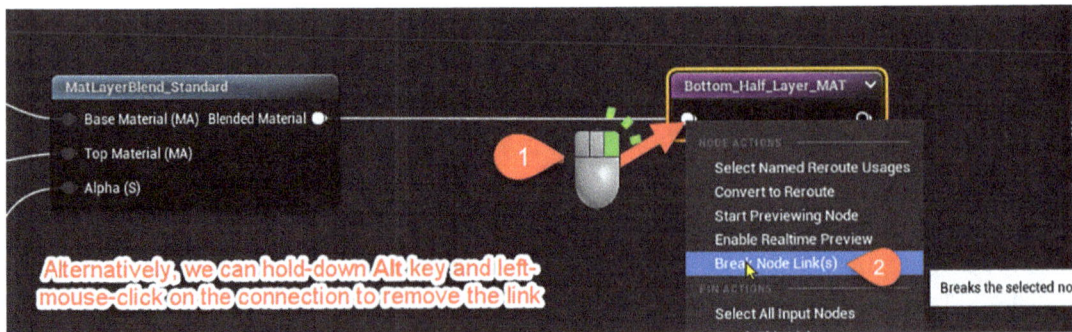

Figure 3-42. *Remove the link between MatLayerBlend_Standard and Bottom_Half_Layer_MAT*

Instead, we will introduce a new set of nodes (as shown in Figure 3-43) that take the slope angle into account.

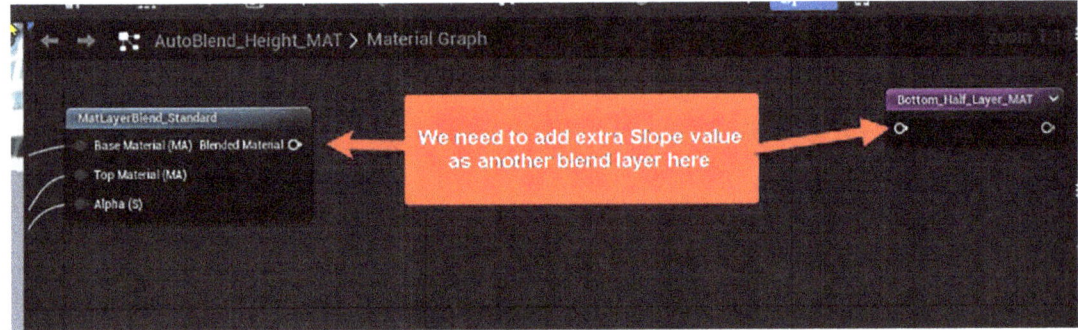

Figure 3-43. *Add extra design nodes between these two connections*

To achieve a landscape that visually responds to the slope's angle, we incorporate a `SlopeMask` node into our material graph, as shown in Figure 3-44. This node is instrumental in blending materials based on the terrain's incline – a steeper slope will result in a different material appearance compared to a gentler incline.

SlopeMask is a technique used to create material effects based on the slope or steepness of a surface. By using a SlopeMask, we can control the blending of textures or materials, making it appear smoother on flat areas and rougher on steep slopes.

CHAPTER 3   AUTO-BLEND LANDSCAPE MATERIALS

In our example, also highlighted in Figure 3-44, the `Middle_MAT` is set as the `Base Material` within the `MatLayerBlend_Standard` node. This setup ensures that areas with the least incline will render the `Middle_MAT`, while areas with the steepest slope will reveal the material that has been blended according to vertical height. The flexibility of this design allows us to substitute the `Base Material` with an alternative of our choosing, facilitating a customized slope-dependent material blend.

For refined control over how the slope affects material blending, we introduce two key parameters: `Contrast` and `FallOff`. These will directly feed into `SlopeMask` and dictate the sharpness of the material transition and the extent of the slope effect, respectively. By exposing these as parameters, they can be easily tweaked from outside the Material Editor, enabling on-the-fly adjustments to perfect the landscape's aesthetic to our exacting standards.

The `Contrast` parameter influences the overall strength of the mask, allowing us to control the intensity of the slope-based effects applied to the terrain.

On the other hand, the `FallOff` parameter determines the smoothness of the transition between sloped and flat regions in the generated mask. A higher `FallOff` value results in a more gradual and subtle blending effect, while a lower value leads to a sharper transition.

In addition to the two parameters required by the `SlopeMask` node, we also need to specify the value of the `TangentNormal` vector that we want to use for this slope calculation. In this case, we are using the up-vector (Z = 1) as the `TangentNormal` vector.

The TangentNormal within the SlopeMask function is used to calculate the gradient or slope of a surface at a given point. The determination of how steep or flat a surface is based on the given TangentNormal vector (i.e., RGB = 0, 0, 1).

Chapter 3   Auto-Blend Landscape Materials

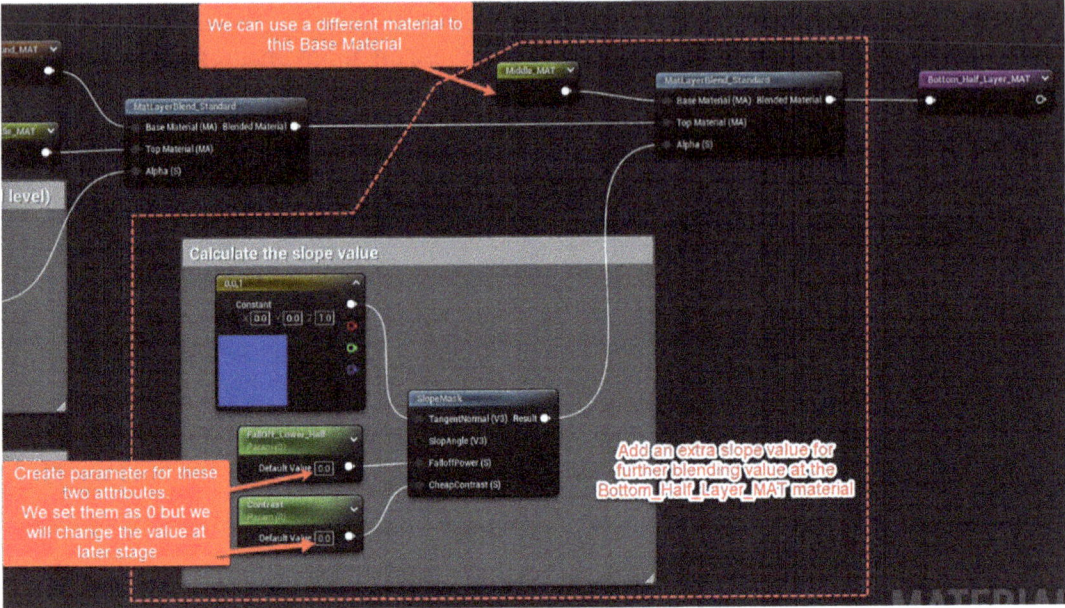

Figure 3-44. *Use SlopeMask to calculate the slope value to allow a further blend of material*

Visualize the Result with Slope Consideration

After incorporating the slope angle into the blending of materials in our landscape, we can observe a much more nuanced gradient on the mountains, particularly in the sloped areas. This results in a more realistic visualization of our landscape. As shown in Figure 3-45, we have the option to adjust the values of two parameters: Contrast and Falloff. By doing so, we can fine-tune the appearance of our landscape according to our preferences.

Be sure to click 'Apply' in the Material Editor to activate the latest changes.

CHAPTER 3   AUTO-BLEND LANDSCAPE MATERIALS

Figure 3-45. *Better blend of material at the slope area, compared to the previous result*

> **EXERCISE: FURTHER IMPROVEMENT**
>
> We have now enhanced the landscape based on height and slope, but this enhancement has only been applied to the bottom half of the landscape.
>
> As an exercise, you should now be able to improve the top half of the landscape. With your improved understanding of auto-blend materials based on height and slope angles, this task should be achievable.

Summary

In this chapter, we explored how to utilize auto-blend materials based on height and slope data in UE5. This technique enables seamless blending of different textures across the landscape, simplifying the workflow for creating expansive and detailed terrains while enhancing the visual appeal of virtual worlds.

Unlike manual hand-painting, which can be time-consuming and lead to inconsistencies in texture application, auto-blend materials elevate the art of virtual world creation. By delving deeper into this chapter, we leveraged these powerful features to craft captivating and realistic landscapes, opening up limitless opportunities for creativity and storytelling in game development and beyond.

CHAPTER 3   AUTO-BLEND LANDSCAPE MATERIALS

This chapter also explained the material setup and essential nodes in material design and showcased a step-by-step process of implementing auto-blend materials with height and slope considerations, resulting in a more immersive and lifelike environment.

In the next chapter, we will learn how to import 2D textures and 3D assets using one of the features in UE5, Quixel Bridge and Megascans. Additionally, we will explore how to implement the Nanite feature in our 3D assets to enhance the performance of our game.

CHAPTER 4

Revitalizing Visuals: Asset Import and Procedural Creation

In this chapter, we will explore the feature of importing assets using the Quixel Bridge Plugin for UE5. Quixel Bridge and Megascans play an integral role in enhancing UE5's visual experience. Quixel Bridge is a powerful asset management tool that serves as a bridge between Quixel's extensive Megascans library and game engines like UE. We will also discuss the quality of Nanite in 3D assets, which is a significant feature introduced in UE5.

As one of the main topics in this chapter, we will begin by examining how to replace the color materials (discussed in the previous chapter of our AutoBlend_Height_MAT landscape material) with the desired textures that we imported from Quixel Bridge. Additionally, we will explore a method to improve the visual appearance of texture tiling.

Finally, we will make use of the plugins available in UE5, particularly the Procedural Content Generation plugin, which allows us to create game contents for our landscape, through customized algorithms and rules rather than relying on time-consuming manual design.

Quixel Bridge and Megascans

With Quixel Bridge, we can effortlessly search, browse, and import an extensive array of photorealistic assets, including 3D models, textures, materials, and more. These assets are meticulously crafted to amplify the visual authenticity of digital environments, thereby granting them a heightened sense of realism and immersion. By furnishing an efficient means of integrating Megascans assets into the UE workflow, Quixel Bridge

CHAPTER 4   REVITALIZING VISUALS: ASSET IMPORT AND PROCEDURAL CREATION

streamlines the process of fabricating breathtaking, lifelike scenes. We can handpick, tailor, and directly import assets into our projects, preserving significant time and effort while upholding a pinnacle of visual quality. Quixel Bridge has now become an inherent component of the UE editor, conveniently accessible through the Window menu, as shown in Figure 4-1.

Before accessing Quixel Bridge and Megascans, as indicated in Figures 4-2 and 4-3, we will need to log in to the Epic server and grant the necessary permissions.

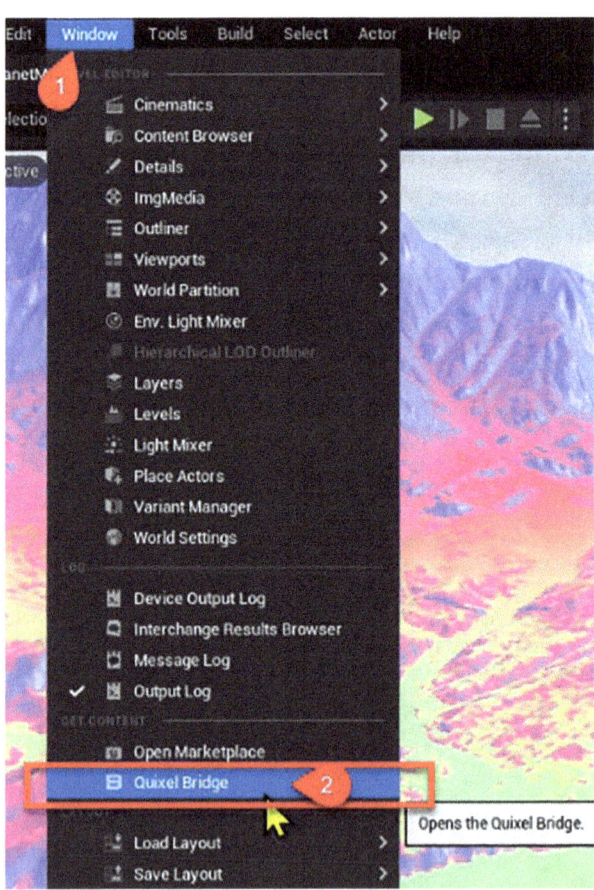

Figure 4-1. Quixel Bridge

CHAPTER 4 REVITALIZING VISUALS: ASSET IMPORT AND PROCEDURAL CREATION

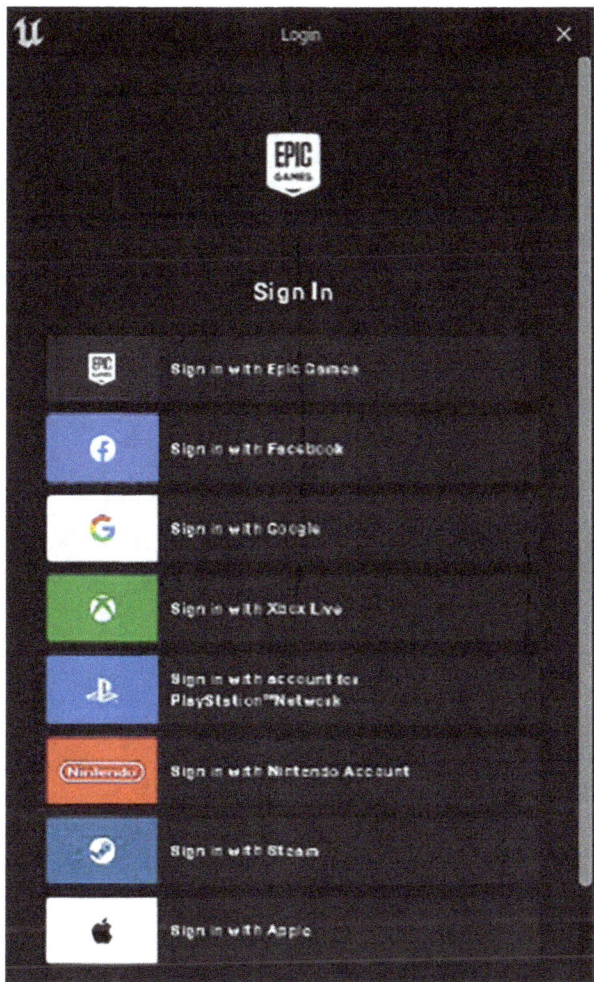

Figure 4-2. *Sign in to Epic*

CHAPTER 4 REVITALIZING VISUALS: ASSET IMPORT AND PROCEDURAL CREATION

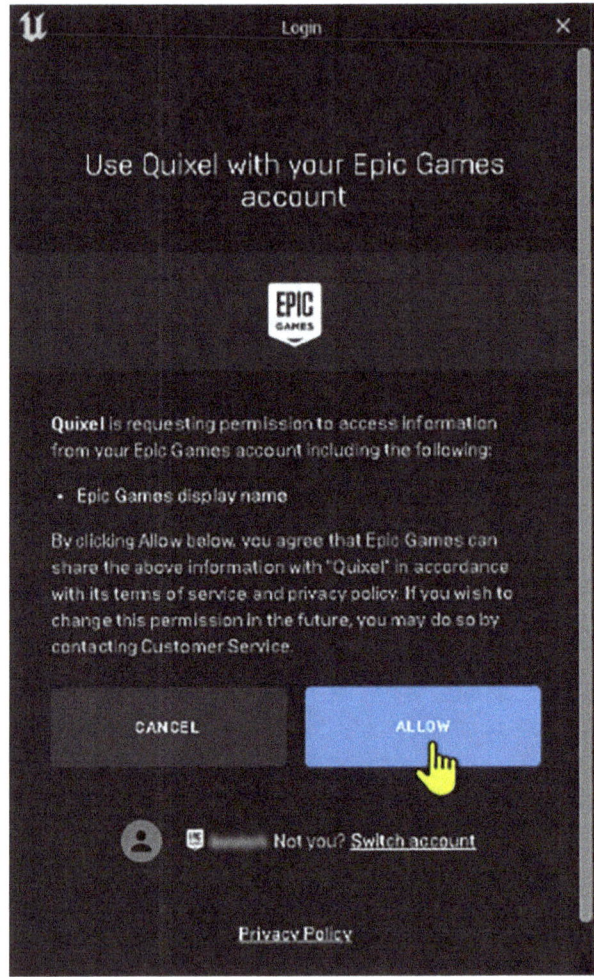

Figure 4-3. Approval of permission for Quixel

Asset ID

In order to make the most of Quixel Bridge and Megascans assets in UE5, we first need to understand the process of using their asset IDs. The procedure begins with opening the Quixel Bridge dialog. From here, we can search for the desired assets directly within the Quixel Bridge. Once we've found the assets we need, we simply copy their asset IDs. This allows for a smooth integration process, as we can then download and add these assets into our project using the copied IDs. This workflow sets clear expectations for the use of Quixel Bridge, ensuring a seamless user experience.

CHAPTER 4 REVITALIZING VISUALS: ASSET IMPORT AND PROCEDURAL CREATION

In the Quixel Bridge dialog, we can paste the given asset ID under the next section, "Surface (Texture) Assets," into the designated search field (as shown in Figure 4-4). UE5 will automatically fetch the asset's details. Confirm the asset import settings such as resolution, LODs, and materials.

Upon importing, UE5 will efficiently download and integrate the Megascans asset into our project, while Quixel Bridge helps in managing updates and variations. This streamlined process enables us to leverage the power of the Megascans library directly within UE5 using asset IDs, thereby enhancing our creative workflow.

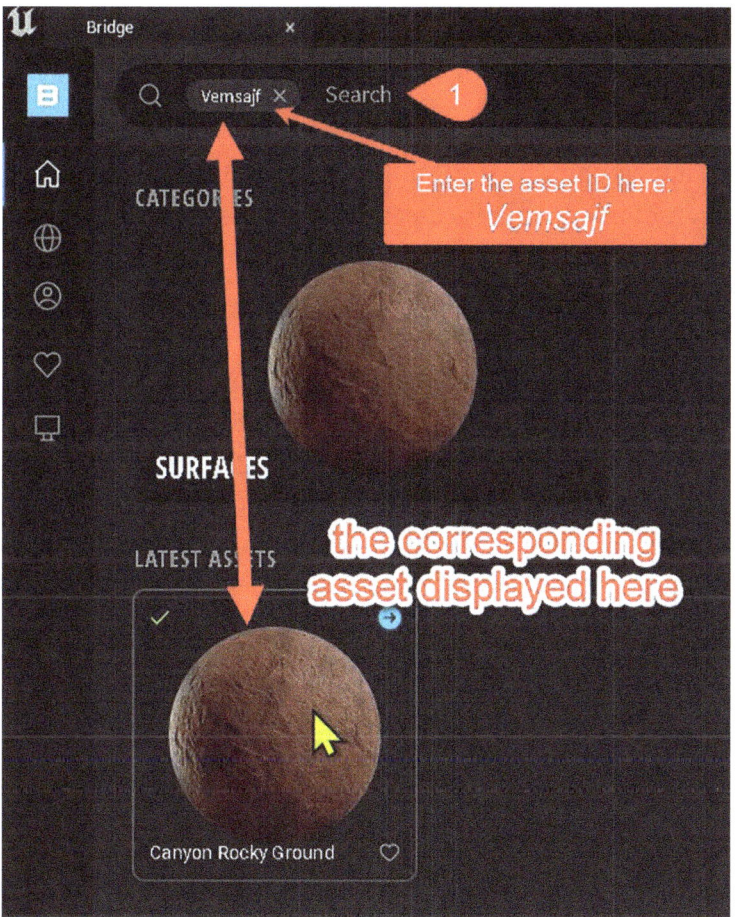

Figure 4-4. *Search with the asset ID*

Surface (Texture) Assets

For this project, we intend to download five specific texture assets (referred to as Surface types) as outlined in the following asset names, each with its corresponding asset ID:

- Canyon Rocky Ground = *vemsajf*
- Beach Cliff = *xdhicfv*
- Rocky Ground = *vjdqcba*
- Rocky Steppe = *uknicjmmw*
- Trampled Snow = *vl3lfddfw*

For our project, we will import medium-quality texture surfaces, striking a balance between visually impeccable and the fluidity of gameplay with protracted load times. The methodology for importing these essential elements is captured in sequential illustrations and steps, ensuring a smooth integration into our project.

Asset Download

1. Begin by navigating to the texture quality settings of your project. As shown in Figure 4-5, select "Medium Quality" from the drop-down menu. This setting is a crucial step, ensuring that the downloaded assets are in harmony with the project's performance parameters.

2. Proceed to download the desired asset by clicking the "Download" icon. It's essential to ensure your selection aligns with the project's thematic needs and performance considerations.

3. Upon completion of the download, the assets will reside in a designated repository within the project's directory. Figure 4-6 illustrates the "Megascan – Surfaces" folder, nested under the "Content" directory, which serves as the vault for these downloaded textures.

CHAPTER 4 REVITALIZING VISUALS: ASSET IMPORT AND PROCEDURAL CREATION

Asset Integration

1. With the assets downloaded, integration is the next phase. Expand the "Content" directory and locate the "Megascan – Surfaces" folder.

2. Here, we'll find a catalog of surface textures, as highlighted in Figure 4-6. These are labeled clearly, such as "Beach_Cliff_xdhicfv" and "Canyon_Rocky_Ground_vemsajf", providing a transparent indication of the texture's essence.

3. Select the desired texture from this list. This is the asset that will be imported into your scene or level, ready for procedural manipulation and placement.

Figure 4-5. *Select the quality to download and add to the project*

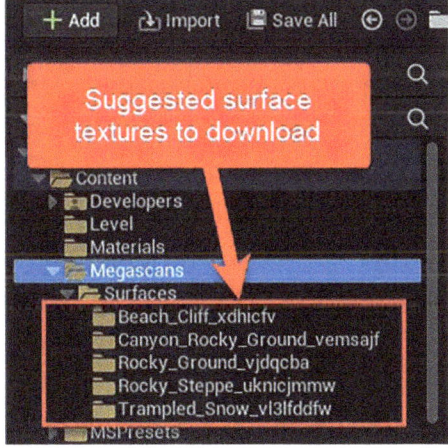

Figure 4-6. *Suggested assets to download*

3D Assets and 3D Plants

In UE, the utilization of 3D assets and lifelike vegetation brings a new dimension of realism and immersion to virtual environments. The incorporation of high-quality 3D assets enables developers to craft intricately detailed worlds with enhanced visual fidelity, resulting in stunningly lifelike scenes that captivate the viewer's imagination.

Furthermore, the integration of 3D plants within UE not only adds an extra layer of authenticity but also contributes to the creation of dynamic ecosystems that respond naturally to environmental changes.

The advanced tools and features in UE5 empower creators to manipulate lighting, shadows, and textures in ways that seamlessly blend these assets into the environment, blurring the line between reality and virtuality.

As a result, UE5 stands as a testament to the evolution of virtual world-building, where 3D assets and botanical elements combine to produce breathtaking, interactive experiences that push the boundaries of visual storytelling.

Nanite

The concept of level of detail (LOD) had been long used to optimize performance by displaying simplified versions of objects at a distance and has evolved to a new level of sophistication. With Nanite, a groundbreaking virtualized geometry system, UE allows for the direct rendering of film-quality assets at runtime, regardless of their complexity.

This innovation eliminates the need for traditional LOD pipelines and empowers us to create intricately detailed assets without concerns about performance. Nanite's real-time geometry processing also enables the portrayal of minute details, even when players are up close, resulting in unprecedented visual fidelity.

The integration of Nanite into the LOD framework of UE5 ensures seamless transitions between different levels of detail, catering to diverse hardware capabilities and providing an immersive and consistent experience across platforms.

Nanite is a virtualized micropolygon geometry system that enables the creation of incredibly detailed and dynamic environments. It empowers us to render highly complex scenes with an unprecedented level of detail, efficiently managing substantial geometric workloads without compromising performance. Through dynamic streaming of the required level of detail based on the player's perspective, Nanite optimizes resource usage and facilitates the development of visually stunning and immersive worlds.

As a result, this virtualization technology allows us to import intricate assets, such as those from Quixel Megascans, while maintaining real-time rendering performance without any compromised effects.

We can find comprehensive information about Nanite, UE5's virtualized geometry system, in the official UE documentation. This includes details on materials supported by Nanite, rendering features, supported platforms, and specifics on fallback mesh and precision settings for the technology.

To access this documentation, do visit the UE5 Documentation page as follows:

https://docs.unrealengine.com/5.0/en-US/nanite-virtualized-geometry-in-unreal-engine/

Import and Enable Nanite on 3D Geometry Assets

We can apply the same procedures to asset IDs that we used for adding surface textures in order to integrate 3D geometry files into our project. We will attempt to download the Nanite quality for the 3D geometry file (if it's available, as shown in Figure 4-7).

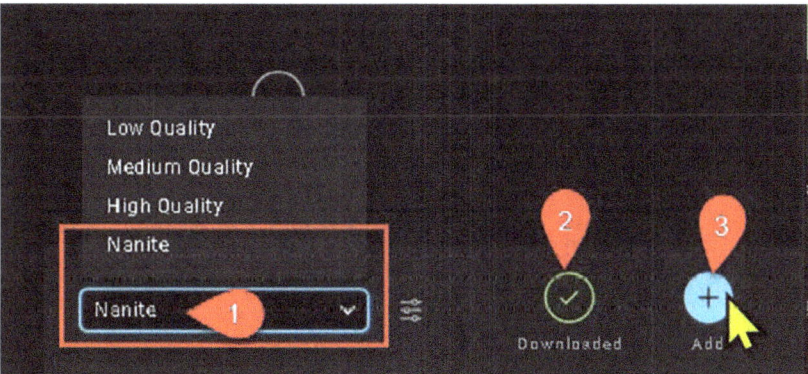

Figure 4-7. *Nanite quality*

If Nanite quality is not available, then we will opt for the Highest Quality option, as demonstrated in Figure 4-8.

CHAPTER 4 REVITALIZING VISUALS: ASSET IMPORT AND PROCEDURAL CREATION

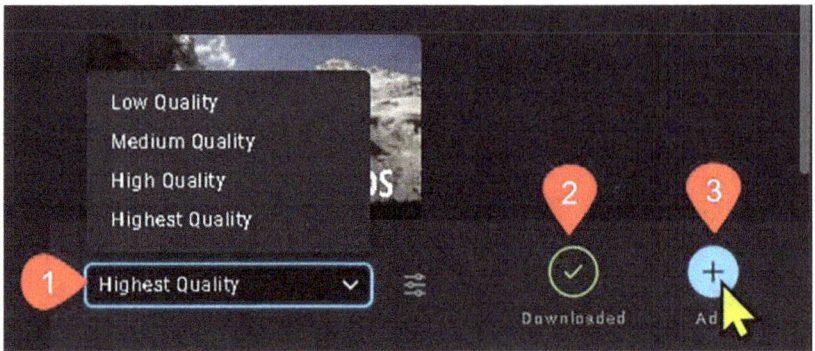

Figure 4-8. *Highest Quality if Nanite is not available*

The five assets that we are going to use in this project, along with their corresponding individual asset IDs, can be found as follows (as shown in Figure 4-9):

- Forest Rock Formation := *wgquaam*
- Small Granite Stone := *wgyxecr*
- Small Sandstone := *vd5scae*
- Small Sandstone Rock := *xdzobhg*
- Broom Creeper := *xfqicgrqx*

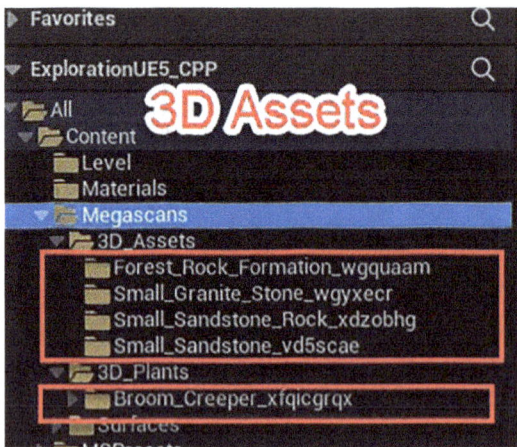

Figure 4-9. *List of 3D assets and plants*

To Enable Nanite on Static Mesh

To enable Nanite on your static meshes within the UE, such as those found in the Broom Creeper package for 3D plants, you'll need to perform a few simple steps:

1. In the UE editor, locate the project's "Content" browser. Within this browser, navigate to the specific package containing our static meshes, as shown in Figure 4-10. For the Broom Creeper package, we will find various static mesh assets named with the prefix "S_Broom_Creeper".

2. Select the static mesh(es) that we wish to enable Nanite on. We can select multiple meshes by holding the "Shift" key and clicking the batch of meshes we want to include.

Figure 4-11 provides a step-by-step process on how to enable Nanite for selected static meshes within the UE editor. Here's a detailed breakdown of each step as shown in the figure:

1. Accessing the Context Menu

    a. After multiple static meshes from the Broom Creeper package in the "Content" browser had been selected, we will right-click one of the selected assets to bring up the context menu.

    b. This action is key to accessing the advanced options for the asset, including Nanite settings.

2. Navigating to Nanite Options

    a. In the context menu, there's a list of various options that can be performed on the selected assets.

    b. We will hover over the "Nanite" option to reveal a side menu.

3. Enabling Nanite

    a. In the side menu that appears upon hovering over "Nanite," we will find an option labeled "Enable Nanite."

    b. Next to this option, there's an indication of the number of meshes that will be affected by this action, in this case, "Enable Nanite (15 Meshes)."

    c. We can click this option to activate Nanite for all selected meshes.

CHAPTER 4 REVITALIZING VISUALS: ASSET IMPORT AND PROCEDURAL CREATION

4. Confirmation and Save

   a. After enabling Nanite, it's good practice to save the changes. The "Save All" button at the bottom of the "Content" browser can be used to save all modifications to the assets.

By following the preceding steps, we can take advantage of Nanite's advanced rendering technology to significantly improve the visual quality and performance of our static meshes in the game engine.

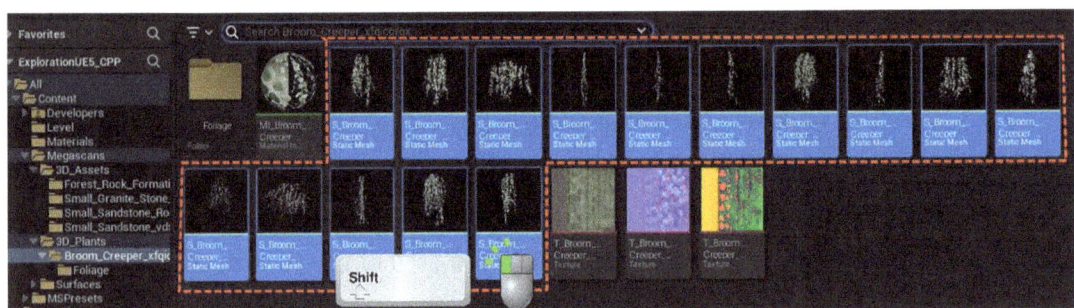

Figure 4-10. Select the static mesh ready for conversion

Figure 4-11. Enable Nanite quality

Visual Improvement on AutoBlend_Height_MAT Landscape Material

Now that we have imported realistic surface textures, we can proceed to replace the colors in our previously created AutoBlend_Height_MAT material with the combination of surface textures (Albedo, Normal, Roughness, and Ambient Occlusion).

Material Function

The primary purpose of Material Functions is to encapsulate a set of material instructions or calculations into a modular and self-contained unit. This modular approach enables developers to create complex shaders and visual effects by combining smaller, manageable elements, streamlining the overall material creation workflow.

Material Functions allow for the creation of consistent and standardized material components that can be reused across various materials, saving time and effort while maintaining visual coherence.

To leverage flexibility and reusability, we intend to implement the concept of Material Function as shown in Figure 4-12:

1. Locate the "Materials" Folder: In the Content Browser, navigate to the folder where we want to create our new Material Function. In this case, we will select the "Materials" folder.

2. Right-click within the editor to bring up the context menu. This menu provides a selection of choices, one of which is the "Material" option.

3. We can hover over the "Material" option, which expands to reveal more specific material-related asset types.

4. Within this expanded "Material" section, we'll find the "Material Function" option.

CHAPTER 4 REVITALIZING VISUALS: ASSET IMPORT AND PROCEDURAL CREATION

Once we've clicked "Material Function" as shown in Figure 4-13, UE will create a new asset and add it to the selected folder in our Content Browser. We can then double-click the new Material Function to open it and start editing in the material editor, where we can define the function's inputs, outputs, and the operations that make up the function itself.

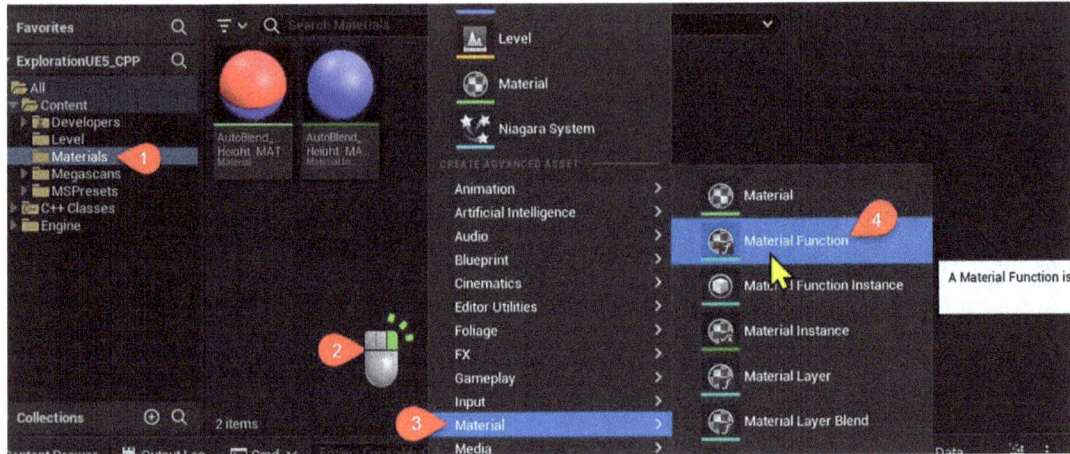

Figure 4-12. *To create Material Function*

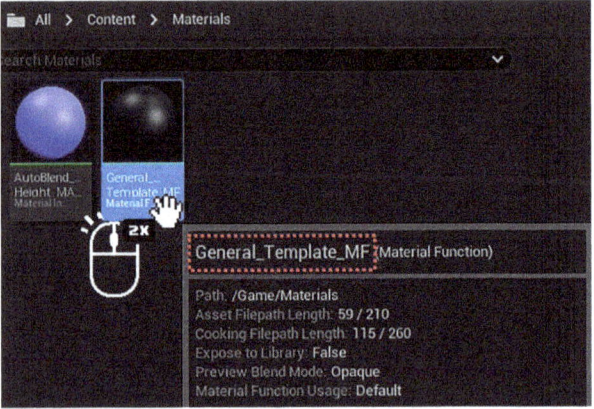

Figure 4-13. *Open up the Material Function*

General_Template_MF

We are creating a Material Function to serve as a general template, enabling us to reuse it for various materials within our AutoBlend_Height_MAT collection (including Ground, Mid, Top, Bottom Slope, and Top Slope materials).

In this Material Function, we are utilizing the preexisting Material Function named `MF_MapAdjustments`, which is accessible from the Megascan Presets function. The process of acquiring these materials can be achieved by following the instructions outlined in Figure 4-14.

MF_MapAdjustments is designed to handle various adjustments to texture maps within a material, such as applying color corrections, brightness and contrast modifications, or even complex transformations like distortion effects.

The presence of multiple entries named "MF_MapAdjustments" in a list of assets, particularly when they are sourced from Megascans, suggests that these are Material Functions that are likely to be very similar, if not identical. In such cases, the redundancy could be due to bulk imports or asset syncing where the same Material Function was downloaded multiple times.

Since they are all from Megascans, it's possible that there is no practical difference between the various "MF_MapAdjustments" entries. This means that choosing any one of them would not significantly impact our project because they all contain the same function or adjustments.

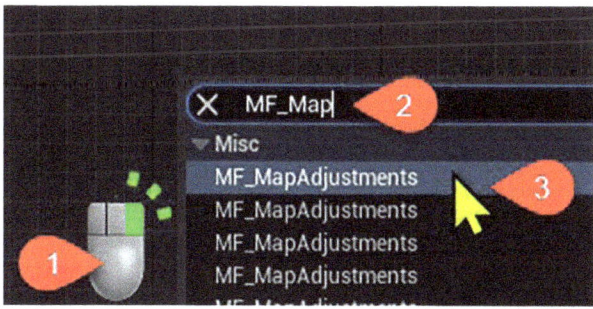

Figure 4-14. *Call out existing Material Function, MF_MapAdjustments*

The "ERROR!" message appearing on the "MF_MapAdjustments" node in the material editor indicates a problem with the node's inputs. In this context, it arises when the node is expecting an external resource (like a texture map or a material parameter) that has not been provided.

To fix this error, we will need to have the inputs of the "MF_MapAdjustments" node connect to missing resources. Once identified, we should link the correct resource or ensure that the missing asset is imported and available in our project.

We will now establish a connection from the Output result of MF_MapAdjustments to the Output Result in our custom Material Function shown in Figure 4-15.

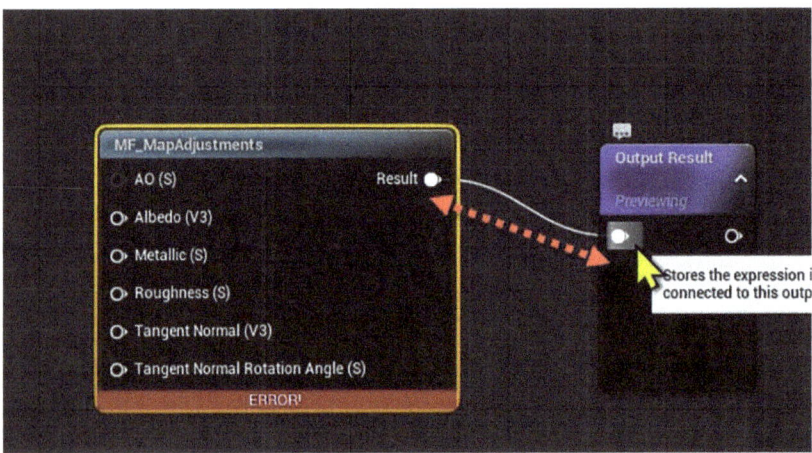

Figure 4-15. *Connecting the result from MF_MapAdjustments to the Output Result of the Material Function*

Set Up the Texture Samplers for Properties

As we are utilizing surface textures downloaded from Quixel Bridge, it's important to note that a majority of these textures are in Virtual Texture format. Consequently, we must ensure that the settings are configured accurately as shown in Figure 4-16. Specifically, we need to set the settings as follows: Shared Wrap for Sampler Source and Virtual Color for the Sampler Type.

In the UE material editor, the "Sampler Source" setting of a Texture Sample node is crucial for optimizing the use of texture samplers within a shader. When set to "Shared: Wrap," it allows multiple texture samplers to use the same sampler, provided they share the same wrapping method. This is essential because GPUs typically have a limit on the number of active texture samplers they can handle, often capped at 16 for many models.

Failing to set the "Sampler Source" to "Shared: Wrap" can lead to exceeding the GPU's texture sampler limit, especially in complex materials with many textures. Once this limit is hit, additional textures may not be sampled correctly, leading to rendering issues or compilation errors. Thus, correctly configuring the Sampler Source is a key step in ensuring material efficiency and avoiding performance pitfalls.

Therefore, neglecting to configure the Sampler Source to Shared Wrap will lead to an issue where we are unable to exceed a total of 16 texture samplers.

As shown in Figure 4-16, we can follow these steps:

1. Select "Shared: Wrap": From the options available in the "Sampler Source" drop-down, select "Shared: Wrap". This ensures that the Texture Sample node shares texture samplers with other nodes that have the same wrap settings, conserving the limited number of samplers available.

2. The selection of "Virtual Color" refers to the setup for Virtual Texturing, a performance optimization technique. By choosing "Virtual Color," we instruct the engine to sample color data from a virtual texture, which efficiently manages texture memory usage by loading only the necessary texture parts. This setup is essential for optimizing materials that use large textures, particularly in expansive environments or games where maintaining performance is crucial.

3. We now select the texture named "DefaultDiffuse" for our Base Color, and be sure that this DefaultDiffuse is a Virtual Texture (VT) sampler type as indicated from the arrow and circle.

CHAPTER 4   REVITALIZING VISUALS: ASSET IMPORT AND PROCEDURAL CREATION

Figure 4-16. *Create a Texture Sampler for the virtual texture node and use it a reroute node named 'BaseColor'*

After setting up the `BaseColour` node to handle the Albedo texture in our material, indicated by the label "`DefaultDiffuse`," we are ready to configure additional texture properties. As per the guidelines in Figure 4-16, we would proceed to set up the nodes for the `Metallic`, Ambient Occlusion combined with Roughness (`AO_R`), and `Normal` maps. As shown in Figure 4-17, the name of each texture used for these properties should correspond to the labels provided in the figure, ensuring that the correct textures are applied to their respective material properties.

CHAPTER 4   REVITALIZING VISUALS: ASSET IMPORT AND PROCEDURAL CREATION

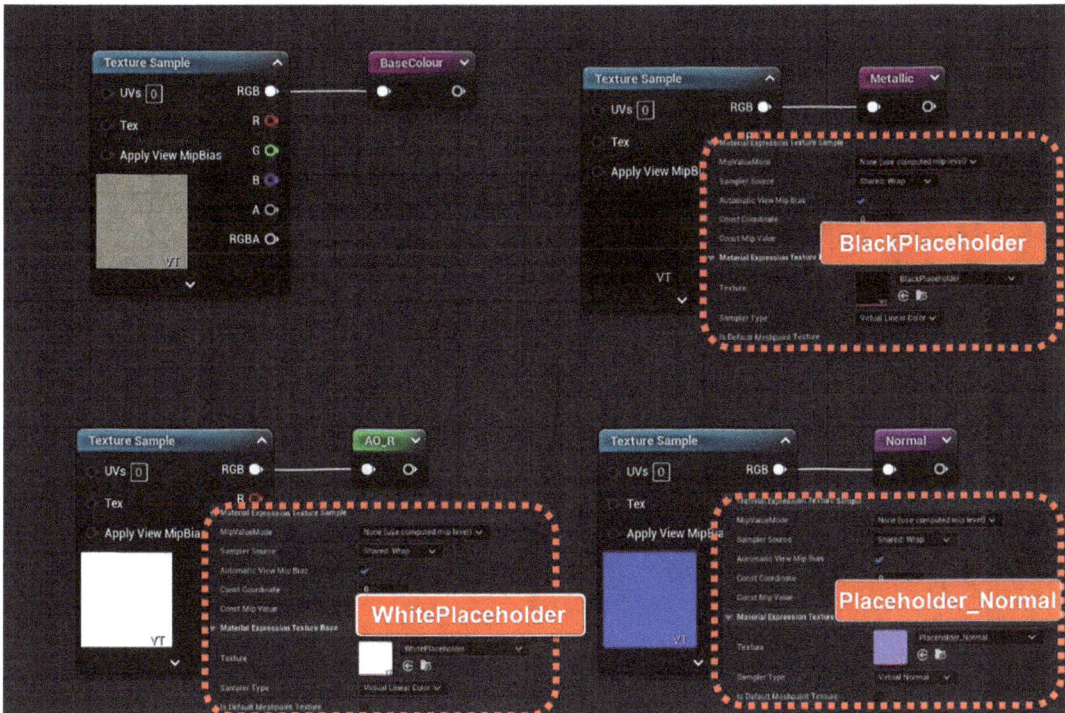

Figure 4-17. *Set up the other three properties for the MF_MapAdjustment Material Function*

Finally, we can establish connections for the nodes we had set up to the MF_MapAdjustment Material Function, incorporating an additional Constant (single value) for the Tangent Normal Rotation. We can see the initial version of our General_Template Material Function as shown in Figure 4-18. After all the preceding settings, we should apply the settings and save our material to ensure that the changes are preserved.

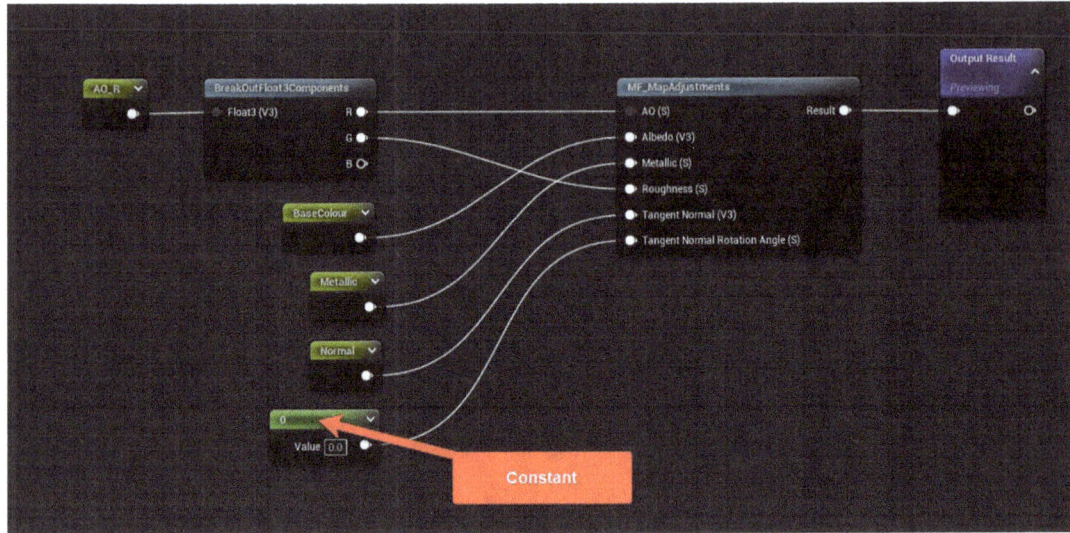

Figure 4-18. *Initial complete version of the Material Function*

Improve Texture Tiling Sampler

Traditional texture tiling based on simple texture coordinates is easily noticeable, resulting in the landscape appearing grid based. In this section, we are creating a new Material Function named `UV_Variation` (as shown in Figure 4-19) specifically designed to address the enhancement of texture tiling.

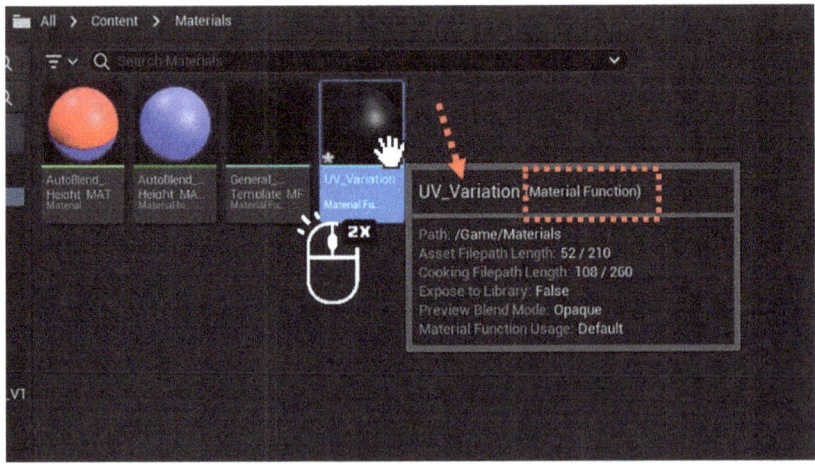

Figure 4-19. *Create and open up a newly created UV_Variation Material Function*

Texture tiling is a technique used in game development to seamlessly repeat a texture across a surface, effectively covering a larger area without increasing the texture's resolution. This is especially important for optimizing memory usage and rendering performance.

Game engines like UE allow us to control the tiling of textures on various surfaces, adjusting parameters such as scale, rotation, and offset to achieve the desired visual effect. Tiling can be crucial in creating immersive and realistic environments by maintaining a consistent texture pattern across objects and terrain.

As shown in Figure 4-20, by applying the Shifted UV algorithm (shown on the right side), we can observe the distinction between the original and modified versions.

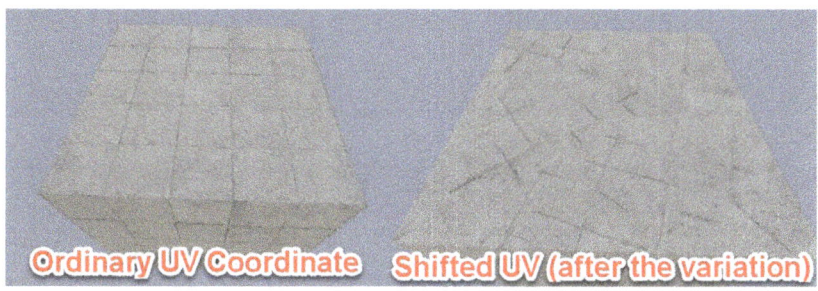

Figure 4-20. *Comparison of ordinary tiling and shifted tiling*

To address the issue of the tiling effect that arises with the use of repetitive textures in a scene, we turn to the Material Function "TextureVariation." As shown in Figure 4-21, this function introduces variations in the texture's UV coordinates, which helps to reduce noticeable repetition. Here are the steps we can follow to set up the figure shown:

1. Create an "Absolute World Position" node to capture the 3D coordinates of each point on your material surface, which we are interested to use the XY output for the division in the next step.

2. Add a "Distance_Away" parameter node, which seems to control the scale of texture variation at a certain distance. Set the default value to 250 or as required for your material.

3. Connect the "Absolute World Position" Z output to the "A" input of the "Divide" node.

CHAPTER 4   REVITALIZING VISUALS: ASSET IMPORT AND PROCEDURAL CREATION

4. Connect the "Distance_Away" output to the "B" input of the "Divide" node.

5. Set up two "Static Bool" parameter nodes, both set to "True," which could be used to toggle features in the material, such as whether the texture variation is influenced by distance.

6. Finally, link all the nodes to the "TextureVariation" node, which appears to have various controls for heightmap, UVs, and other texture properties.

In the material editor of UE, the preview color you see can sometimes vary unexpectedly, displaying as solid yellow, a mix of yellow-green, or solid green.

This discrepancy in color preview could be indicative of bugs within the material editor, where the rendering of material previews may not accurately reflect the intended outcomes.

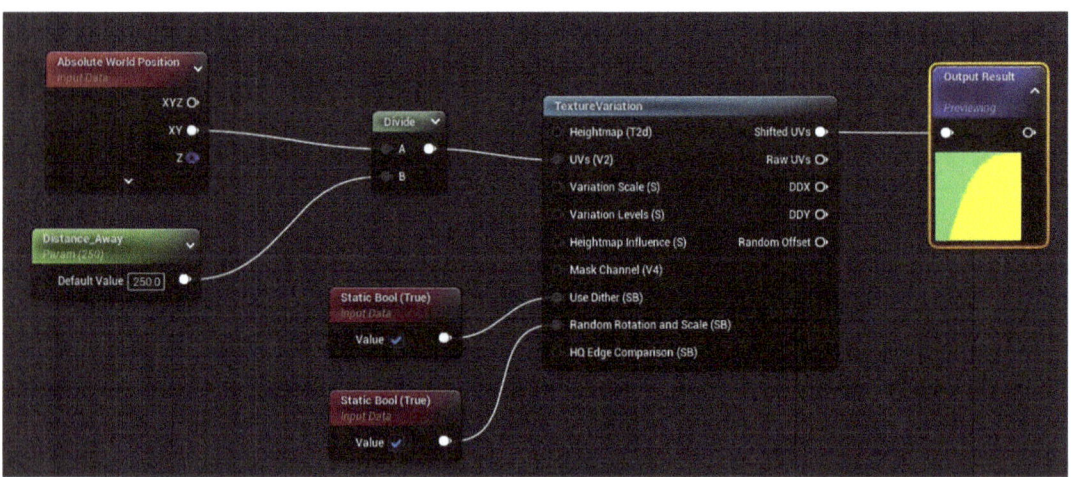

Figure 4-21. Simple setup for the UV_variation Material Function

The "Static Bool" node, as implied by the screenshot, is a type of node in UE's material editor that allows us to set a boolean value. In material creation, this could be used to conditionally control the flow of operations or enable/disable certain features within the material.

In the context of this screenshot, these two nodes are controlling whether the features of "Use Dither" or "Random Rotation and Scale" within the "TextureVariation" node should be active.

Static Bool nodes are an essential part of creating dynamic and adaptable materials, as they can be used to quickly toggle features without the need to recompile or adjust the underlying shader code. This can streamline the workflow and make it easier to experiment with different material properties and effects within the UE.

To further refine the appearance and control the scale of the texture based on the viewer's distance, we introduce a scalar parameter named "Distance_Away". The effectiveness of these adjustments can be evaluated in Figure 4-20, which we like to achieve the resulting visual enhancement.

To avoid the repetitiveness often seen with texture tiling, we typically employ basic techniques like rotating and scaling the texture's UV coordinates.

However, the "TextureVariation" function offers a more sophisticated approach. This method not only provides UV variation but also allows for the integration of displacement maps, enhancing the material's realism by creating height-based intersections.

Subsequently, a temporal dither is applied, which further refines the material's appearance by smoothing out the transitions between the displacement map's varying heights. This advanced technique results in a more natural and less grid-like pattern on applied surfaces.

Incorporate UV_Variation into General_Template_MF

After setting up the UV_Variation Material Function, we can now integrate it into our initial setup of General_Template_MF by initiating the Material Function call, as shown in Figure 4-22. Before we start to the integration, be sure that we double-click "General_Template_MF" from the Content Browser to ensure that we are working on the correct Material Function.

1. We will right-click the editor to activate the context menu, and we can type in "MaterialFunctionCall" to see this selection from the drop-down list. Notice that we can find this "MaterialFunctionCall" under the "Functions" category.

2. We can now select the "MaterialFunctionCall" from the list, and we can see this node in our material graph canvas to be ready for us to use.

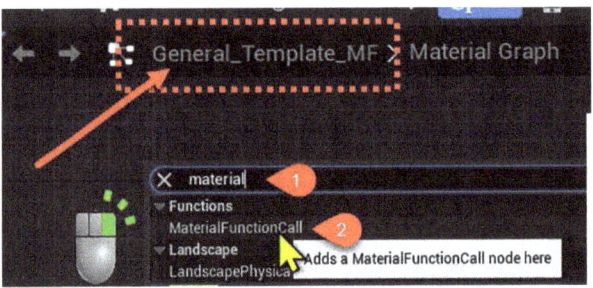

Figure 4-22. Initiate a Material Function call inside General_Template_MF

In Figure 4-23, we can set up the link to the UV_Variation as the function for this Material Function node. This allows the variation effects processed by UV_Variation to be applied across the texture samplers in the material.

Figure 4-23. Calling UV_Variation as the Material Function

CHAPTER 4 REVITALIZING VISUALS: ASSET IMPORT AND PROCEDURAL CREATION

Figure 4-24 illustrates the implementation showing UV_Variation being fed into various texture sampler nodes to enhance the textures with UV coordinate variations. However, this does not include the metallic texture sampler, as it's indicated that there is no metallic texture to be used, since in physically based rendering, metallic properties are consistent across a surface and do not require the same UV coordinate variations as other textures. This selective application ensures that only relevant texture samplers receive the UV variation effect, optimizing the material's performance and visual output.

UV variations are typically applied to maps like albedo or normal to reduce visible tiling, but since metallic surfaces are uniform in nature, adding such variation to the metalness of a material would be unnecessary and could lead to nonphysical results.

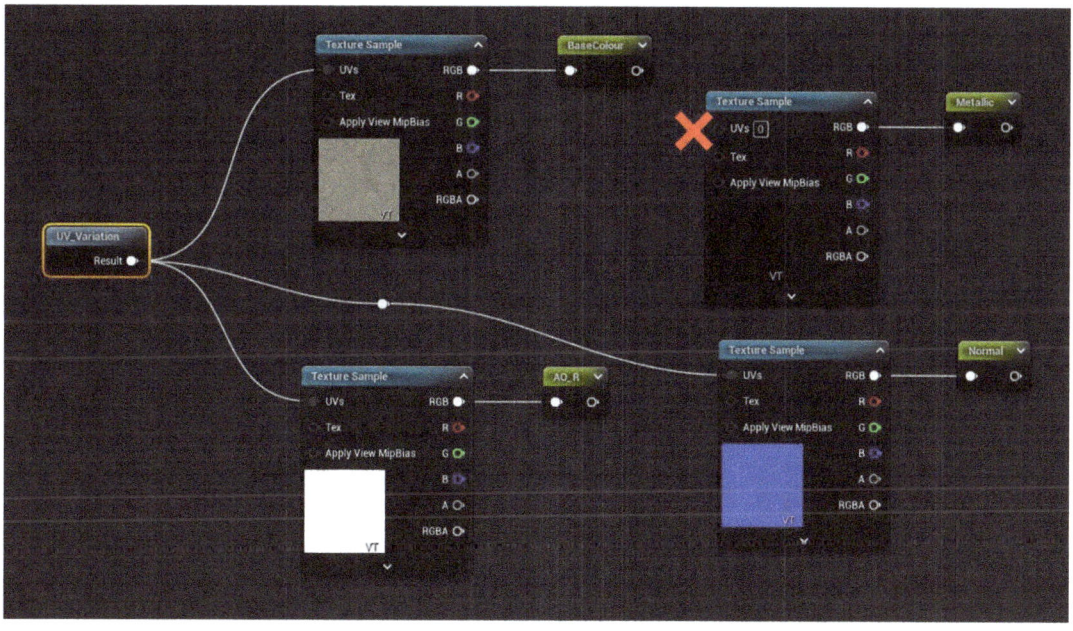

Figure 4-24. UV_Variation connects to the Texture Sampler

Complete the AutoBlend_Height_MAT Landscape Material

After setting up the General_Template_MF, we can proceed to duplicate this Material Function as shown in the steps in Figure 4-25.

CHAPTER 4   REVITALIZING VISUALS: ASSET IMPORT AND PROCEDURAL CREATION

After setting up the General_Template_MF, which serves as a master blueprint for Material Functions, duplicating it for specific use cases is a strategic approach.

This practice safeguards the original template from unintended alterations during project iterations and customizations. It allows for the creation of bespoke Material Functions tailored to unique aspects of the game environment or objects while maintaining a clean and consistent base to revert to or reference.

By duplicating the master template, developers can experiment and expand their material library efficiently, ensuring that the core design principles embedded within the General_Template_MF remain intact and universally applicable across the project.

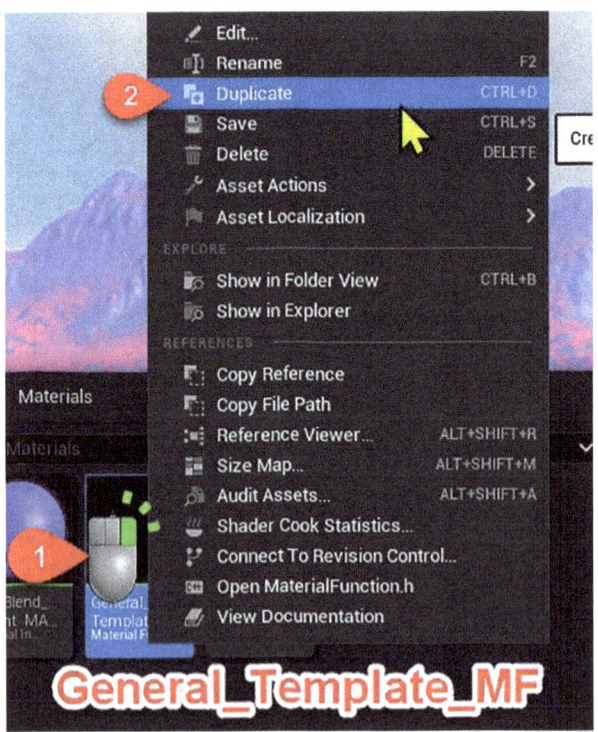

Figure 4-25. Duplicate the General_Template_MF

CHAPTER 4   REVITALIZING VISUALS: ASSET IMPORT AND PROCEDURAL CREATION

We will duplicate it into five distinct Material Functions as shown in Figure 4-26:

- Ground_MF
- Middle_MF
- Top_MF
- Lower_Slope_MF
- Upper_Slope_MF

Figure 4-26. *Each of the duplicated Material Function*

To isolate each texture sampler for the distinct Material Function we duplicated earlier, we will establish distinct parameter names for the texture samplers within the Material Function, as shown in Figure 4-27, by performing the following steps.

Additionally, inside each of the Material Functions, we will assign unique group names to each texture sampler:

1. Connect UV Variation: The "UV_Variation" node is connected to the "UVs" input of the "Ground_BaseColor" node to provide UV coordinate variations.

2. Paramcter Details – Ground_BaseColor: We will now parameterize the Texture Sampler that leads to Base_Color, and we named this parameter "Ground_BaseColor".

3. Assign Group Name: Here, we can type the group name into the "Group" field to categorize this parameter. We named this category "Ground" for this particular node.

127

CHAPTER 4   REVITALIZING VISUALS: ASSET IMPORT AND PROCEDURAL CREATION

A group is set up to categorize related parameters within a material, facilitating easier navigation and adjustment.

Assigning a group name, such as "Ground" for ground-related texture parameters, consolidates these settings in the material instance, streamlining the interface for users and enhancing the efficiency of the material modification process.

This approach helps to organize our texture samplers when employing these materials outside the material editor.

4. Connect UV Variation to AO_R: We will now parameterize the Texture Sampler that leads to AO_R.

5. Set Parameter Name: The "Parameter Name" is set to "Ground_AO_R" to identify the purpose of this parameter.

6. Assign Group Name to AO_R: Similar to step 3, the "Group" field for the "Ground_AO_R" node is set to "Ground."

7. Connect UV Variation to Normal: We will now parameterize the Texture Sampler that leads to Normal.

8. Set Parameter Name for Normal: The "Parameter Name" for the normal map is set to "Ground_Normal".

9. Assign Group Name to Normal: The "Ground_Normal" node's "Group" field is set to "Ground," keeping all related parameters organized under the same category.

Be sure to set the settings as follows: Shared Wrap for Sampler Source and Virtual Color for the Sampler Type.

These steps show the importance of proper organization and naming conventions in material setup, which can streamline the workflow and make it easier to manage complex materials.

CHAPTER 4   REVITALIZING VISUALS: ASSET IMPORT AND PROCEDURAL CREATION

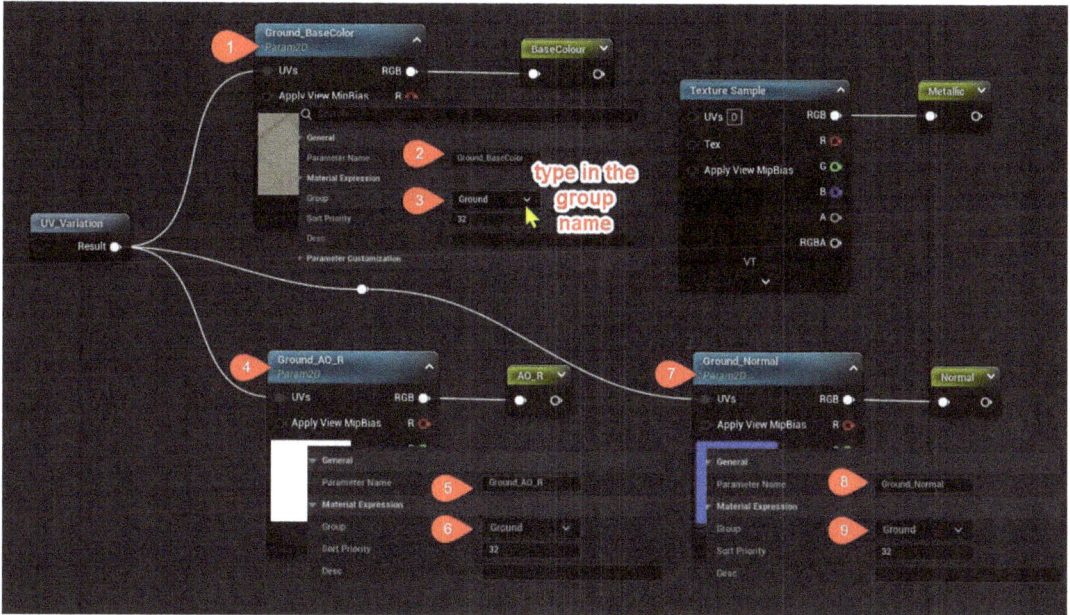

Figure 4-27. *Set up an individual parameter name and group name for each Material Function. This example is for the Ground Material Function*

Incorporating the Material Function "Ground_MF" into our "AutoBlend_Height_MAT" material involves a specific adjustment as described here:

1. To incorporate the "Ground_MF" Material Function into the "AutoBlend_Height_MAT" material in UE, we can search for and add the "Ground_MF" Material Function by either dragging it from the Content Browser or right-clicking in the graph area and selecting it from the list.

2. Once placed, connect the outputs of "Ground_MF" to the appropriate inputs in the "AutoBlend_Height_MAT" material, ensuring that the function's attributes such as textures, normals, and any additional parameters blend seamlessly with the existing material setup.

CHAPTER 4   REVITALIZING VISUALS: ASSET IMPORT AND PROCEDURAL CREATION

3. The existing connections to the "Ground_MAT" named reroute are intentionally disconnected. This step is taken to integrate the "Ground_MF" function, which subsequently alters the material color from green to the desired "Ground_MAT" material. The reason for not removing the previously used elements immediately may be to allow for easy reversion or comparison during the transition phase.

It's important to ensure that the material editor visually reflects this replacement process to avoid confusion, as highlighted by the discrepancy between the described text and the image in Figure 4-28.

As seen in the figure, we also label the node as "Bottom layer materials" to document these nodes in the material editor and keep this section as a reference.

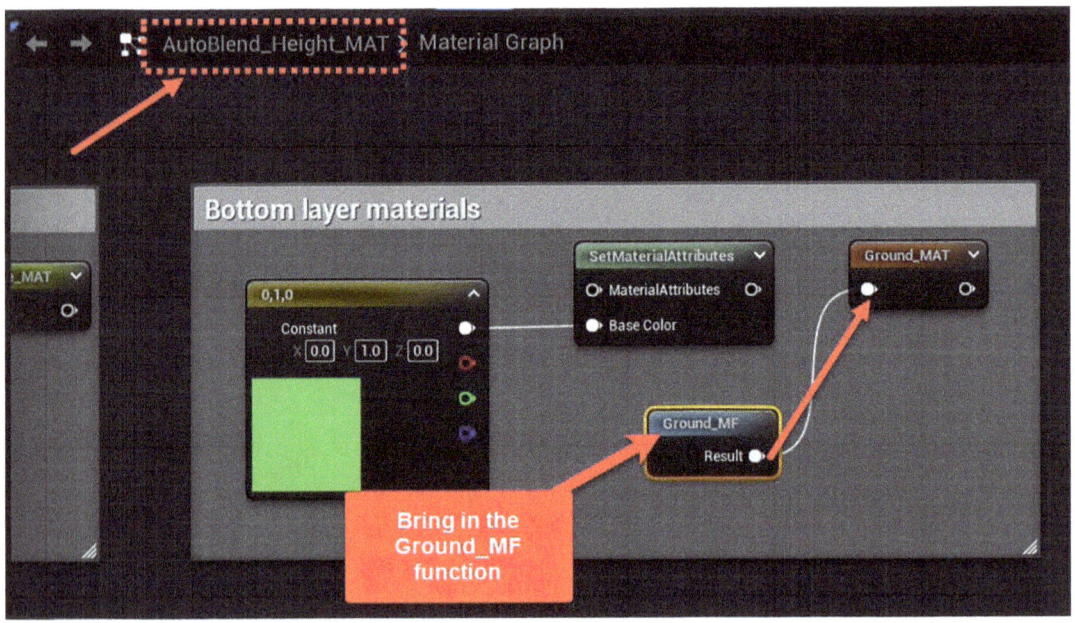

Figure 4-28. *Replace the green color connection from Ground_MF to Ground_MAT*

130

We can now observe that there will be a distinct category designed specifically for us to associate our desired textures of the Ground in AutoBlend_Height_MAT_INST material instance, as shown in Figure 4-29 and described in the following steps:

1. Material Instance Selection: Be sure that we open up the "AutoBlend_Height_MAT_Inst" material instance for editing.

2. Parameter Grouping: Under the "Details" panel, expand the "Ground" group. Ensure that each property in this group is enabled by clicking on them.

3. Texture Assignment: Within the "Ground" group, individual textures are assigned to parameters such as "Ground_AO_R", "Ground_BaseColor", and "Ground_Normal" correspondingly as follows:

    a. Ground_AO_R: T_CanyonRockyGround_vemsajf_2K_ORDp

    b. Ground_BaseColor: T_Canyon_Rocky_Ground_vemsajf_2K_D

    c. Ground_Normal: T_Canyon_Rocky_Ground_vemsajf_2K_N

4. Preview Thumbnails: The thumbnail previews of the textures provide a quick visual reference for us to ensure that the correct textures are being applied to each parameter.

You are welcome to adjust these parameters, such as to change the material's properties, like color, roughness, and normal mapping, which can be previewed in real time on the material applied to a mesh in the scene. This setup is typically used to fine-tune the appearance of materials without altering the underlying material blueprint.

Encountering an issue where textures fail to assign to a Material Instance can be puzzling. In this case, the problem persisted with the textures not attaching as expected.

The resolution came through a workaround: switching out the default virtual textures for their non-virtual counterparts. By doing so, the non-virtual textures, which are standard texture assets within UE, were successfully assigned where the virtual textures faced compatibility or assignment issues.

CHAPTER 4 REVITALIZING VISUALS: ASSET IMPORT AND PROCEDURAL CREATION

This solution underscores the importance of flexibility and alternative approaches when troubleshooting material and texture assignments within UE, ensuring that the material setup remains functional and the development process can continue unimpeded.

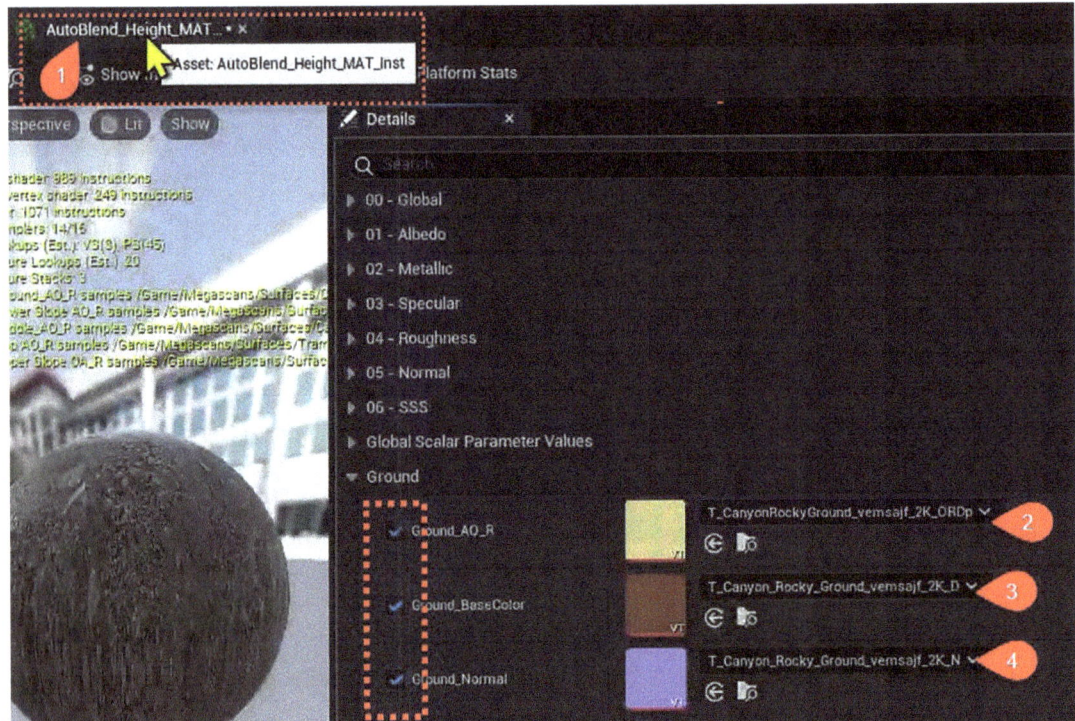

Figure 4-29. *Attach the relevant texture that belongs to the Ground textures (in this case, we use textures for CanyonRockyGround)*

EXERCISE

We can now repeat the same process as we did for the "Ground_MAT" for the other four Material Functions: `Middle_MF`, `Top_MF`, `Lower_Slope_MF`, and `Upper_Slope_MF`.

You can use the following suggested texture sampler for each property of the Material Function:

Middle_MF with Group_Name: **Middle**
    AO_R : T_RockyGround_vjdqcba_2K_ORDp
    Base Color : T_RockyGround_vjdqcba_2K_D
    Normal : T_RockyGround_vjdqcba_2K_N
Top_MF with Group_Name: **Top**
    AO_R : T_Trampled_Snow_vl3lfddfw_2K_ORDp
    Base Color : T_Trampled_Snow_vl3lfddfw_2K_D
    Normal : T_Trampled_Snow_vl3lfddfw_2K_N
Lower_Slope_MF with Group_Name: **Lower_Slope**
    AO_R : T_BeachCliff_xdhicfv_2K_ORDp
    Base Color : T_BeachCliff_xdhicfv_2K_D
    Normal : T_Beach_Cliff_xdhicfv_2K_N
Upper_Slope_MF with Group_Name: **Upper_Slope**
    AO_R : T_Rocky_Steepe_uknicjmmw_2K_ORDp
    Base Color : T_Rocky_Steepe_uknicjmmw_2K_D
    Normal : T_Rocky_Steepe_uknicjmmw_2K_N

With the preceding Material Function set up with the appropriate texture samplers, we can now substitute the relevant input to each of the named reroute nodes as shown in Figures 4-30 to 4-32.

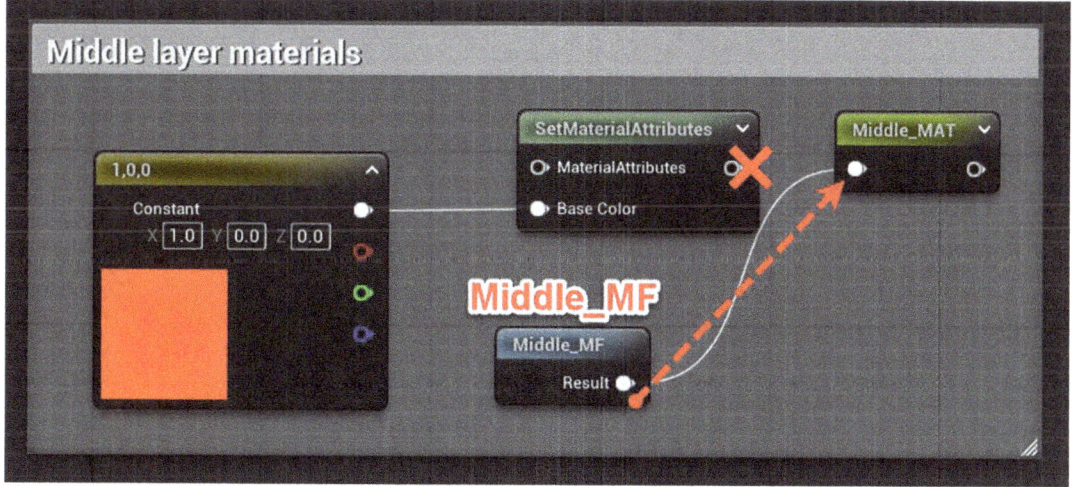

Figure 4-30. *Substitute the initial "red" base color for the Middle layer material with the output of the Middle_MF material*

CHAPTER 4   REVITALIZING VISUALS: ASSET IMPORT AND PROCEDURAL CREATION

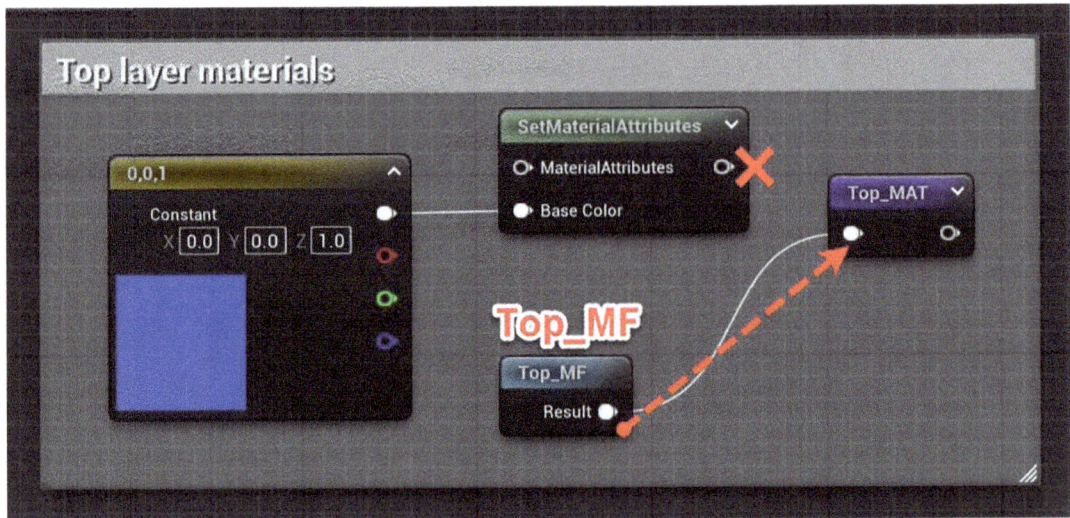

Figure 4-31. *Substitute the initial "blue" base color for the Top layer material with the output of the Top_MF material*

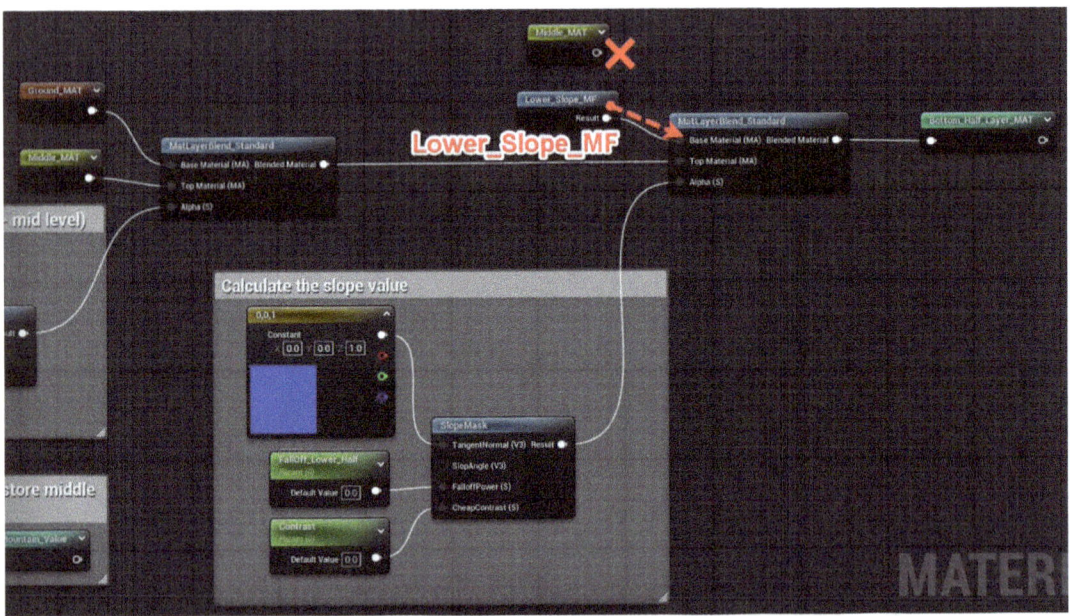

Figure 4-32. *Substitute the Middle_MAT for the Bottom_Half_Layer_MAT with the Lower_Slope_MF material*

CHAPTER 4   REVITALIZING VISUALS: ASSET IMPORT AND PROCEDURAL CREATION

Figure 4-33 showcases the resolution proposed for the exercise in Chapter 3, serving as a potential answer to the exercises outlined there.

Your approach to material blending may differ; however, the core concept introduced in Chapter 3 revolves around leveraging "Upper_Slope_MF" as the foundational material for the top half of the terrain in the "AutoBlend_Height_MAT".

This technique is pivotal for creating a cohesive transition on steeper inclines, ensuring that the upper sections of your landscape exhibit a blended, realistic integration of textures and materials.

This illustrates how Material Functions can be layered to simulate environmental features, such as the gradation seen on a slope, by blending different textures based on height data.

As a reminder, be sure to set the settings as follows: Shared Wrap for Sampler Source and Virtual Color for the Sampler Type.

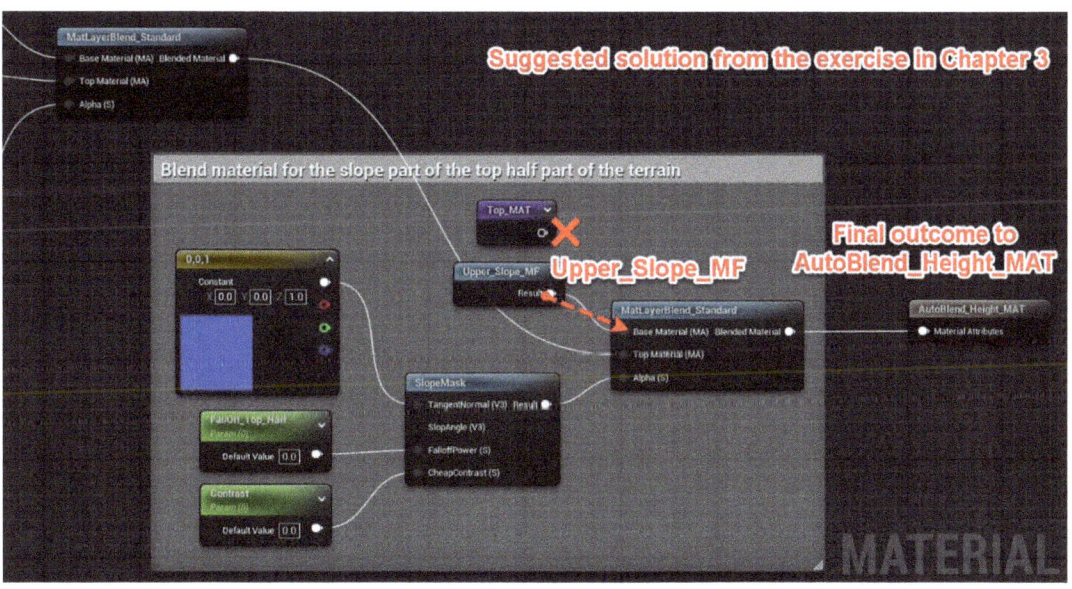

Figure 4-33. Substitute the Top_MAT for the final outcome AutoBlend_Height_MAT with the Upper_Slope_MF material

Finally, we can also adjust the remaining parameters, such as `Distance_Away`, `Ground_Minimum`, `Middle_Mountain`, and `Top_Mountain`, to achieve the visual appearance we desire as shown in Figure 4-34.

Further information about these parameters in the Global Scalar Parameters can be read as follows. The purpose of these parameters is to control various aspects of a material's appearance within our landscape in the level.

The numbers associated with each parameter, such as "Distance_Away", "Falloff_Lower_Half", "Ground-Minimum", "Middle_Mountain", "Random Stripe Offset", and "Top-Maximum", represent specific values that have been set to define certain characteristics. For example:

- "Distance_Away" is to determine how texture or effects change with distance from the camera or a certain point.

- The "Falloff" parameter is to control the rate of change or transition between two states or areas, possibly affecting how textures blend or the intensity of an effect from one area to another.

- "Ground-Minimum" and "Top-Maximum" define the lower and upper bounds for the elevation of our landscape.

- "Middle_Mountain" specifies a midpoint value for elevation changes or another specific attribute related to our mountainous terrain.

- "Random Stripe Offset" is a value for introducing randomness to patterns or stripes, which might be part of a procedural texturing technique.

The specific negative or positive numbers indicate the magnitude or direction of the effect or value. The reason for these exact numbers would be based on the desired visual outcome or the technical requirements of the material within the game or scene. Designers adjust these values through experimentation and visual feedback to achieve the right look and performance.

The values assigned to the material parameters are based on the author's artistic discretion and do not follow a specific formula. They have been fine-tuned to the creator's aesthetic preference, highlighting the subjective nature of visual design.

CHAPTER 4   REVITALIZING VISUALS: ASSET IMPORT AND PROCEDURAL CREATION

You are encouraged to experiment with different values to tailor the material's appearance to your own artistic vision and project needs, as there is no one-size-fits-all set of numbers in the creative process.

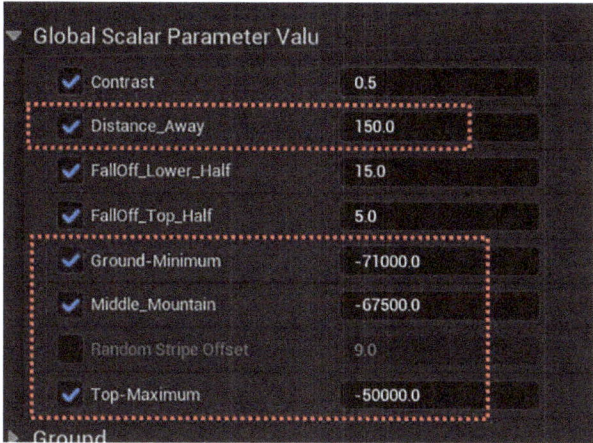

Figure 4-34. *Adjust the relevant values to suit the desired visual*

In UE5, a notable transition is the deprecation of both the tessellation feature and the utilization of the displacement map function in material creation. This move marks a significant shift in the rendering pipeline and material workflow.

Tessellation, which once played a crucial role in adding intricate geometric detail to surfaces, has been phased out in favor of using Nanite to display 3D geometry assets in the level that provides similar intricate detail with improved efficiency.

Procedural Content Generation (PCG)

In UE, PCG can offer significant advantages by enabling developers to generate diverse and expansive virtual worlds efficiently. This could involve generating terrain, architecture, vegetation, and even quests or challenges in games. PCG within UE5 might leverage advanced technologies like Nanite virtualized geometry and Lumen global illumination to create stunning, detailed, and dynamic procedural content.

Ensure that you have activated the Procedural Content Generation plugin, as indicated in Chapter 1 and as shown in Figure 1-20.

Set Up the PCG Volume

To incorporate a PCG volume into our scene, we can locate the PCG volume actor within the "Add Actor" toolbar, as shown in Figure 4-35 with the following steps.

The screenshot shows the interface of UE and outlines the steps to navigate to a specific feature within the engine:

1. Accessing the Place Actors Panel: Here, we are using the Add Actor icon at the top toolbar.

2. Volumes Category: In the "Place Actors" panel, the "Volumes" category has been expanded, showing a list of different volume types that can be placed in the level. These volumes are special actors used to define spaces within the world that can modify or control behavior and properties of objects within their bounds.

3. Procedural Content Generation Volume: Here, we can locate the "Procedural Content Generation" volume, which we can use to facilitate the dynamic creation or alteration of content at runtime.

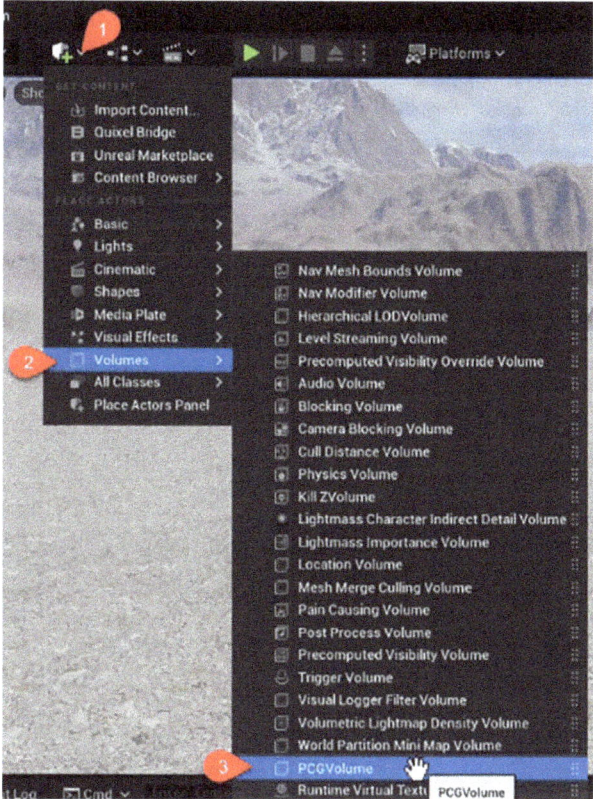

Figure 4-35. *Add a PCG volume*

To properly situate the Procedural Content Generation (PCG) volume within our landscape, we utilize the 2D views available in UE's editor for precise positioning and sizing. Through the top 2D view, we can adjust the volume's XY coordinates and its width and length dimensions, ensuring it spans the full extent of the landscape horizontally. This is illustrated in Figures 4-36 to 4-38, where we fine-tune these settings.

Furthermore, the Front 2D view allows us to manipulate the YZ position and the volume's height, as shown in Figures 4-39 and 4-40. This vertical adjustment is crucial for encompassing the entire landscape's elevation range within the volume's bounds. By employing these views and making careful adjustments, we align the PCG volume with our landscape's dimensions, which is essential for the accurate generation of content across the entire terrain.

The general description of the steps is typically involved in configuring a Procedural Content Generation (PCG) volume in UE, as per the process outlined:

- Top View Configuration (Figures 4-36 to 4-38): Access the Top 2D view in UE to oversee the landscape from above. Here, adjust the PCG volume's X and Y position to align it with the landscape's horizontal dimensions. Resize the volume's width (X-axis) and length (Y-axis) to ensure it encompasses the entire area where procedural content will be generated.

- Front View Configuration (Figures 4-39 and 4-40): Switch to the Front 2D view to observe the landscape from the side. This perspective allows you to adjust the PCG volume's Y position (depth) and Z position (height). Modify the volume's depth to match the landscape's range and its height to cover the full vertical extent of the terrain features.

Through these steps, using the reference bounding box size as a guide, you can accurately position and scale the PCG volume to match the full area of the landscape, ensuring that any procedural content generated fits seamlessly within the game world's existing topography.

In our project, we need to configure the PCG volume to fit our landscape. To achieve this, we will employ the Top 2D view to fine-tune the XY position and XY bounding size (as shown in Figures 4-36 to 4-38). Additionally, we will utilize the Front 2D view to adjust the YZ position and Z bounding size (as shown in Figures 4-39 and 4-40). This step ensures that the generated contents align with the entire landscape, using the reference bounding box size.

CHAPTER 4 REVITALIZING VISUALS: ASSET IMPORT AND PROCEDURAL CREATION

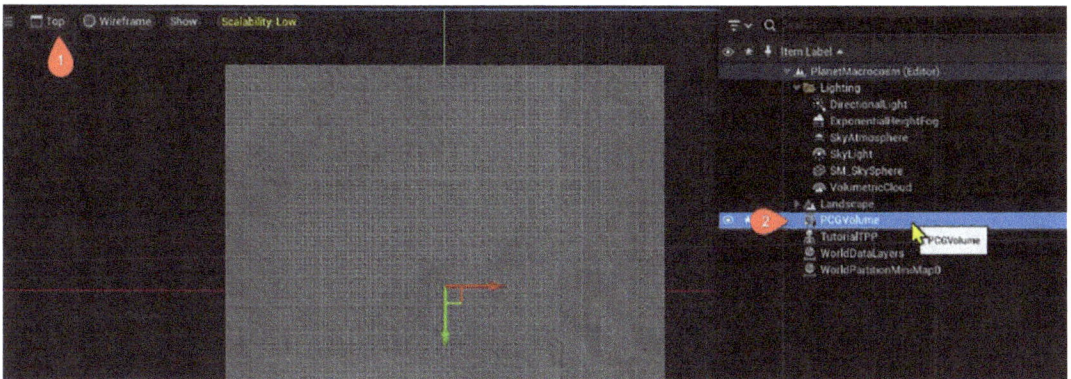

Figure 4-36. *Arrange the PCG volume using the 2D view (Top-view perspective)*

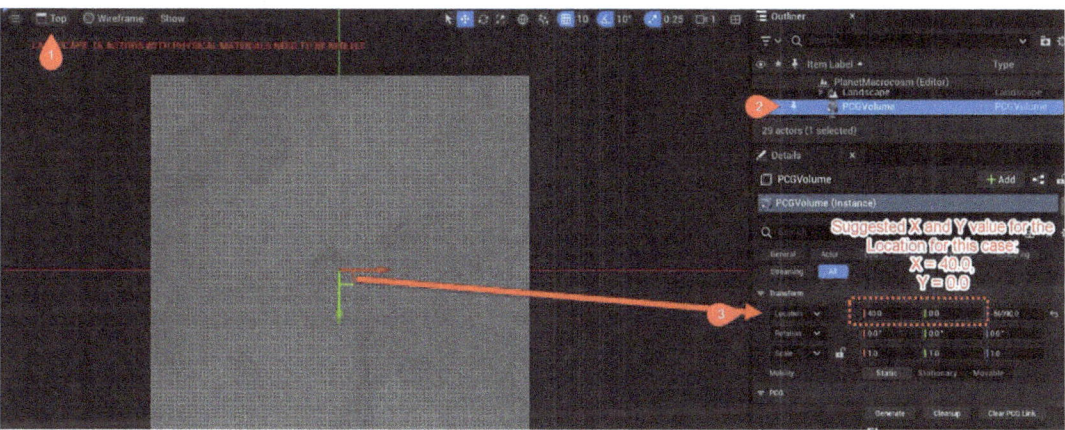

Figure 4-37. *Rearrange the location of the PCG volume in the center of the landscape*

CHAPTER 4   REVITALIZING VISUALS: ASSET IMPORT AND PROCEDURAL CREATION

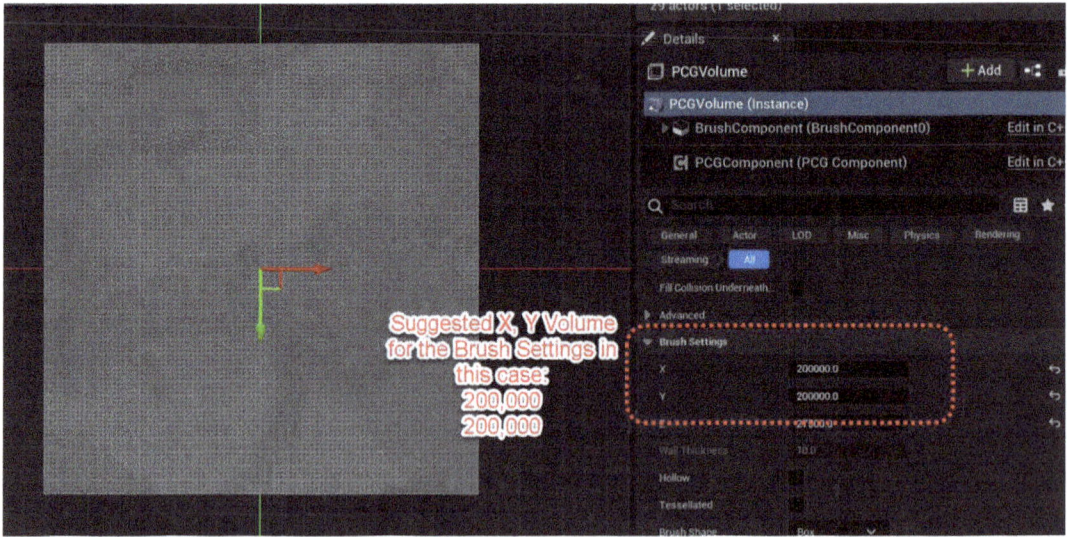

Figure 4-38. *Extend the X and Y volume for Brush Settings to fit the whole landscape in top view*

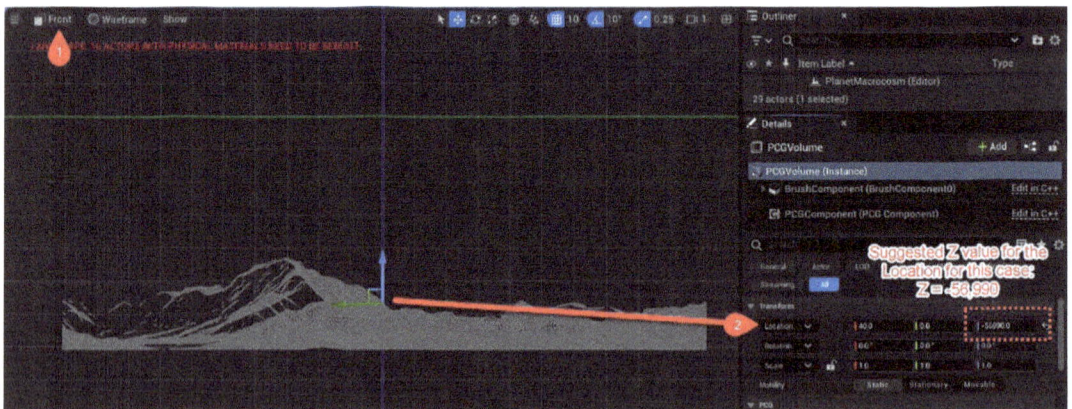

Figure 4-39. *Switch to front view, so the PCG volume can be arranged in the center in Z value*

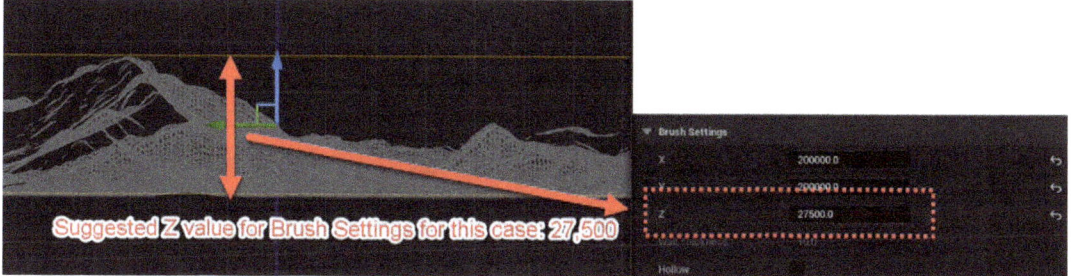

Figure 4-40. *Rearrange the Z bound size and position Z to fit the whole landscape*

The PCG volume requires an algorithm for generating the required assets at specific positions. This is accomplished by creating our own PCG graph (as shown in Figure 4-41) and attaching it to the graph instance, as shown in Figures 4-42 and 4-43 with the following steps for creating and configuring a PCG in UE:

- Initiating PCG Graph Creation: As shown in Figure 4-41, we create a new folder named Volume, and we activate the context menu, before we navigate to and select the "PCG" category, which is dedicated to procedural content tools.

- PCG Graph Asset Creation: Upon selecting "PCG Graph" within the "PCG" category, a new PCG graph asset is created, as shown in Figure 4-42. This new asset, named "Content_Volume", appears in Figure 4-43 and is specifically tagged as a "PCG graph," indicating its role in generating procedural content.

- Assigning the PCG Graph to a PCG Volume: In Figure 4-43, we can configure a PCG volume within the level. In the details panel of the selected PCG volume, we can find the "PCG" section, which includes a "Graph Instance" subsection. Here, we assign the "Content_Volume" PCG graph to the volume by selecting it from the drop-down menu. This links the PCG graph's rules and parameters to the volume, enabling the PCG system to generate content based on the Graph's logic.

- Finalizing PCG Graph Integration: By completing the drop-down selection, the PCG volume in the level is now set to generate content dynamically. This content will conform to the design and constraints specified within the "Content_Volume" PCG graph.

CHAPTER 4   REVITALIZING VISUALS: ASSET IMPORT AND PROCEDURAL CREATION

This detailed process sets the foundation for leveraging UE's procedural generation capabilities, allowing for a more efficient and automated approach to content creation in the game development workflow.

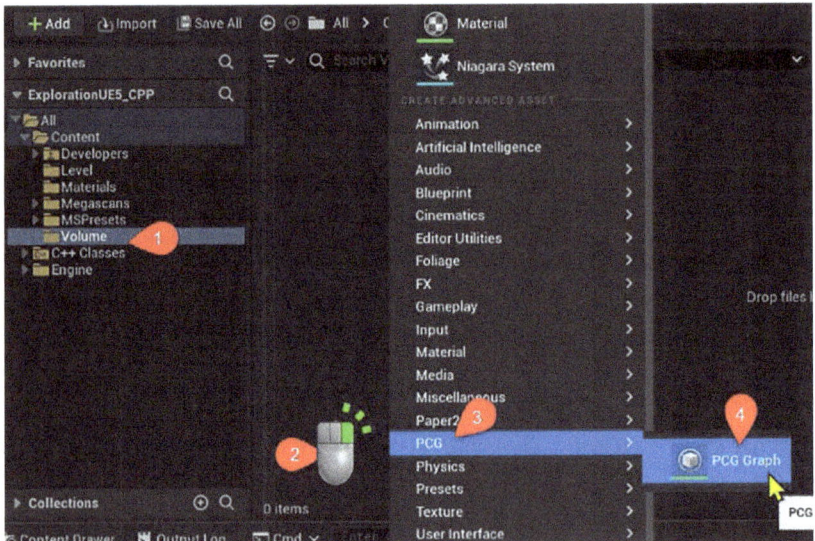

Figure 4-41.   *Create a PCG graph for the PCG volume*

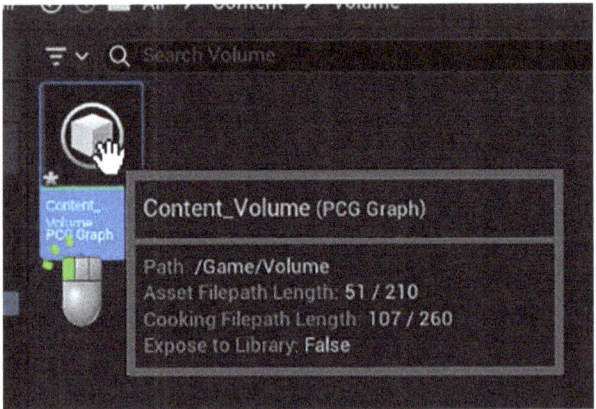

Figure 4-42.   *Single-click to select the PCG graph*

Figure 4-43. *Attach the created PCG graph to the PCG volume's graph instance*

Important PCG Nodes to Look Into

Before delving into the complete design of our Procedural Content Generation to set up our 3D assets in our game level, it is important for us to examine some of the nodes that we are going to use in this design. This examination will enable us to gain a deeper understanding, and it will also provide us with the necessary skills to further manipulate the showcased design, based on what we have learned.

Input Node

The Landscape output and Landscape Height output obtained from the Input node (as depicted in Figure 4-44) within the PCG graph editor of UE5 constitute two distinct methods of acquiring information about the landscape within our level:

- The Landscape output furnishes a series of points that correspond to the vertices of the landscape mesh.

- The Landscape Height output furnishes a scalar value denoting the elevation of the landscape at a given location.

These outputs can be utilized for sampling, alteration, or the generation of points on the landscape, contingent upon our requirements.

To illustrate, we can employ the Landscape output for generating a point cloud that emulates the landscape's contour. Subsequently, we can utilize other nodes to refine or manipulate these points. Furthermore, the Landscape Height output can be employed

CHAPTER 4   REVITALIZING VISUALS: ASSET IMPORT AND PROCEDURAL CREATION

to fine-tune the elevation of points based on the landscape's topography or to devise a mask that omits points exceeding certain height thresholds. Moreover, by amalgamating these outputs with other inputs like splines or meshes, we can engender more intricate and diverse procedural content.

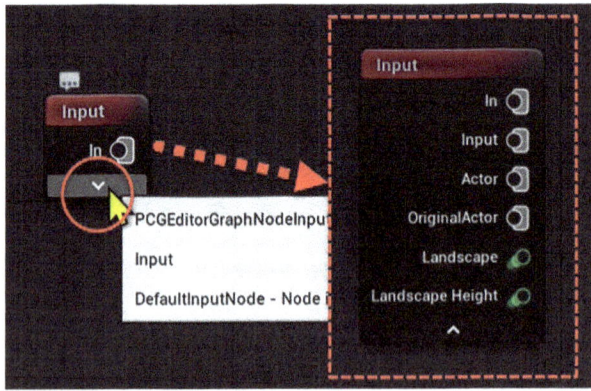

Figure 4-44. First node as Input

Surface Sampler

The Surface Sampler node enables us to generate procedural content based on the surface properties of any mesh within the scene. As depicted in Figure 4-45, we have the ability to regulate the sampling process by adjusting parameters such as points per square meter, point extension, and looseness.

"Points per square meter" determines the quantity of samples extracted from the mesh for each unit area. "Point extension" defines the dimensions of the bounding box around each sample point. "Looseness" governs the extent to which the sample points can deviate from their original positions on the mesh. By modifying these parameters, various levels of detail and variation can be achieved for the generated content.

Figure 4-45. Surface Sampler node and its settings

Point Filter

The Point Filter node refines point sets by filtering them based on criteria such as distance, density, angle, and tag. It takes an input point set and produces an output set of filtered points, allowing customization of properties like proximity thresholds, angle limits, and tag matching.

This node proves useful for performance optimization by removing distant points, ensuring a uniform distribution by adjusting density, creating varied shapes by considering angles, and achieving scene coherence by filtering points based on tags.

As shown in Figure 4-46, we can use the Operator to indicate what kind of modification or transformation we want to apply to the Target Attribute. The Target Attribute is the portion of the sentence that identifies the attribute or property of the points that we want to modify or transform.

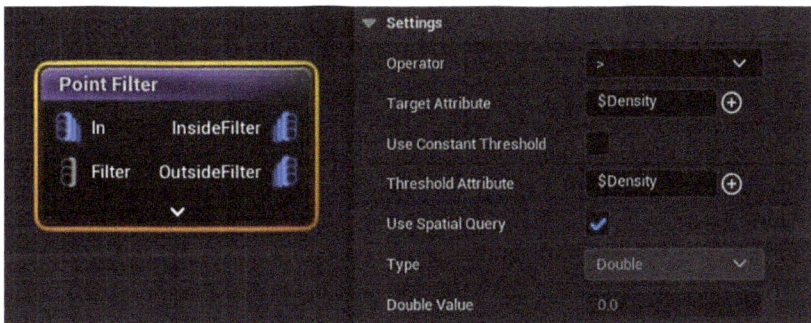

Figure 4-46. Point Filter and its settings

Transform Points

The Transform Point node is versatile and serves multiple purposes, including point relocation, rotation, scaling, and skewing. It can also aid in aligning points with additional objects or surfaces like meshes or splines.

As depicted in the settings presented in Figure 4-47, the Transform Point node allows us to displace the samplers with randomized offsets in position, rotation, and scaling. Furthermore, this node enables the generation of procedural patterns or shapes by applying diverse transformations to points.

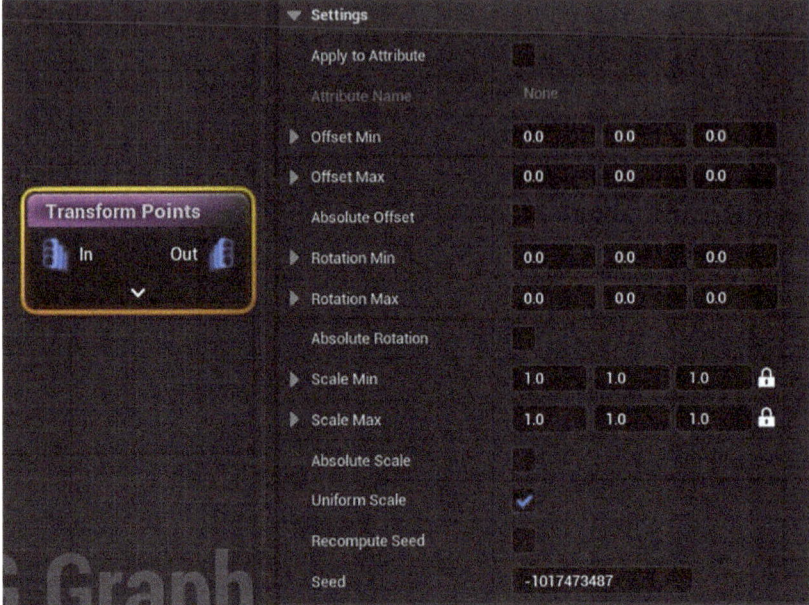

Figure 4-47. Transform Points and its settings

Projection

The Projection node is a node that projects samplers onto a surface. As depicted in Figure 4-48, the Projection node takes a Projection Target as an input and a projection mode as parameters. The Projection Target is an input that enables us to specify a target surface for the projection. By default, the projection is applied to the entire surface of the object that utilizes the PCG graph as a material. However, if we intend to project the texture solely onto a specific part of the object, we can utilize the Projection Target input to establish a mask for the projection.

CHAPTER 4 REVITALIZING VISUALS: ASSET IMPORT AND PROCEDURAL CREATION

Figure 4-48. *Projection node and its settings*

Normal to Density

The Normal to Density node converts the normal vector of a point into a density value. This density value can then be utilized to govern the spawning of meshes or other assets at that particular point. For instance, it can be applied to generate grass on flat surfaces while avoiding steep ones. As shown in Figure 4-49, adjustments can be made to configure the preferred normal vector. Additionally, there is a parameter known as Density Mode that dictates the method by which the density value is derived from the dot product value.

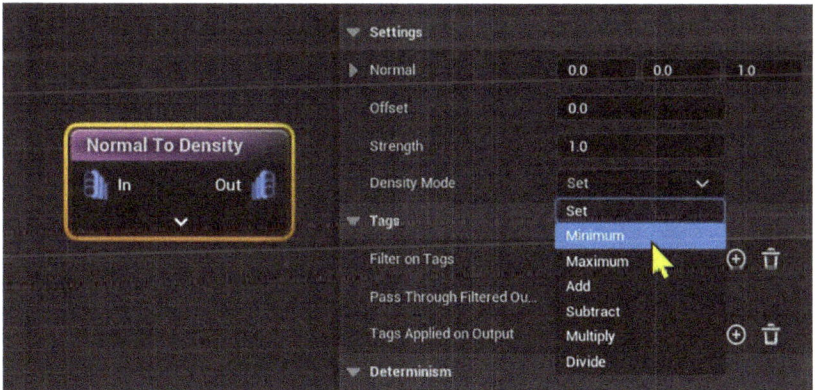

Figure 4-49. *Normal to Density and its settings*

Density Filter

The Density Filter node filters points based on their density value. The density value is a measure of how crowded a point is by other points in its vicinity. As shown in Figure 4-50, it has two parameters: lower bound and upper bound. These parameters define the range of density values that the node will accept. Points with density values outside this range will be discarded.

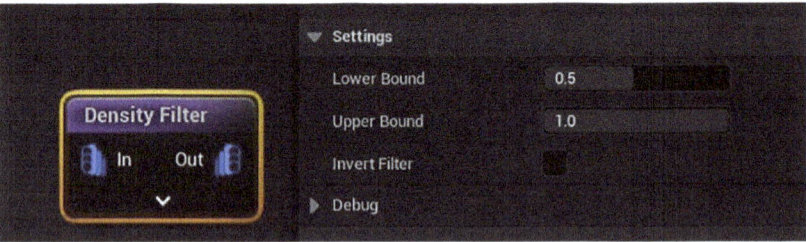

Figure 4-50. *Density Filter and its settings*

Self-Pruning

The Self-Pruning node can be used to remove overlapping points in a point cloud based on their bounds. The bounds are the dimensions of the meshes or actors that are spawned on the points. As shown in Figure 4-51, the Self-Pruning node has a Pruning Type parameter that offers different pruning options.

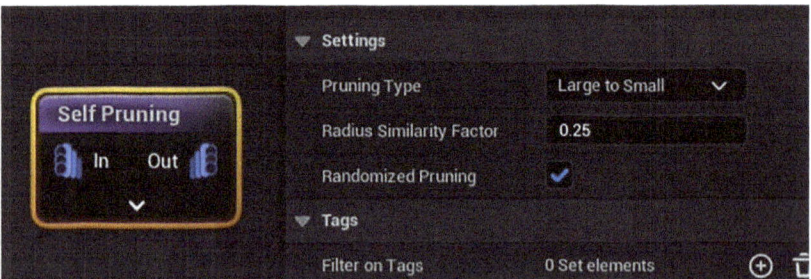

Figure 4-51. *Self-Pruning and its settings*

Bounds Modifier

The Bounds Modifier node in UE5 enables adjustments to a point cloud's boundaries. These boundaries, depicted in Figure 4-52, establish the point cloud's spatial range using minimum and maximum coordinates within the world space. These boundaries play a role in point generation, system performance, and memory usage in the PCG system.

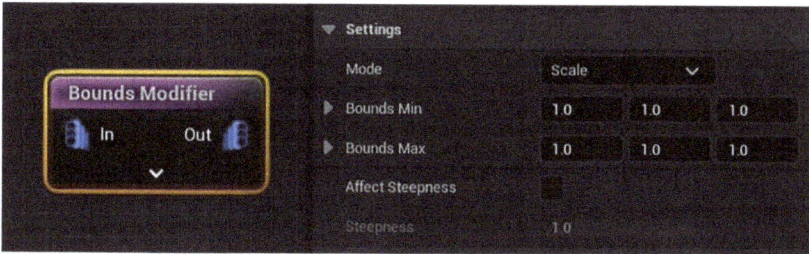

Figure 4-52. Bounds Modifier and its settings

Difference

The Difference node is a node that takes two inputs: a set of points and a geometry. The node outputs the points that are outside the geometry, effectively subtracting the geometry from the point set. As shown in Figure 4-53, the node also has a Density Function input, which is a custom function that can be used to modify the density of the points based on their position, distance, or other attributes.

The Density Function can be used to create interesting effects, such as hollowing out the geometry, creating holes or patterns, or adding noise or variation to the point distribution. We can also use a Density Function to create a terrain-like geometry by subtracting a set of samplers from the source of the sampler points.

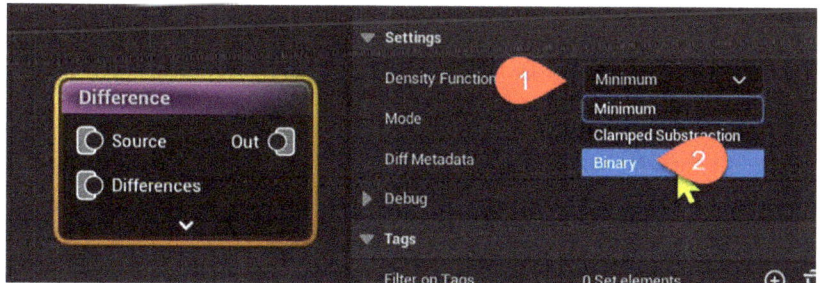

Figure 4-53. Difference and its settings

Static Mesh Spawner

The Static Mesh Spawner node is a tool for generating static mesh assets within a scene. It takes a collection of input points and generates corresponding Static Mesh Fragments. These fragments are segments of geometry that can be further manipulated using other nodes. As shown in Figure 4-54, the node includes a Mesh Entries input, an array of static mesh assets utilized for spawning the geometry.

When multiple mesh entries are present, the node randomly chooses one asset from the array for each point (or the randomness can also be based on the weight of individual assets). This selected asset is then spawned at the designated point's position, orientation, and scale. Various parameters within the node offer control over spawning behavior, including options for random rotation, random scale, alignment mode, and collision mode.

The Static Mesh Spawner node is particularly valuable for constructing intricate scenes that encompass diverse variations of static meshes, such as structures, boulders, vegetation, or objects.

Optimization is key in certain situations, and in the individual mesh entries, we have the flexibility to modify collision modes, enabling options like no collision or shadows for very small spawned meshes. We can further control the rendering distance (culling) for these meshes, ensuring that only nearby objects are rendered. This optimization significantly improves gameplay performance and efficient resource usage.

Here are the steps and descriptions that we can follow to add new static mesh into this Static Mesh Spawner:

1. Mesh Entries Setup: As shown in Step 1, we can add additional "Mesh Entries" elements, where we can define the array of static meshes to spawn. Each index within this array corresponds to a different mesh that can be procedurally placed in the scene as shown in Step 2.

CHAPTER 4   REVITALIZING VISUALS: ASSET IMPORT AND PROCEDURAL CREATION

2. Component Class and Descriptor: Step 3 shows that within the "Mesh Entries," there's a "Descriptor" field, and a "Component Class" drop-down is set to "Hierarchical." This specifies the type of component that will be used for spawning, with "Hierarchical" suggesting the use of a hierarchical instancing system for more efficient rendering of multiple instances of the same mesh.

3. Static Mesh Selection: Step 4 shows that we can select the specific static mesh asset we want to be procedurally generated in the world, though the figure is showing "None" which indicates that there is no static mesh selected.

This node is part of a larger procedural system that automates the placement of static mesh assets in a scene, and these configurations are crucial in determining how the assets will be distributed, which assets will be used, and how they will be rendered for performance optimization.

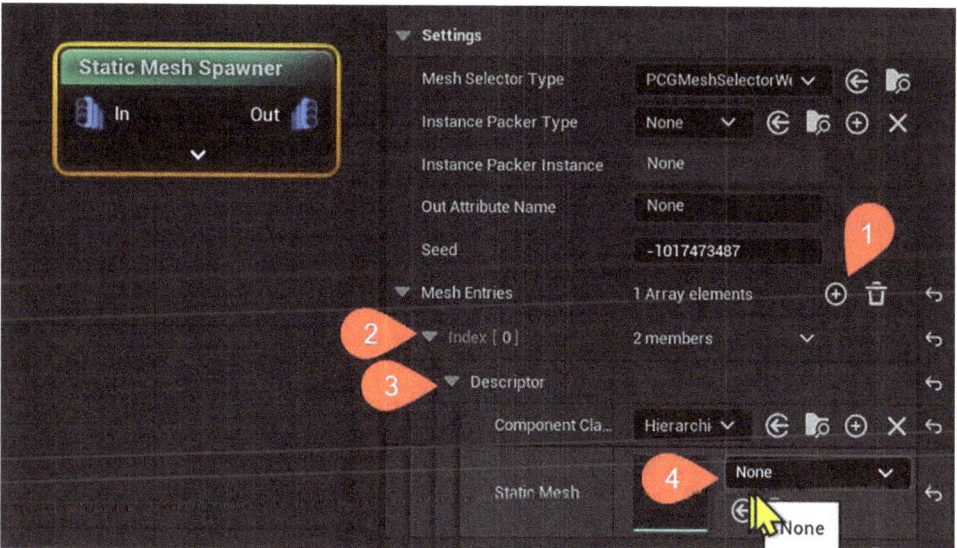

Figure 4-54. *Static Mesh Spawner and its settings about Mesh Entries*

Force Regen

The Force Regen button in the PCG graph editor (as shown in Figure 4-55) is a useful feature that allows us to manually trigger the regeneration of the procedural content. This can be helpful when we want to see the changes in our graph without having to exit and reenter the editor or when we want to test different random seeds for our content.

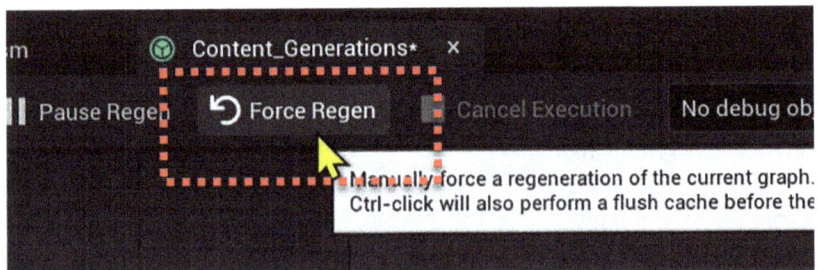

Figure 4-55. *Force Regen button at the top of the PCG graph editor*

When we click it, the PCG graph will execute and update the content in the viewport. We can also use the keyboard shortcut Ctrl+R to force regen our content.

Design of Content Volume PCG Graph

While the behavior of the PCG graph may differ from that of a material editor, the fundamental concept of utilizing the graph closely resembles that of a material editor. To access the graph editor, simply double-click the PCG graph asset (as shown in Figure 4-56).

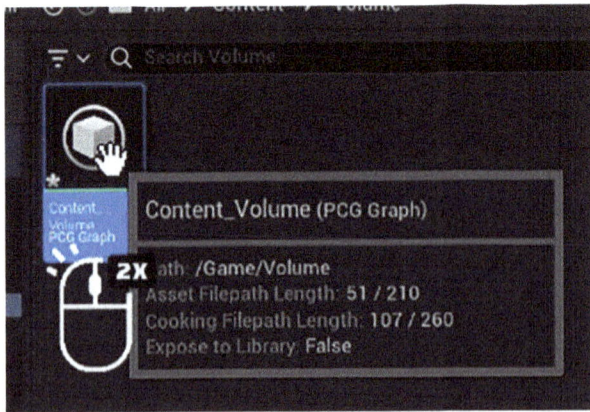

Figure 4-56. *Double-click to open up the Content_Volume PCG graph*

Procedural Generate Large Rocks

The initial stage of our Procedural Content Generation graph involves spawning a large rock, as shown in Figure 4-57. However, our intention is for these substantial rocks to be visible only below a certain level of the landscape, rather than across the entire landscape.

This constraint can be achieved by configuring the Surface Sampler, which is connected to the Point Filter. Within this setup, the Point Filter can be adjusted using the operator <= for a specific $Position.Z value (as the attribute).

Subsequently, we can connect the Transform Points node to enable the random rotation and scaling of these large rocks and also apply Normal Project with specific bounds of the Normal to Density value.

Before we spawn the large rock, we will make a final adjustment to the Density Filter to ensure that the samplers will not be too densely packed together.

Finally, we will set up the spawner static mesh and its settings, as shown in Figure 4-61, using the downloaded static mesh from Megascans as demonstrated in the earlier section of this chapter.

The details of each process can be seen in the figures following Figure 4-57.

Figure 4-57. *Generate large rocks*

Figure 4-58 shows a portion of a PCG graph, including the following elements and their corresponding values:

1. Input Node – Landscape Height: The input node, labeled "Landscape Height," is where the graph receives data about the terrain's elevation. This data is then used to inform the rest of the procedural generation process.

2. Surface Sampler Node: The "Surface Sampler" node is connected to the "Landscape Height" input and appears to be responsible for sampling the landscape's surface based on the height data. Its output will provide surface information necessary for the placement of objects.

3. Point Filter Node: The "Point Filter" node is where certain conditions are defined to filter points on the landscape's surface, as determined by the "Surface Sampler." This node may be used to exclude or include areas of the landscape for object placement based on various criteria.

4. Settings – Point Filter: The settings for the "Point Filter" node are shown in the details panel, where parameters like "Points Per Squared Meter," "Looseness," and "Seed" for randomness are configured. The specific value is "Double Value" set to –63,008, which represents a threshold within the filter's operation within the PCG graph.

These components collectively define how the procedural system samples the landscape's surface and decides where to place content based on the height information and filtering criteria. The numbered settings can be fine-tuned to achieve your desired distribution and density of procedurally generated content.

CHAPTER 4   REVITALIZING VISUALS: ASSET IMPORT AND PROCEDURAL CREATION

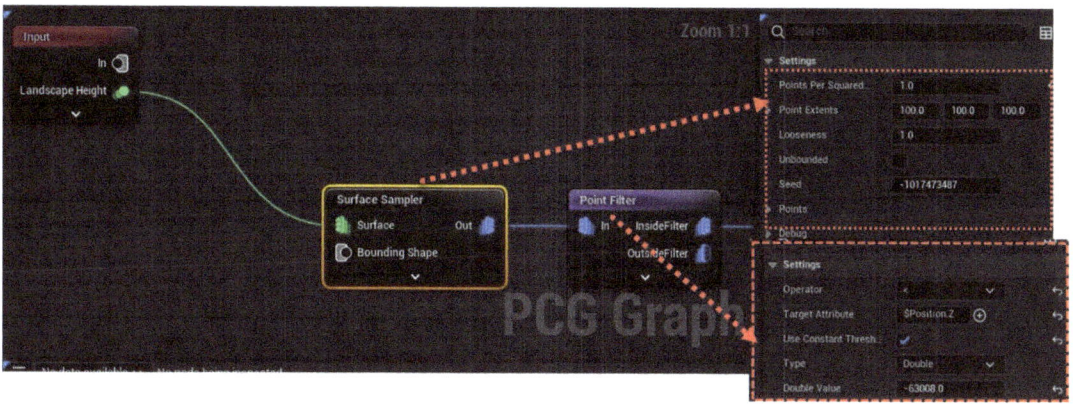

Figure 4-58. *Surface Sampler connecting to Point Filter*

The second portion of this PCG graph connection is now carried on by Figure 4-59, after the Point Filter connecting to Transform Points and so on, as described in the following:

1. Transform Points Node: The "Transform Points" node, connected to the "Inside filter" parameter from the "Point Filter" node, is used to manipulate the position, rotation, and scale of the points based on the input data. This allows for adjustments to the placement of generated content in relation to the landscape features.

2. Projection Node: After the points are transformed, they are passed through a "Projection" node, which projects them onto a target, such as the terrain, ensuring that the generated content conforms to the landscape's contours. In this case, we use the input data from the Landscape parameter in the Input node.

3. Normal to Density Node: The "Normal to Density" node takes the projected points and adjusts their density based on the normal of the projection target. This can be used to control the distribution of procedural content, such as vegetation, so it's denser on flat areas and sparser on steep slopes.

157

CHAPTER 4 REVITALIZING VISUALS: ASSET IMPORT AND PROCEDURAL CREATION

4. The numbers on the details panel correspond to specific settings within the "Normal to Density" node:

   a. Offset Name: A label for identifying the offset in a system where multiple offsets might be used.

   b. Offset Max and Min: Define the range of values for the procedural offset.

   c. Rotation Min and Max: Specify the range for the procedural rotation of content, with a visual indicator set at 360 degrees.

   d. Scale Min and Max: Set the limits for scaling procedural content, ensuring it stays within desired size bounds.

The editor provides a visual interface for setting up complex rules that determine the placement of procedural content, with the ultimate goal of creating a more dynamic and varied game environment.

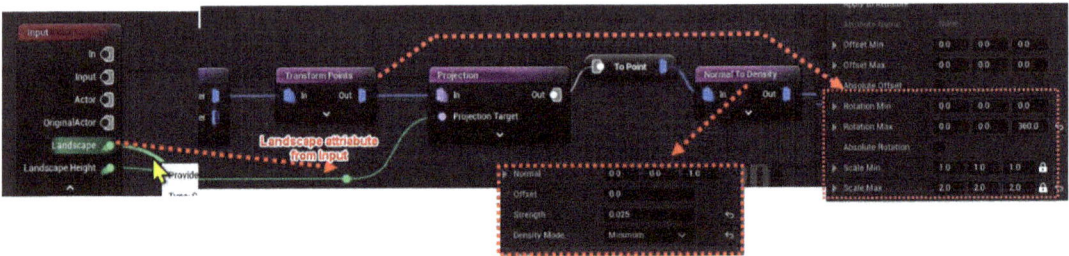

Figure 4-59. *Use of the Landscape input for the Projection node, after the Transform Point, and connect to the Normal to Density*

Figure 4-60 features the final stages of this set of connections:

1. Normal to Density Node: This node is responsible for converting normal map information into density values. This is used to control the density of procedurally placed objects based on the angle of the terrain's surface, affecting how many objects are placed on slopes vs. flat areas.

2. Density Filter Node: Following the conversion, the "Density Filter" node applies a filter based on the density values determined by the "Normal to Density" node. This is used to fine-tune the placement of objects further, ensuring they are distributed according to the desired density across the terrain.

CHAPTER 4   REVITALIZING VISUALS: ASSET IMPORT AND PROCEDURAL CREATION

   a. Settings Panel: The panel shows settings related to the "Density Filter" or "Static Mesh Spawner" node. The range of bound value (0.9–1.0) corresponds to a limit or threshold value that is being set for the procedural system. This might define a cap on the maximum number of objects to be spawned or a minimum required density for spawning to occur.

3. Static Mesh Spawner Node: The "Static Mesh Spawner" node is where the actual instantiation of the static meshes occurs, based on the filtered density values. This node controls the spawning of 3D objects onto the terrain, placing them where they are needed according to the procedural rules defined in the graph.

These nodes work together to dictate where and how often certain objects appear within a game environment, based on terrain data and specified rules. The final part of the connection solidifies the procedural logic, culminating in the placement of static meshes to enhance the game's visual complexity and realism.

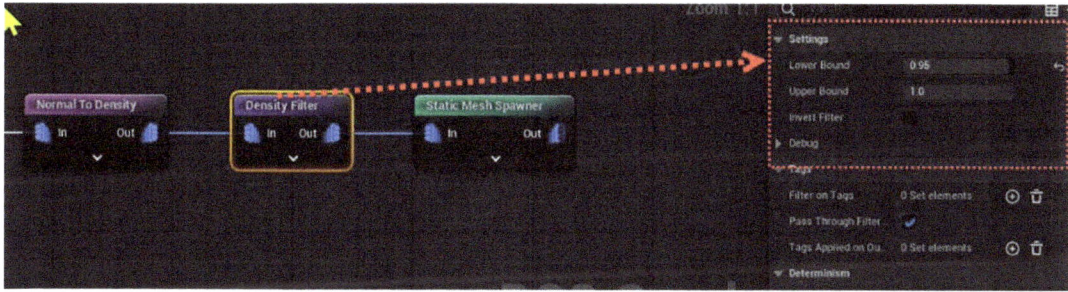

Figure 4-60. *The density value for this large rock sampler*

Figure 4-61 shows the steps and settings that we can set for the "Static Mesh Spawner" node:

1. Static Mesh Assignment: The "Static Mesh" field is where we select the specific mesh that will be procedurally placed in the scene. In this case, we use a static mesh named "S_ForestRock_Formation_vqauv" to be used by the spawner.

2. Mobility Setting: The "Mobility" setting determines whether the spawned meshes are static, stationary, or movable. We set the "Mobility" option to "Static," meaning the instances of this mesh will not be expected to move or change in game, which can be beneficial for performance.

159

3. Collision Presets: The "Collision Presets" dropdown is set to "BlockAll" ensuring that the spawned meshes use a collision preset that blocks all types of collisions, which is important for gameplay and physics interactions.

4. Instance End Cull Distance: We set the "Instance End Cull Distance" with the value of 75,000 to indicate the distance from the camera at which the meshes will no longer be rendered. This is a performance optimization setting that helps to manage the rendering load by culling distant objects from the scene.

Each step here in this setting configures how the static meshes will be generated and interact within the game world, affecting both visual fidelity and performance. The "`Static Mesh Spawner`" node is a key component in controlling the procedural placement of assets in the environment.

CHAPTER 4   REVITALIZING VISUALS: ASSET IMPORT AND PROCEDURAL CREATION

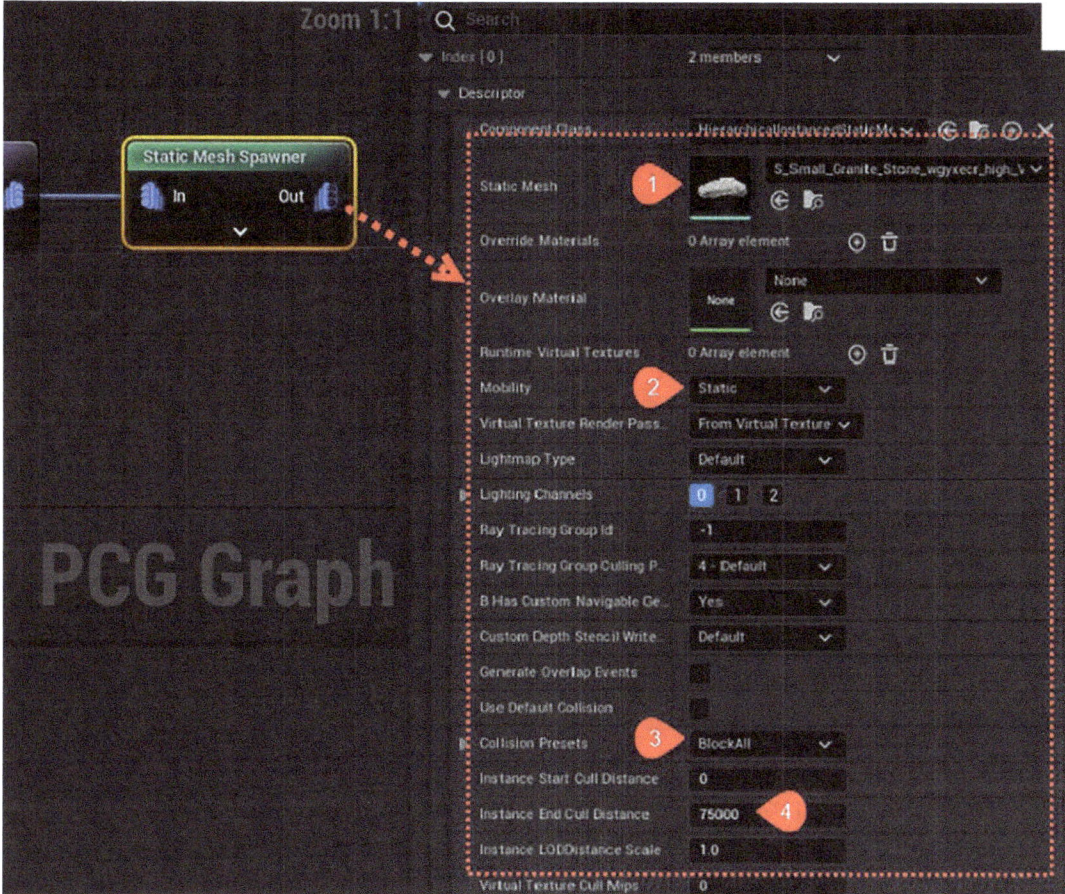

Figure 4-61. Setting up the spawner for the large rock and its attribute value

Procedural Generate Gravel Stones

The approach to procedurally generating gravel stones is quite similar to that of large rocks. However, in this case, we aim to generate them in larger quantities and at closer distances to each other. This is illustrated in Figure 4-62. An essential aspect of this design is ensuring that gravel stones do not spawn within the bounding positions already occupied by the large rocks (given that the large rocks are static). This optimization prevents unnecessary resource usage.

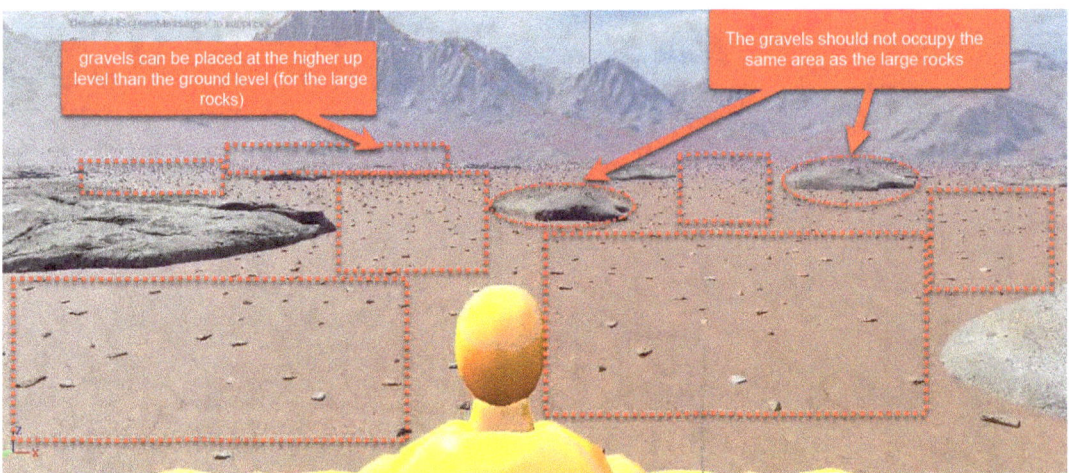

Figure 4-62. *Procedurally generate the gravel stones*

In our terrain, gravel stones are procedurally placed at higher elevations than large rocks, using a distinct Z position in the Point Filter for targeted elevation control. To ground the stones and avoid a floating appearance, their Z offset is slightly reduced, ensuring they nestle into the landscape. Moreover, we will also apply the `Difference` node that ensures the gravel does not spawn in large rock areas, maintaining separation and enhancing realism, while the gravel's varied static meshes have no collision enabled to optimize performance, relying on culling distance for efficient rendering.

Figure 4-63 shows the beginning of the configuration within a procedural generation system for the gravel stones:

1. Surface Sampler Node: This node takes in "Landscape Height" data from the "OriginalActor" input and samples the surface based on this information. The "Bounding Shape" output defines the area within which sampling occurs.

    a. Settings – Point Filter: The settings panel for the "Point Filter" shows configurations such as "Points Per Squared Meter," "Point Extents," and "Looseness," which affect the density and distribution of the points. The auto-generated seed value is sufficient to be used to maintain consistent randomness in procedural generation.

CHAPTER 4   REVITALIZING VISUALS: ASSET IMPORT AND PROCEDURAL CREATION

2. Point Filter Node: The "Point Filter" is connected to the "Surface Sampler" and filters points on the landscape's surface according to certain criteria. The "InsideFilter" and "OutsideFilter" outputs indicate whether a corresponding point is categorized as either inside or outside under a specified condition.

   a. Simultaneously, we desire a higher quantity of gravel stones to be spawned at an elevation above any of the large rocks. This necessitates the use of a separate Surface Sampler with distinct settings, including a different $Position.Z value for the Point Filter.

   b. Tag Settings: Within the settings panel, under "Tags," further adjustments are made with an "Operator" as <= and a "Target Attribute" specified as "$Position.Z". The "Use Constant Threshold" is enabled, and a "Float Value" is provided. This value, denoted by the number –68,500, sets a vertical threshold, which can be used to filter points based on their height in the landscape.

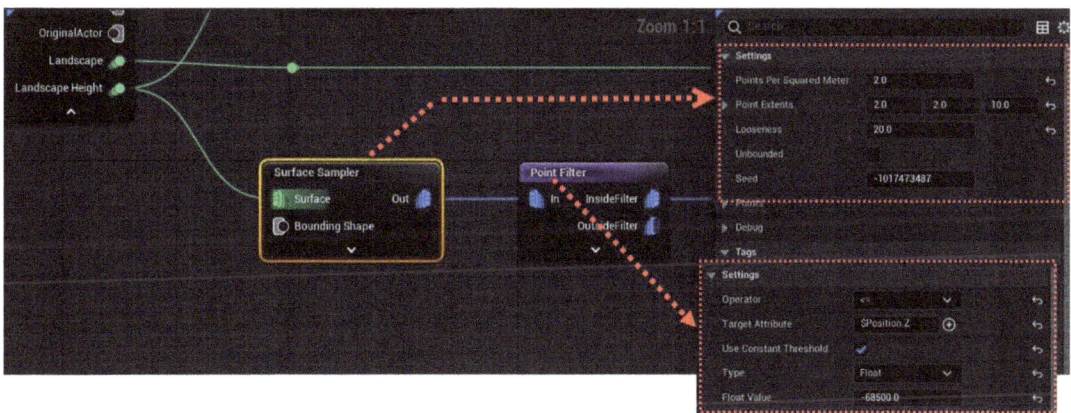

Figure 4-63. *Surface Sampler and Point Filter for generating gravel stones*

Figure 4-64 shows the follow-on of the node graph (from Point Filter) where the height of objects relative to the landscape is being manipulated:

1. Transform Points Node: This node adjusts the position, rotation, and scale of points that have been generated on the landscape. Manipulating these points is essential to determine where objects will be procedurally placed in relation to the terrain's topography.

163

a. Offset Min and Max: Here, specifically with a value of –1.0 pointed by the number, it indicates the range within which the object's height can be offset relative to the ground. A negative offset value is to allow us to position the generated contents to bury slightly within the landscape, which is to ensure they don't appear to float above the ground.

2. Projection Node: Points that have been transformed are then projected onto a target, which is typically the landscape itself. This step ensures that the objects adhere to the landscape's contours and height variations.

3. Normal to Density Node: The "Normal to Density" node is used to convert the surface normals of the landscape into a density map. This map can dictate where objects are more or less likely to be placed, based on the inclination of the terrain. We can leave the default settings for this Normal to Density node.

We should remember that manipulating height without understanding the source of ground elevation can be challenging. We are retrieving the height data that is from the landscape's elevation information, which informs how objects should be placed relative to the surface. Without this context, procedural content cannot integrate seamlessly with the terrain.

The offset parameter, which we intentionally set to a negative value to embed objects into the landscape, is a technique to make sure objects, such as this gravel stones, are grounded and do not appear to be floating due to slight inconsistencies or variations in the terrain's surface. This technique can enhance realism, as objects are rarely perfectly placed on the terrain in real life; they often intersect slightly with the ground.

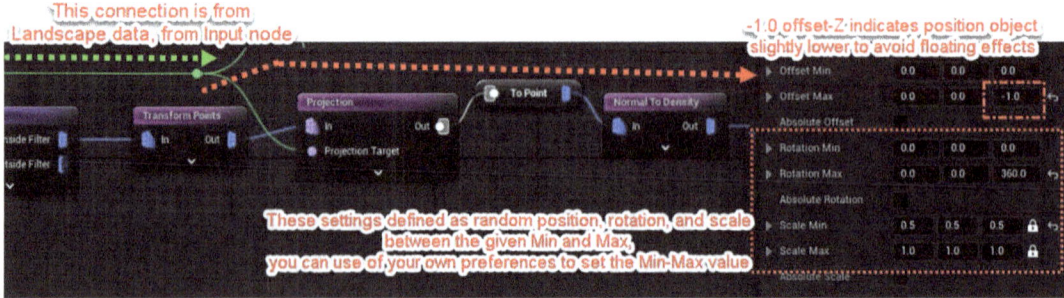

Figure 4-64. *Set the Offset Max.Z value to "sink" down the sampler a little*

CHAPTER 4   REVITALIZING VISUALS: ASSET IMPORT AND PROCEDURAL CREATION

The RotationMax.Z value is set to 360 degrees, which typically allows for a full rotational range in the Z-axis. This means that any object or particle affected by this setting can rotate a complete circle, giving it freedom to face any direction. This is useful in simulations where objects need to appear randomly oriented, as it would in nature, contributing to a more organic and varied visual output.

The Scale Min value, which is set to 0.5, specifies the smallest size that the objects or particles can scale down to, which is half of their original scale. By not allowing the scale to go down to zero, we ensure that the objects are always visible and contribute to the scene while still allowing for significant variation in size. This can prevent the objects from becoming too small to be noticeable or from disappearing entirely, which could be important depending on the visual effect or interaction you aim to achieve.

Figure 4-65 shows the next segment of the PCG, continuing from the Normal To Density connection to the Density Filter and Self Pruning nodes to generate the gravel stones. Firstly, we will set up the necessary nodes to organize the gravel stones itself, and we can set up the associated settings:

1. Density Filter Node: The "Density Filter" node adjusts the distribution of points based on density values. The settings displayed show that the filter has been configured with a `lower bound of 0.5 and an upper bound of 1.0`, establishing a range within which the density values will be considered valid for the placement of objects.

2. Self-Pruning Node: The "Self-Pruning" node is configured to refine the distribution of points further. The settings reveal that the "Pruning Type" is set from "Large to Small," with a "`Radius Similarity Factor`" of 0.25. This suggests that points too close to each other, based on this factor, will be pruned to prevent crowding, with preference given to keeping smaller objects over larger ones within close proximity.

The numbers provided in the settings are critical for the procedural generation logic. They determine the density of the points that will be used for spawning objects and how densely these objects will be allowed to cluster together.

These settings help control the procedural placement of objects to achieve a desired level of natural randomness and distribution across the game environment.

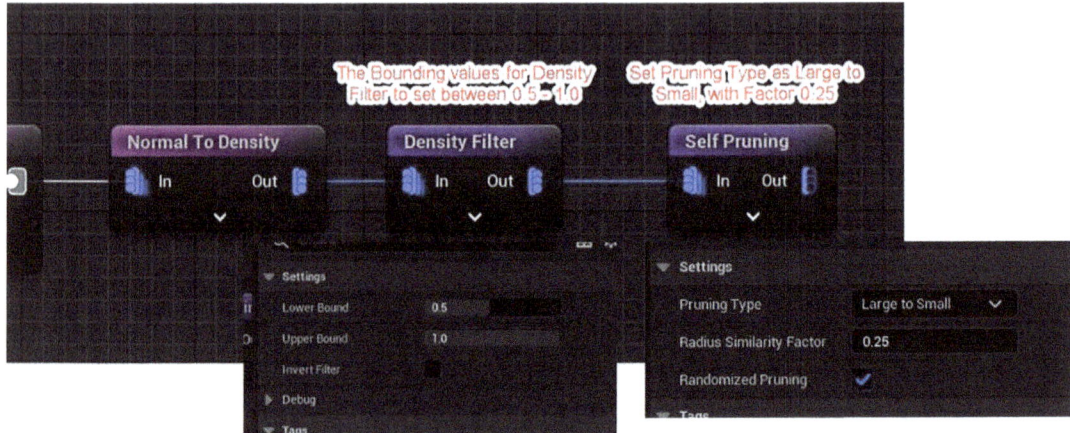

Figure 4-65. *Continue to set up the next segment to generate a gravel stone*

This segment that is outlined in Figure 4-66 shows the process of spawning static meshes in relation to the terrain and other objects, specifically ensuring that new objects like gravel stones do not overlap with existing large rock formations (we can see the configuration settings after this section):

1. Connect the Density Filter Node from above PCG Large Rock to the Bounds Modifier.

2. Bounds Modifier Node: The "Bounds Modifier" node adjusts the boundaries within which the points are valid, possibly refining the area for spawning further.

3. Self-Pruning Node: The "Self-Pruning" node eliminates points that are too close to each other to prevent overlapping meshes.

4. Union Node: This node combines multiple sets of points, incorporating the modifications made by the "Self-Pruning" and "Bounds Modifier" from the previous section into a single set of points to be used by the "Difference" node.

5. Difference Node: The "Difference" node subtracts one set of points from another, here indicated to be used for ensuring that the points where gravel stones will be generated do not overlap with the area under large rocks. The "Difference" node ensures that procedural generation respects the existing large rock formations by not placing new objects in the same space.

6. Static Mesh Spawner Node: This node takes the filtered points and uses them to spawn static meshes within the game world. The placement of these meshes is based on the data provided by the "Density Filter."

Each node is a step in the procedural generation process, with its own role in determining where and how objects are placed in relation to the landscape and other features. The final outcome is a landscape that dynamically spawns objects like gravel stones in a realistic and visually appealing manner without intersecting with large rock formations.

Lastly, as shown in Figure 4-66, the Difference node can be employed to ensure that gravel stones are not spawned within the area occupied by the large rock. This is accomplished by utilizing the large rock's bounds as the source for the Difference from the gravel stone area. Furthermore, three distinct gravel stone variations are configured as static mesh entries, with their collision set to None due to their status as small objects within the level. By establishing an appropriate culling distance, the rendering resources can be efficiently optimized.

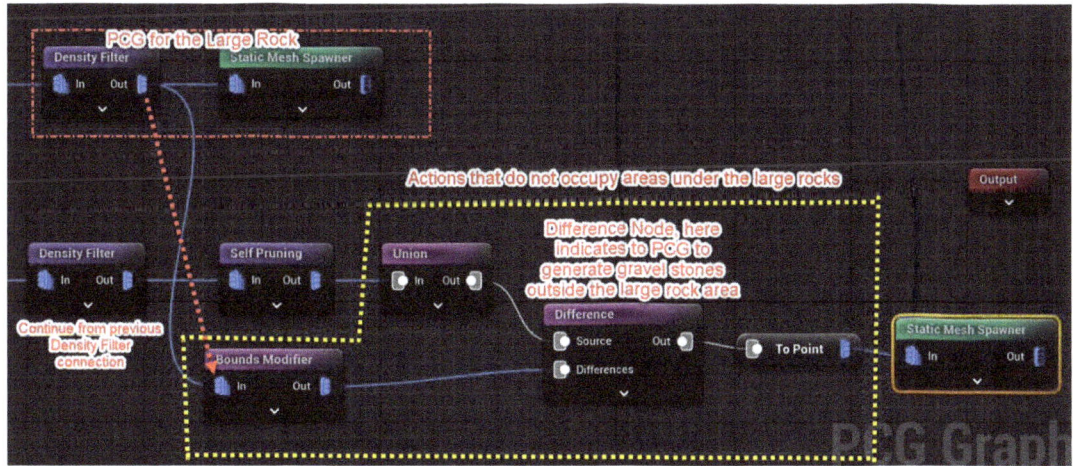

Figure 4-66. Settings that do not occupy the large rock area, and set three different gravel stones for the static mesh entries

CHAPTER 4   REVITALIZING VISUALS: ASSET IMPORT AND PROCEDURAL CREATION

Figure 4-67 shows the configuration of a PCG for the segment from before:

1. Bounds Modifier Node: The "Bounds Modifier" node is configured with specific minimum and maximum scale bounds, set to 1.5 and 1.55, respectively, as indicated by the red dashed box. This adjustment defines the scaling limits for objects being spawned, ensuring they fit within certain size parameters.

2. Difference Node: The "Difference" node, configured with a "Density Function" set as "Binary" and "Mode" as "Inferred," subtracts one set of points from another. This setup ensures that objects (e.g., gravel stones) are not placed in areas where other objects (e.g., large rocks) exist, as per the procedural rules.

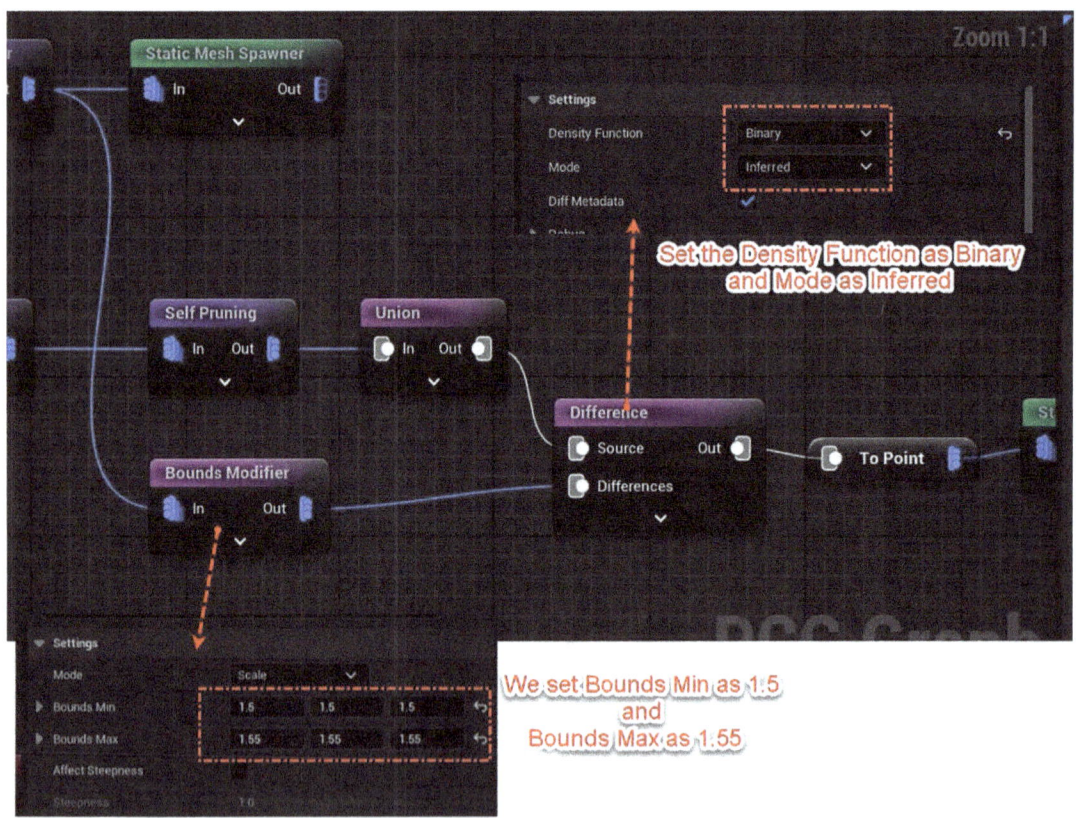

Figure 4-67. *Settings for the Bounds Modifier and Differences*

CHAPTER 4   REVITALIZING VISUALS: ASSET IMPORT AND PROCEDURAL CREATION

Figure 4-68 shows the configuration panel of a "Static Mesh Spawner" node from before detailing how static meshes are set up for procedural spawning the gravel stones:

1. Mesh Entries Configuration: The "Mesh Entries" are set up with three different static meshes for procedural spawning. The specific meshes are named "Small_Granite_Stone_wgyxexr", "Small_Sandstone_Rock_xdzobhg", and "Small_Sandstone_vd5scae", which are the variants of rocks that will be used in the PCG process.

2. Mobility Setting: The "Mobility" attribute for each mesh is set to "Static," meaning once placed, these meshes will not move or be subject to dynamic changes during gameplay.

3. Collision Preset: The "Collision Preset" is set to "NoCollision," indicating that these meshes will not interact with other game physics or collisions. This is typical for small objects where collision detection is not necessary or could be performance-intensive.

4. Instance End Cull Distance: The "Instance End Cull Distance" is set to "75,000", which likely represents the distance from the camera at which the engine will stop rendering these objects. This is a performance optimization to prevent rendering objects that are too far away to be seen by the player.

These settings in the PCG graph are crucial for optimizing the game's performance while ensuring that the environmental details, such as rocks, are generated where desired in the game world. The numbers provided for each attribute help define the behavior of each static mesh within the environment, contributing to the overall realism and efficiency of the game.

CHAPTER 4   REVITALIZING VISUALS: ASSET IMPORT AND PROCEDURAL CREATION

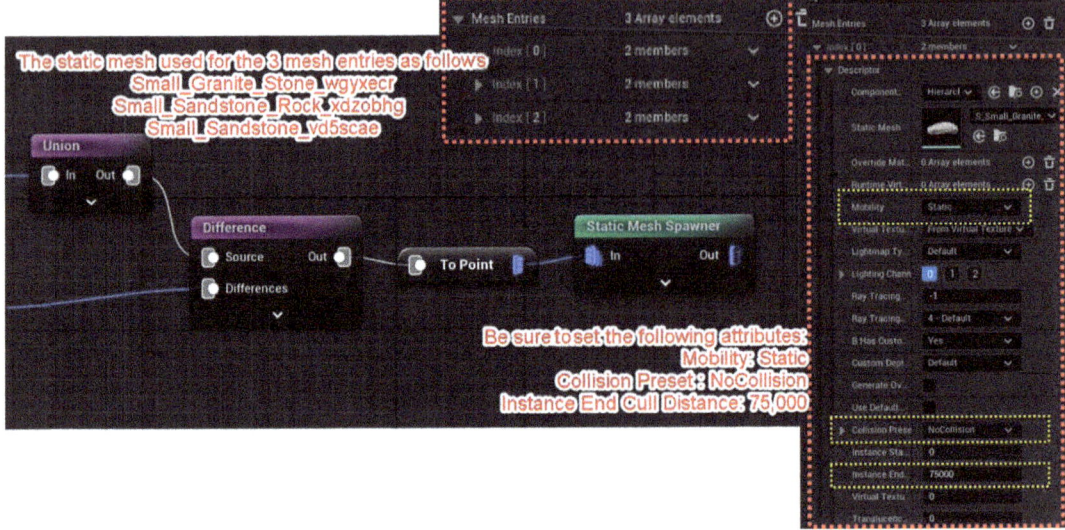

Figure 4-68. Settings for the Static Mesh Spawner node

Procedural Generate Vegetations

We would like to generate another asset: vegetation within the level. Specifically, we aim to generate vegetation up the mountain where no gravel stones are being spawned, as shown in Figure 4-69.

Figure 4-69. Procedurally generated vegetation up the mountain

The procedural setup shown in Figure 4-69 can be restructured to facilitate the placement of vegetation in relation to the small granite areas. To accomplish this, the existing "`Point Filter`" nodes (*the lowest bottom Point Filter in the editor*) used for

170

gravel stones are repurposed. Contrary to the original configuration which utilizes the "Inside Filter" to designate placement within the gravel stone areas, the "Outside Filter" is employed instead. This strategic adjustment dictates that vegetation will only be generated **outside** the predefined small granite zones, ensuring no overlap occurs between the vegetation and gravel stone placements, as shown in Figure 4-70.

The intentional use of the gravel stone reference in the filter is to guarantee that vegetation spawns only at elevations higher than those of the gravel stones, thereby maintaining a clear stratification between these two types of environmental features.

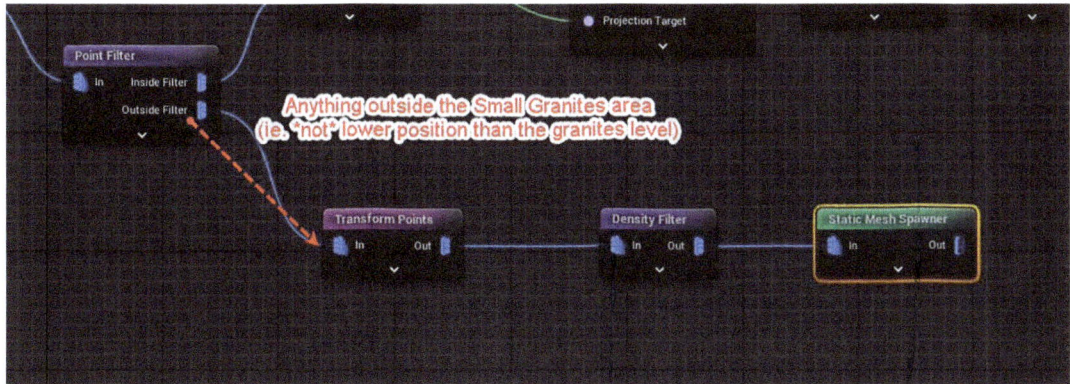

Figure 4-70. *PCG nodes to generate the vegetations (above the gravel stone level)*

Figure 4-71 shows the "Transform Points" node settings as follows:

1. Transform Points Node: This node is responsible for modifying the properties of points in 3D space, which will dictate where and how objects are placed within the game world. The points could represent locations for object spawning, and the transformations adjust their final position, rotation, and scale.

2. Settings for Transformation

   a. Rotation Settings: The "Rotation Min" and "Rotation Max" are set at 0 and 360 degrees, respectively, indicating that the objects can be rotated randomly along the vertical axis, allowing for a full 360-degree range of possible orientations.

CHAPTER 4   REVITALIZING VISUALS: ASSET IMPORT AND PROCEDURAL CREATION

b. Scale Settings: The "Scale Min" and "Scale Max" settings are set at 5 and 10, respectively. These values define the minimum and maximum bounds for scaling objects, allowing for a range of sizes from half the original size up to double the original size.

c. Uniform Scale: The checkbox for "Uniform Scale" is marked so that the scaling transformation will maintain the object's proportions, scaling equally along all axes.

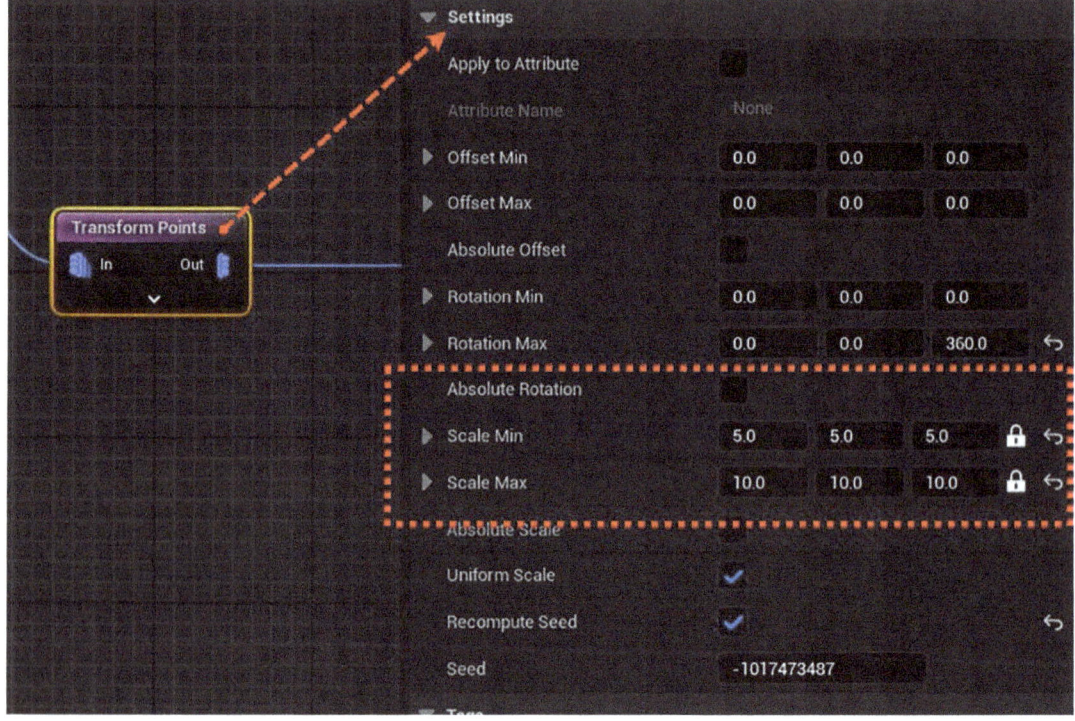

Figure 4-71. *Settings for the Transform Points (vegetations)*

Figure 4-72 shows the "Density Filter" node and its settings within a Procedural Content Generation (PCG) graph:

1. Density Filter Node: This node is utilized to control the density of procedural points that are used to spawn objects. The points passing through this node will be filtered based on the density settings defined in the right-hand settings panel.

CHAPTER 4   REVITALIZING VISUALS: ASSET IMPORT AND PROCEDURAL CREATION

2. Settings for Density Filter

    a. Lower Bound: The lower bound of the density filter is set to 0.9, which specifies the minimum threshold of density that points must meet to be considered for spawning objects.

    b. Upper Bound: The upper bound is set to 1.0, establishing the maximum density threshold. Points with a density value within the 0.9 to 1.0 range will pass through the filter for further processing in the PCG graph.

The numbers in the settings determine the range of point density that is allowed by the filter, directly affecting where and how densely objects will be placed in the game environment. A higher lower bound means fewer points will qualify, potentially resulting in less cluttered object placement. Conversely, a higher upper bound could allow more points to pass through, possibly leading to a more densely populated area with procedural objects. The "Density Filter" is a crucial tool for managing the distribution of procedurally generated content, ensuring that the game world remains visually balanced and performs well.

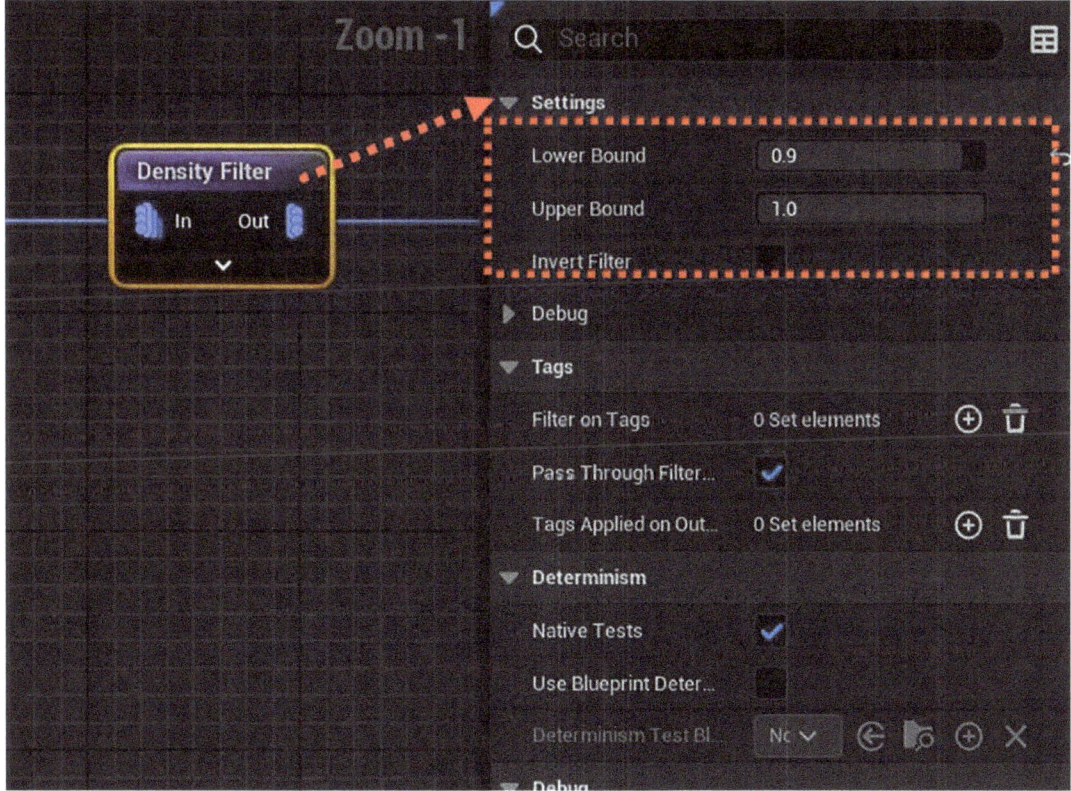

Figure 4-72. Settings for the Density Filter (vegetation)

CHAPTER 4 REVITALIZING VISUALS: ASSET IMPORT AND PROCEDURAL CREATION

Figure 4-73 shows a "Static Mesh Spawner" node within a Procedural Content Generation system. The node and its settings are configured for spawning static meshes as follows:

1. Static Mesh Spawner Node: This node is used to place static mesh objects into the game world based on the input points it receives. The objects to be spawned are defined in the "Mesh Entries" section of the settings.

2. Mesh Entries: The "Mesh Entries" setting shows an array with three elements, indicating that there are three different types of meshes that this spawner can place. Each index in the array likely corresponds to a different mesh that will be procedurally generated.

3. Descriptor Settings

    a. The "Descriptor" section within each index specifies the details of the static meshes, such as their "Component Class" and the specific "Static Mesh" to be used. In this case, the "Component Class" is set to "Hierarchical Instanced Static Mesh Component," which is efficient for rendering multiple instances of the same mesh.

    b. In this case, we will use the following static mesh as examples; however, you can use any other alternative 3D plants: "S_Broom_Creeper_xficgrqx_0_Var5_lod0", "S_Broom_Creeper_xficgrqx_0_Var8_lod0", and "S_Broom_Creeper_xficgrqx_0_Var11_lod0".

    c. Mobility: The "Mobility" setting is configured to "Static," meaning that the placed meshes will not move or change post-placement.

    d. Collision Presets: The "Collision Presets" are set to "NoCollision," specifying that these meshes will not interact with the game's physics engine for collision detection.

    e. Instance End Cull Distance: The "Instance End Cull Distance" is set to 75,000. This value represents the distance from the camera at which the engine will stop rendering these objects, a common optimization technique in game design to improve performance.

CHAPTER 4   REVITALIZING VISUALS: ASSET IMPORT AND PROCEDURAL CREATION

The numbers and settings in this node define how and where the static meshes will appear in the game world. The "Static Mesh Spawner" is essential for adding complexity and detail to the environment without manually placing each object, streamlining the level design process.

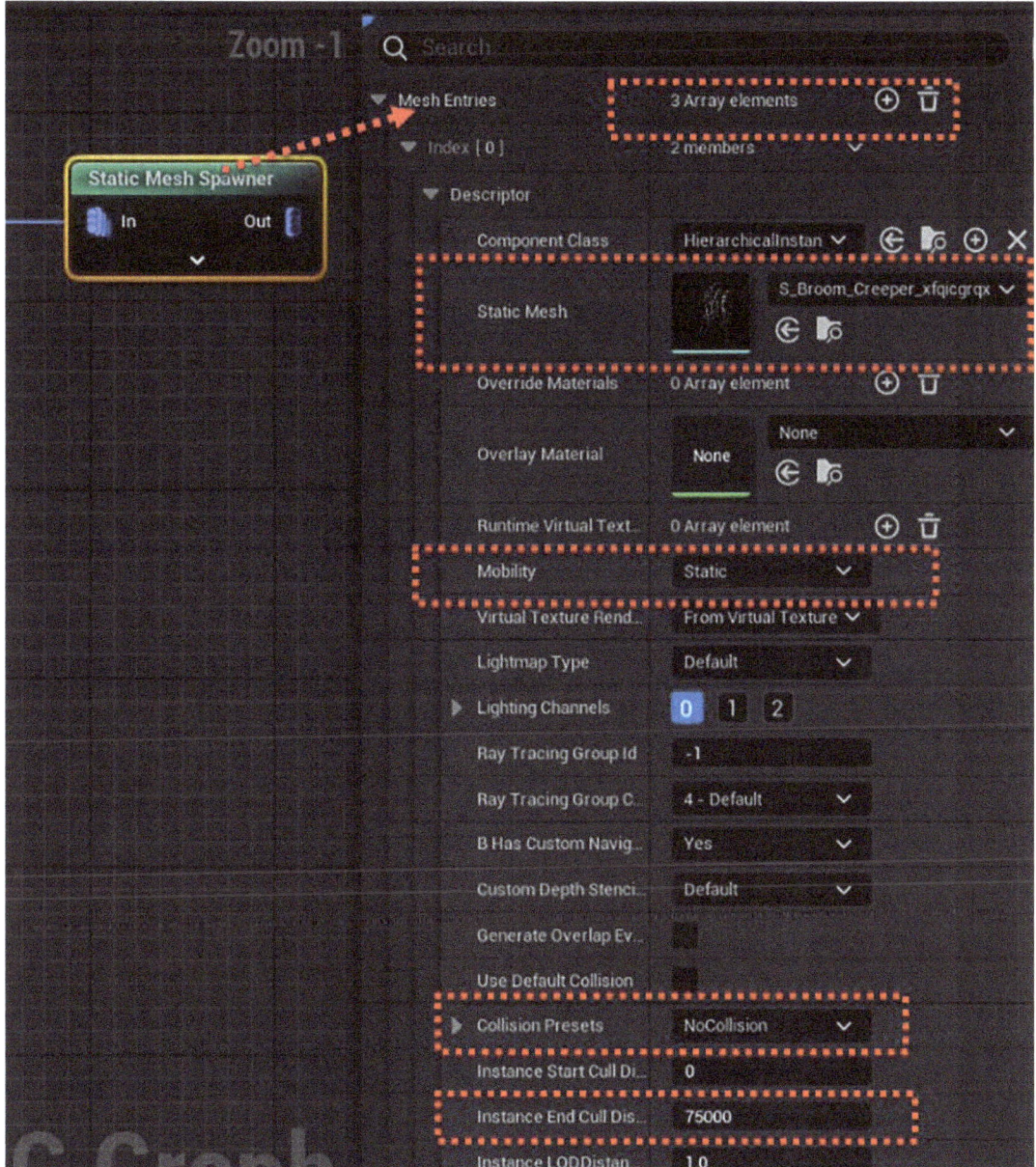

Figure 4-73. *Settings for the Static Mesh Spawner to generate the vegetations*

The settings and usage of the nodes in this PCG graph are merely suggestions. Therefore, you are welcome to experiment and try out various nodes and settings to create a visually appealing landscape as a whole.

Final Visualization Result

Figure 4-74 showcases the final result of a carefully designed Procedural Content Generation (PCG) process within a virtual landscape. It vividly illustrates diverse elements such as PCG vegetations, which are densely scattered across the terrain's higher elevations, contributing to the ecosystem's realism.

In the lower area of the terrains, we observe PCG gravel stones peppered across the surface, offering a natural variation to the terrain. Dominating the landscape are the PCG large rocks, strategically placed to provide focal points and enhance the rugged feel of the environment.

This harmonious blend of PCG elements results in a dynamic and immersive world, demonstrating the powerful capabilities of PCG in crafting detailed and lifelike game environments.

CHAPTER 4   REVITALIZING VISUALS: ASSET IMPORT AND PROCEDURAL CREATION

Figure 4-74. *Final result of PCG large rocks, gravel stones, and vegetations*

When faced with issues where PCG is not functioning as expected, a methodical approach to debugging is essential.

Start by isolating the problem: verify that all assets are correctly referenced and that there are no broken links or missing files. Examine the PCG logic for errors in node connections or misconfigured parameters that could be disrupting the procedural workflow.

Additionally, check for compatibility issues that may have arisen from software updates or changes in the project settings. Utilize the debugging tools such as pressing the D key when selecting a node and observe if there's any cube generated, observing the generation at each stage to pinpoint where it diverges from expected behavior.

Review the console for any error messages or warnings that could offer clues, and consider simplifying the PCG setup to its most basic form to ensure fundamental processes are operational before gradually reintroducing complexity.

Through meticulous testing and validation of each component within the PCG system, you can identify the root cause and apply the necessary fixes to get your procedural content up and running smoothly.

Also, be sure to have the PCG volume setup and configuration with the correct PCG graph asset (refer to the beginning section from Figures 4-35 until 4-43).

Summary

In UE5, this chapter extensively explored enhancing visual prowess through Quixel Bridge integration and Procedural Content Generation (PCG). Quixel Bridge acts as a transformative bridge, linking UE5 with Megascans and facilitating effortless asset import. This synergy empowers creators to seamlessly integrate photorealistic 3D models, textures, and materials, fundamentally elevating the visual authenticity and immersive potential of digital environments. Additionally, Nanite technology is introduced, revolutionizing asset rendering by enabling real-time display of film-quality assets regardless of complexity. This innovative approach eliminates traditional LOD constraints, yielding an unparalleled level of visual detail and realism, especially in close-ups.

Delving into Procedural Content Generation, the chapter unveiled the PCG graph editor as a powerful toolset. This dynamic environment offers a plethora of nodes and techniques to manipulate points and assets systematically. Practical examples showcased the potential of PCG, illustrating the generation of diverse elements such as large rocks, gravel stones, and vegetations. By merging Quixel Bridge's integration, Nanite's rendering prowess, and the creative possibilities of PCG, the chapter painted a vivid picture of UE5 as a platform poised to redefine digital landscapes, marrying intricacy and immersion in an unprecedented visual storytelling experience.

In the upcoming chapter, we will enhance the visual details of material blending between the large rock and the landscape. As shown in Figure 4-75, there exists a noticeable edge separation between the large rock and the terrain, which lacks realism. We will delve into learning how to utilize the Runtime Virtual Texture feature to seamlessly blend materials from the static mesh and the landscape materials.

CHAPTER 4 REVITALIZING VISUALS: ASSET IMPORT AND PROCEDURAL CREATION

Figure 4-75. *Obvious edge separation between the large rock and the terrain*

CHAPTER 5

Enhancing Visual Realism with Runtime Virtual Textures and Material Blending

This chapter will continue from the previous chapter where we will try to sort out the issues with continuous blending between the material from the 3D assets and the landscape materials. We will explore using the feature of Runtime Virtual Textures that allow an advanced rendering technique that enhances the efficiency and quality of texture streaming in large and detailed virtual environments.

Runtime Virtual Texture

Runtime Virtual Texture (RVT) is a technology within UE that addresses the challenge of rendering large and highly detailed environments with high-quality textures without consuming excessive memory or causing performance issues.

In traditional texture streaming methods, lower-resolution versions of textures, called mipmaps, are loaded as the player approaches an object or surface. However, in scenes with a vast number of high-resolution textures, this can still lead to memory overload and visible texture popping.

In UE5, the RVT dynamically generates texture data through Runtime GPU processing. This mechanism bears resemblance to the conventional approach of texture mapping. The RVT efficiently stores shading information across expansive regions, proving particularly advantageous for landscape shading. This is especially true when incorporating materials resembling decals and utilizing splines that seamlessly adapt to the natural lay of the land.

Color Map in RVT

RVT takes a different approach. It allows textures to be streamed directly from storage rather than preloading all textures into memory. This is achieved by creating a virtual texture that represents a massive texture atlas (the RGB color). This atlas contains different levels of detail, like mipmaps, but these levels are generated dynamically as needed, instead of being precomputed and stored in memory.

When a player approaches an object or surface, UE's RVT system calculates which sections of the virtual texture need to be loaded at the required level of detail and streams only that information from storage. This significantly reduces memory consumption while maintaining high-quality textures up close.

For example, in an open-world game with intricate landscapes, the terrain textures can be incredibly detailed. Instead of loading all the terrain textures into memory, RVT generates the necessary texture data on the fly, providing a balance between realism and performance.

Increase Framework and Optimization with RVT

As RVT dynamically generates texture mipmaps and streams the textures directly from storage, RVT enables the display of high-resolution textures without overloading memory resources. This technology is particularly beneficial for expansive open worlds and highly detailed scenes where traditional texture streaming methods might lead to performance bottlenecks. UE5's RVT system aims to provide a seamless and immersive visual experience by efficiently managing texture data, contributing to the overall realism and quality of interactive virtual worlds.

As a groundbreaking feature in UE, RVT presents a transformative solution that significantly elevates both framework robustness and optimization within the realm of game development. Consider a sprawling open-world game where players traverse diverse landscapes, each demanding intricate textures and fine details.

Traditionally, the challenge lies in accommodating these demands without burdening the system's memory. This is where RVT steps in. As players navigate through the environment, the technology dynamically loads only the necessary high-resolution textures, seamlessly streaming them from the disk. This targeted approach slashes memory usage while upholding visual fidelity. For instance, imagine a forest in the game with an array of unique flora and fauna. RVT would allow each tree, leaf, and creature to boast lifelike textures, without overwhelming the hardware. This advancement ensures that UE5 not only enriches the visual aspects but also empowers developers to craft expansive worlds that run fluidly and remain immersive.

World Height in RVT

The relationship between world height and Runtime Virtual Texture in UE5 is a crucial factor in achieving immersive and detailed virtual environments. World height, often represented through heightmaps or 3D elevation data, plays a pivotal role in determining the level of terrain detail and complexity.

RVT, a cutting-edge rendering technique, allows for efficient memory usage by streaming high-resolution texture data only where needed, based on the viewer's proximity to the surfaces. The accuracy of world height data directly impacts how textures are applied to different elevations, influencing the visual fidelity and realism of landscapes. UE5 utilizes this synergy between world height and virtual texturing to enhance the visual quality of expansive game worlds, providing an unparalleled sense of depth and intricacy to the virtual spaces.

Imagine a vast open-world game set in a mountainous region. The terrain consists of rolling hills, towering peaks, and deep valleys. To create this environment, developers utilize world height data, which represents the elevation of each point on the terrain. This data is often encoded in heightmaps, where brighter values represent higher points and darker values indicate lower points.

Instead of loading entire high-resolution textures for the entire terrain, virtual texturing enables the engine to dynamically stream texture data based on the player's viewpoint. As the player moves through the world, only the textures needed for the visible surfaces are loaded, saving memory and improving performance.

Now, the relationship between world height and virtual texturing becomes evident. Let's say the player stands at the base of a towering mountain. The height data determines how the textures are applied to the mountain's sides, taking into account the steepness and elevation changes. As the player climbs, the engine dynamically loads higher-resolution textures to ensure that the close-up surfaces are as detailed and realistic as possible.

Conversely, as the player looks across a wide valley, the height data dictates the terrain's shape, which in turn affects how textures are stretched or compressed. RVT ensures that even the distant parts of the valley have appropriate textures, maintaining a consistent level of quality throughout the player's field of view without overloading memory.

In this scenario, the synergy between world height and Runtime Virtual Texture creates a breathtaking experience. The detailed representation of elevation in the heightmaps allows the engine to accurately apply textures to various surfaces, while virtual texturing optimizes memory usage and maintains performance, resulting in a seamless and immersive open-world environment.

Overall, UE5's integration of world height and RVT empowers developers to craft expansive and visually stunning game worlds, where terrain intricacies are matched by efficient rendering processes, ultimately enhancing the overall player experience.

Material Blending and RVT in UE5

In UE, the concepts of material blending and RVT are related in the sense that they can be used together to enhance the visual quality and efficiency of a game's graphics. When using RVTs, we can apply material blending techniques to the high-resolution textures that are being streamed in real time.

For instance, blending materials between a 3D asset and the landscape is a technique that enables the creation of seamless transitions between various objects and the terrain in our scene. This technique can enhance the realism and visual quality of our environment by eliminating harsh edges and unnatural gaps.

As RVT is a feature introduced in UE5, it facilitates material blending by storing the shading outcomes of intricate materials on a virtual texture. Subsequently, this texture can be utilized by other materials and actors to seamlessly blend with the landscape. This texture is used to generate RVT.

For example, imagine a scenario where we have a terrain in an open-world game. With RVTs, we can stream high-resolution textures for the terrain's surface, allowing for intricate details even at a distance. Material blending can then be used to layer different textures on the terrain based on factors like slope, altitude, or masks generated procedurally.

This creates a more realistic and visually appealing terrain, with the blending occurring on the high-resolution textures that are being streamed using RVTs.

CHAPTER 5   ENHANCING VISUAL REALISM WITH RUNTIME VIRTUAL TEXTURES AND MATERIAL BLENDING

Set Up an RVT Asset in UE5

Before we can utilize RVT in our level, we need to include an RVT as an asset in our content folder. For the purpose of this project, we will generate the RVT asset within a folder we have created named `RuntimeVT`, as shown in Figure 5-1.

1. Create a new folder (in this case, the author named the folder RuntimeVT).

2. Right-click the blank space on the Content to activate the selection of new asset creation.

3. Hover over the `Texture` selection to activate the submenu of the textures.

4. Then, click the `Runtime Virtual Texture` to create a new RVT asset.

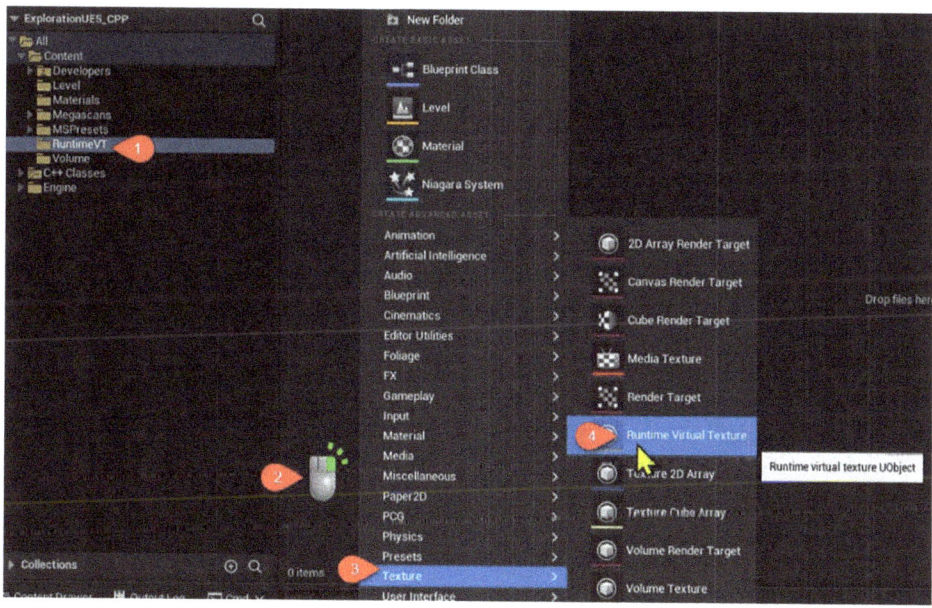

Figure 5-1. *Add the Runtime Virtual Texture asset*

While the default settings for the RVT are sufficient for the purpose of our Color (RGB) RVT, it is wise for us to explore the settings for the type of virtual texture contents. We can access the RVT editor by double-clicking the `Colour_RVT` asset, as shown in Figure 5-2.

CHAPTER 5   ENHANCING VISUAL REALISM WITH RUNTIME VIRTUAL TEXTURES AND MATERIAL BLENDING

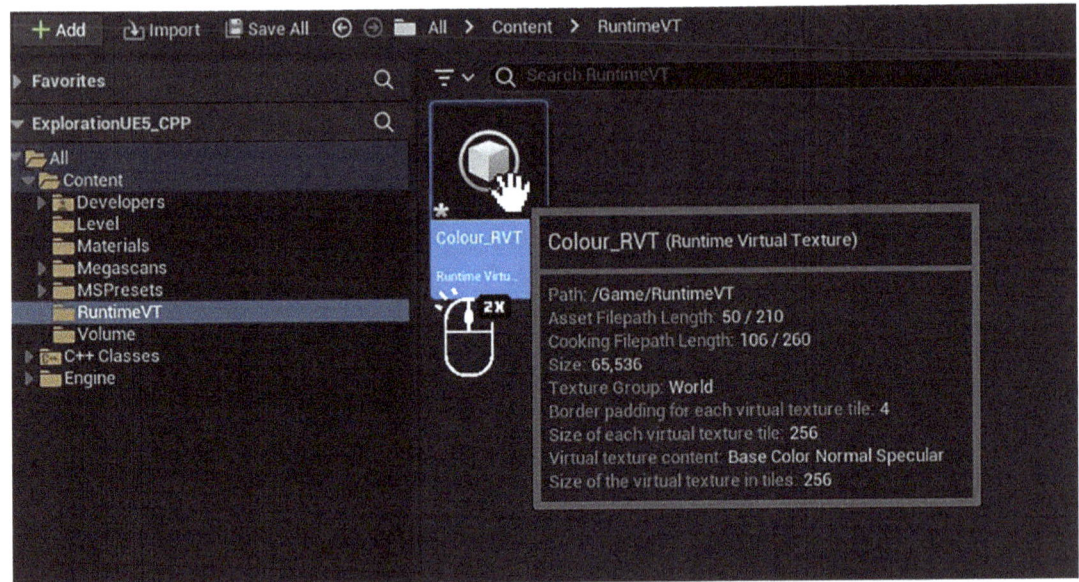

Figure 5-2. *Double-click to open up the RVT asset*

As shown in Figure 5-3, we need to ensure that the `Virtual Texture Content` setting is configured as `Base Color`, `Normal`, `Roughness`, and `Specular`.

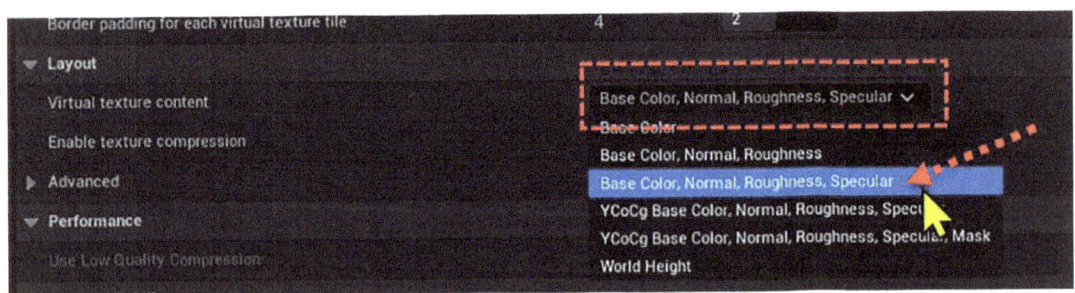

Figure 5-3. *Set the virtual texture content as Base Color, Normal, Roughness, Specular*

Including the Specular attribute in the RVT settings is crucial, even if we are not utilizing Specular in our project materials. The UE project might crash if this Color RVT does not have the Specular attribute as part of the virtual texture content settings.

As a component of our material blending pipeline in UE5, we will need an additional RVT asset that utilizes the Virtual Height as the content for the virtual texture. This can be accomplished by following these steps to add another RVT asset and configuring the Virtual Height settings, as shown in Figure 5-4, and be sure that this Virtual Height RVT is set as world height as the virtual texture content as shown in Figure 5-5.

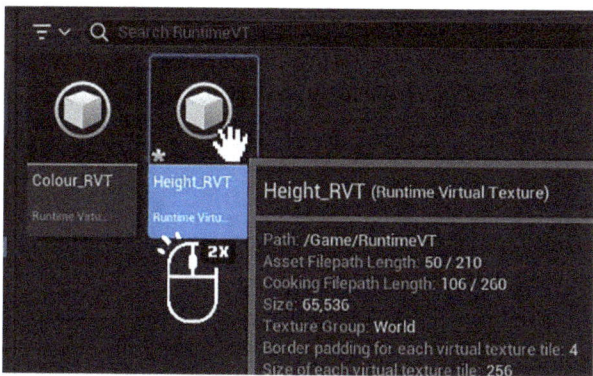

Figure 5-4. Add another RVT asset

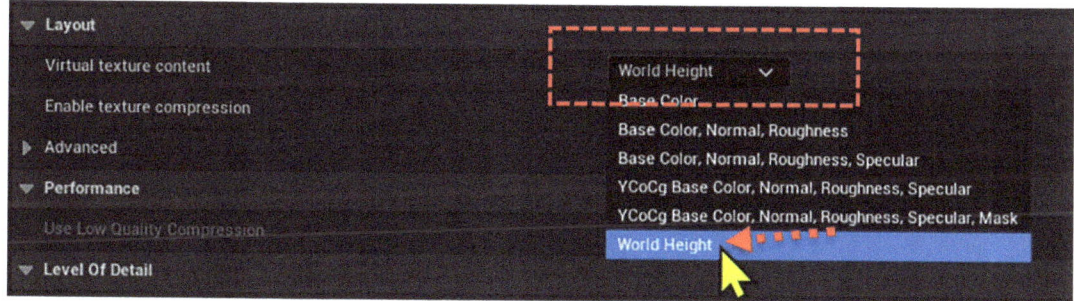

Figure 5-5. World height as the virtual texture content

RVT Volume

An RVT volume is a special region within the level where textures are streamed and displayed dynamically at runtime, based on the camera's view and the player's position. The purpose of RVT volumes is to provide a way to efficiently handle large and detailed textures without consuming excessive memory or causing performance issues.

The Setup of RVT Volume

To add an RVT volume, follow a similar concept as adding the aforementioned PCG volume in Chapter 4, as shown in Figure 5-6, with the steps described as follows:

1. Step 1 – Accessing Volumes: Within the "Place Actors" panel, the user has navigated to the "Volumes" category, which is a collection of special actor types that define areas in the world with specific behaviors or properties.

2. Step 2 – Expanding Volume List: The "Volumes" category has been expanded to reveal a list of different volume types available for use within the game level. These volumes can control various aspects of gameplay and environmental interaction.

3. Step 3 – Selecting a Specific Volume: The user is about to select or has selected the "Runtime Virtual Texture Volume" from the list. This volume type is typically used to enhance the visual fidelity of textures in the game environment by allowing textures to be updated at runtime based on certain conditions or interactions.

CHAPTER 5   ENHANCING VISUAL REALISM WITH RUNTIME VIRTUAL TEXTURES AND MATERIAL BLENDING

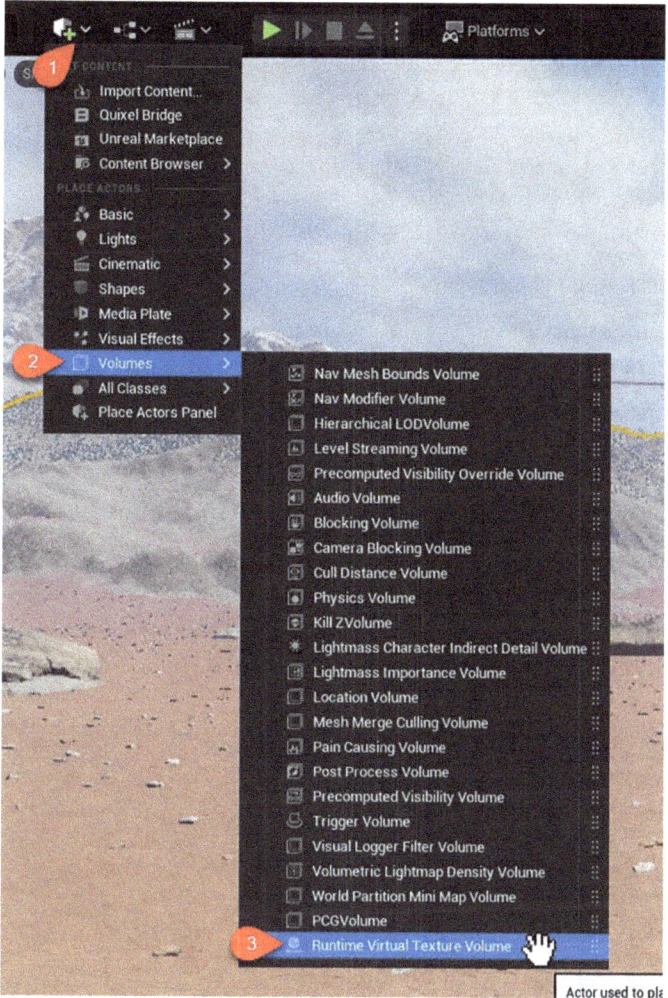

Figure 5-6. *Add an RVT volume in the level*

Since we require two sets of RVT volumes, it's best to name them appropriately, as demonstrated in Figure 5-7.

CHAPTER 5   ENHANCING VISUAL REALISM WITH RUNTIME VIRTUAL TEXTURES AND MATERIAL BLENDING

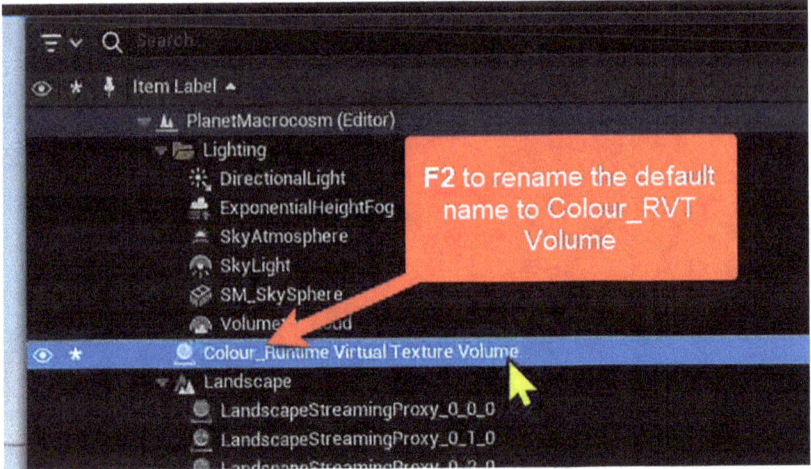

Figure 5-7. Rename the volume as a color-based volume

The RVT volume is responsible for providing the texture atlas map to the RVT asset. To achieve this, we need to associate the prepared RVT asset with the Virtual Texture property, as shown in Figure 5-8.

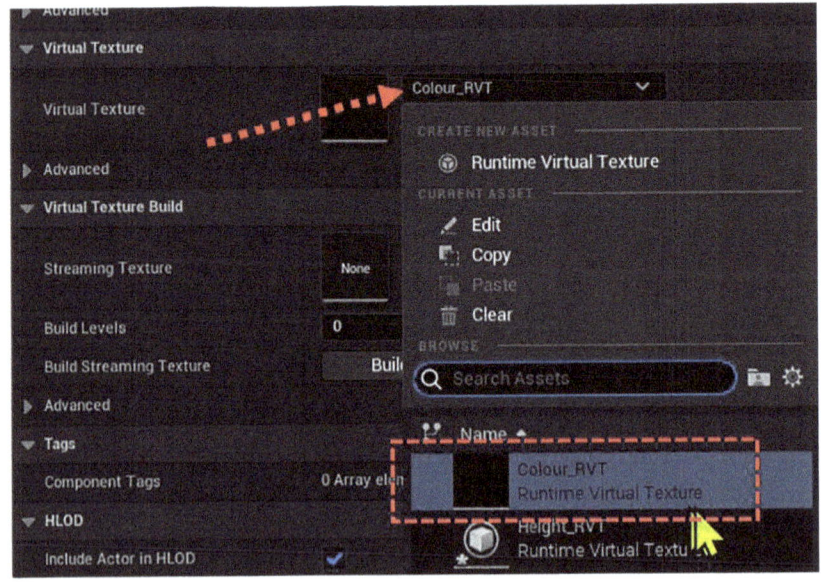

Figure 5-8. Set the appropriate RVT asset (Colour_RVT in this case)

190

CHAPTER 5   ENHANCING VISUAL REALISM WITH RUNTIME VIRTUAL TEXTURES AND MATERIAL BLENDING

It's crucial to configure the RVT volume to cover the entire landscape volume. The recommended approach is to utilize the Top perspective and Front perspective to adjust the position and size, as shown in Figures 5-9 and 5-10, as follows:

- Viewport: In the upper left, we see the editor's viewport set to "Top" view to provide a bird's-eye view of the landscape or level onto which a virtual texture will be applied.

- Transform Settings: The transform settings for the "Runtime Virtual Texture Volume," including the "Location" and "Scale."

- The Location values are set to -101,240 in X and Y coordinates, positioning the volume within the world.

- The Scale is uniformly set to 202,500 in X and Y dimensions, indicating the size of the volume that will influence the landscape.

- Viewport Display: The viewport is set to "Front" view and "Wireframe" mode, providing a side perspective of the landscape. This view can be particularly useful for adjusting the vertical (Z-axis) placement and scale of volumes in relation to the terrain's elevation.

- Transform Settings for Volume: On the right, the Transform settings for the selected volume are visible.

    - The "Location" values indicate that the volume's position in the world is set to -72,050 on the Z-axis, which positions the volume in three-dimensional space relative to the origin point of the level.

    - The "Scale" values are set to 30,000 on the Z-axis, which determines the size of the volume and, consequently, the area of the landscape that the virtual texture will affect.

The numbers in the Transform settings are essential for correctly placing the volume within the landscape, which will influence the rendering and performance of the virtual textures applied to the terrain. Proper configuration ensures that virtual textures are mapped accurately and optimized for the game's visual and performance requirements.

CHAPTER 5   ENHANCING VISUAL REALISM WITH RUNTIME VIRTUAL TEXTURES AND MATERIAL BLENDING

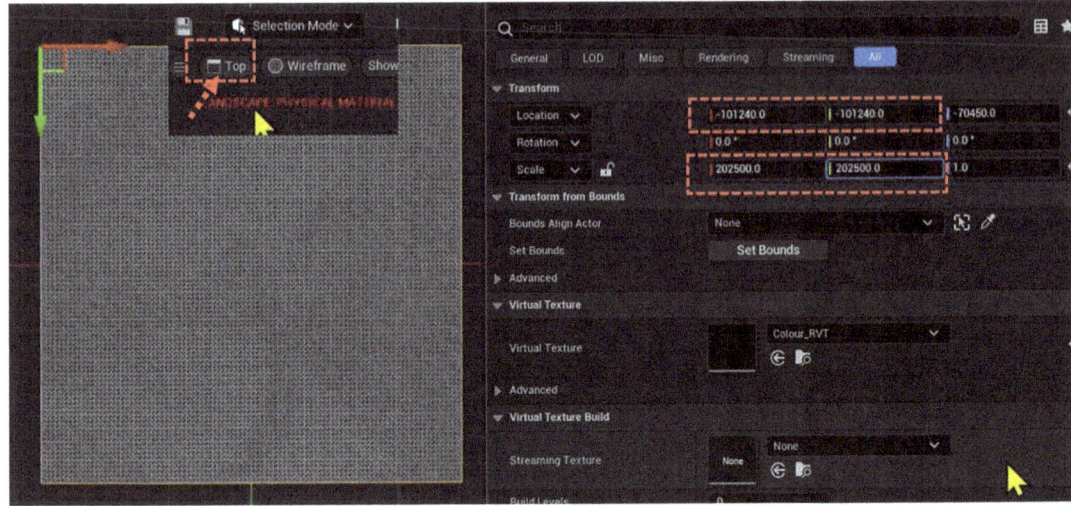

Figure 5-9. Reposition and rescale the volume (from the Top perspective)

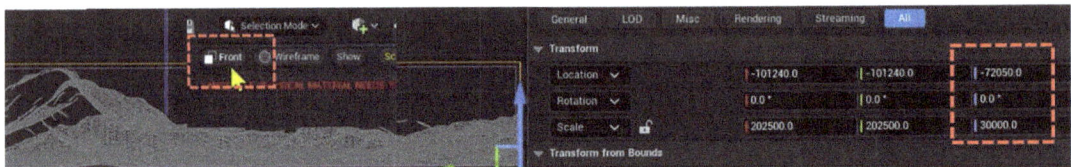

Figure 5-10. Reposition and rescale the volume (from the Front perspective)

Lastly, we will attach the bounding property to the Landscape actor, as shown in Figure 5-11. This ensures that we can appropriately sample the Landscape's RGB and Height to the RVT assets at a later stage, as follows:

1. Bounds Align Actor: The "Bounds Align Actor" drop-down is set to "Landscape," indicating that the bounds for the virtual texture volume will be aligned to the landscape actor within the scene. This is important for ensuring that the virtual texture is properly mapped over the terrain.

2. Current Actor Selection: Figure 5-11 shows the current actor selected in the editor, which is the "Landscape." This could be the target to which the virtual texture volume's bounds should be aligned.

CHAPTER 5   ENHANCING VISUAL REALISM WITH RUNTIME VIRTUAL TEXTURES AND MATERIAL BLENDING

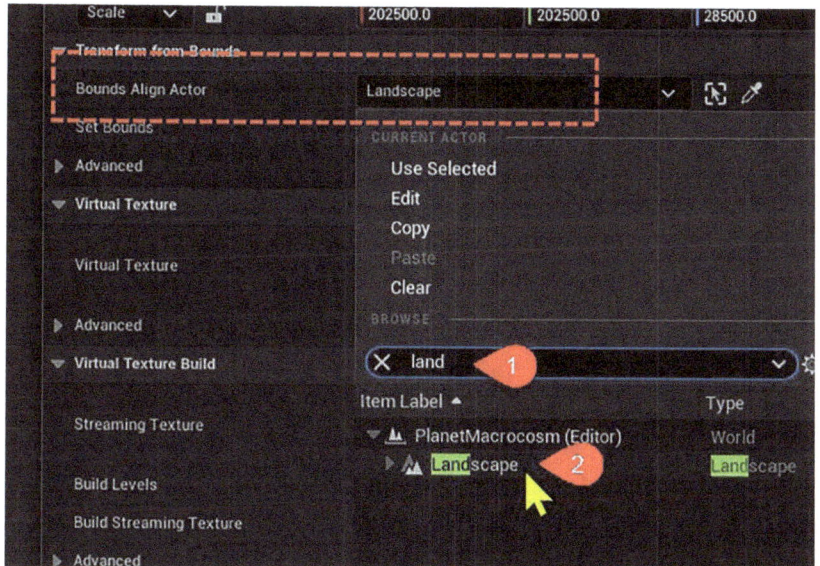

Figure 5-11. *Attach the Landscape as the bound align actor*

To streamline the process of adjusting the position and size for a new set of RVT volumes within the project, the following steps can be taken:

1. Duplicate the existing Colour_RVT_Volume to create a new RVT volume, as shown in Figure 5-12.

CHAPTER 5   ENHANCING VISUAL REALISM WITH RUNTIME VIRTUAL TEXTURES AND MATERIAL BLENDING

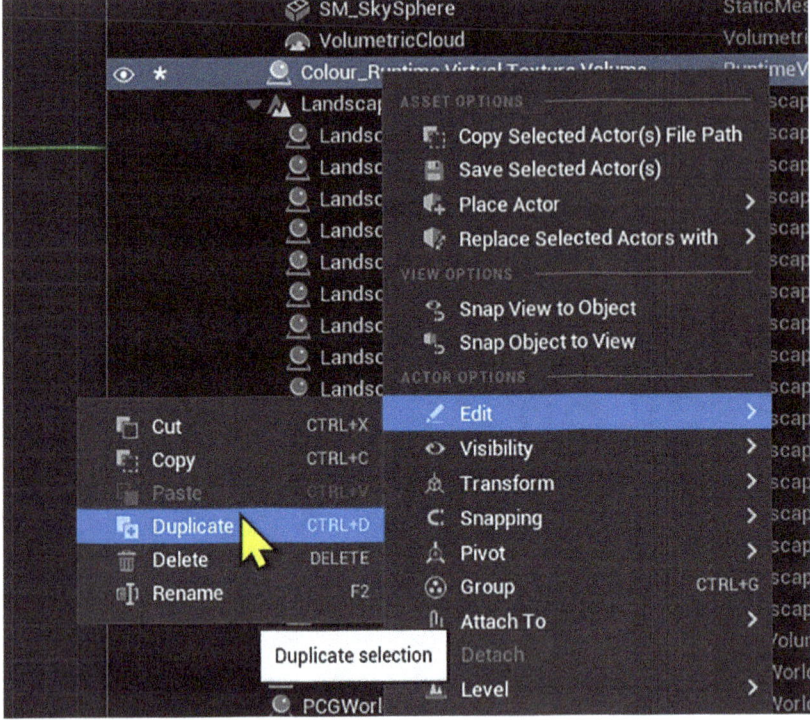

Figure 5-12. *Duplicate the current RVT volume*

2. Rename the duplicated volume to Height_RVT_Volume as shown in Figure 5-13.

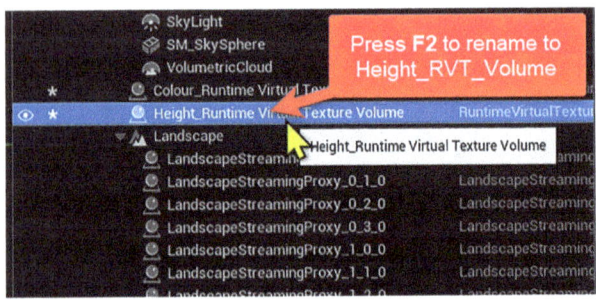

Figure 5-13. *Rename the new volume as Height_RVT_Volume*

3. Configure the virtual texture settings to utilize the Height_RVT asset, as shown in Figure 5-14.

CHAPTER 5 ENHANCING VISUAL REALISM WITH RUNTIME VIRTUAL TEXTURES AND MATERIAL BLENDING

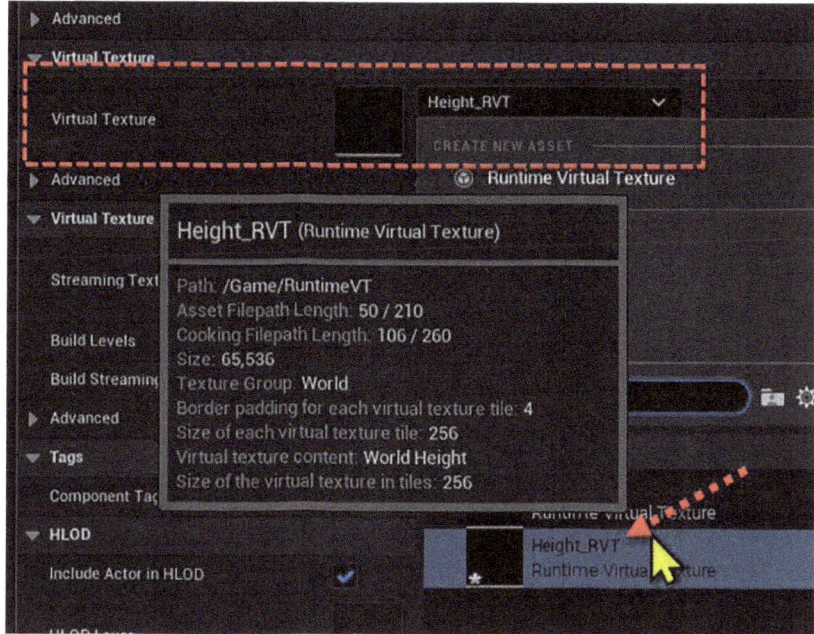

Figure 5-14. *Attach the Height_RVT asset to the virtual texture property*

The preceding steps aim to simplify the task of adapting the same position and size settings for `Height_RVT_Volume` in the level.

Set RVT Volumes Inside Landscape Partitions

It is important to note that we were utilizing Open World Partitions in UE. The original landscape has been divided into these partitions, and it's crucial to consider that each partition possesses its own distinct properties. For instance, we are now required to associate all partitions with the designated RVT asset as part of the setup process.

As shown in Figure 5-15, we must

1. Select the parent Landscape entity itself.

CHAPTER 5   ENHANCING VISUAL REALISM WITH RUNTIME VIRTUAL TEXTURES AND MATERIAL BLENDING

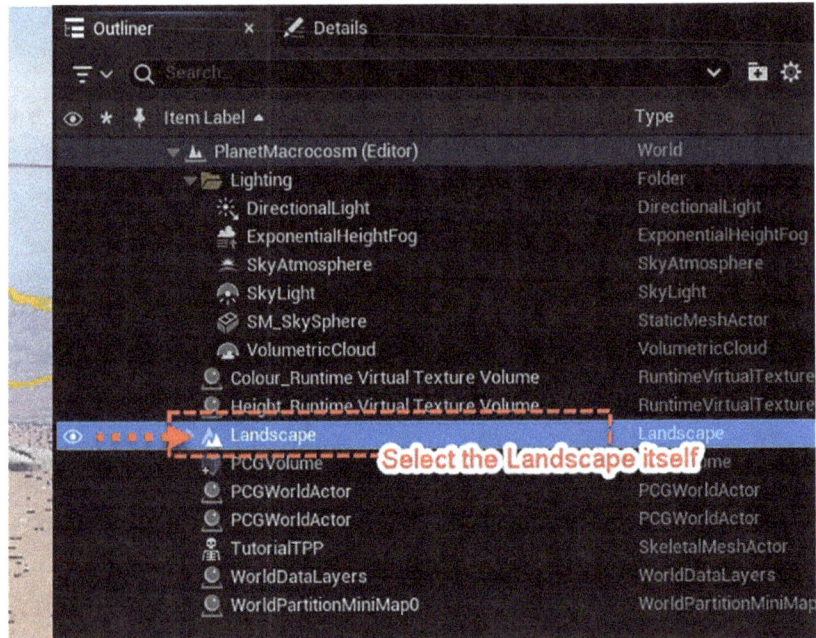

Figure 5-15. *Select all partitions from the open world landscape*

2. Then, we perform the following steps to attach both RVT assets to the Draw Virtual Texture under the Virtual Texture sections of the landscape partitions, as shown in Figure 5-16:

   a. Under the selected Landscape entity.

   b. Under the Rendering category.

   c. As Step 1 indicates, click to add two array of Virtual Texture elements.

   d. In Steps 2 and 3, attach both Colour_RVT and Height_RVT in the corresponding Virtual Texture element.

There is no specific order for adding RVT assets to the list of RVTs in the Draw Virtual Texture property in the landscape partition actor.

CHAPTER 5   ENHANCING VISUAL REALISM WITH RUNTIME VIRTUAL TEXTURES AND MATERIAL BLENDING

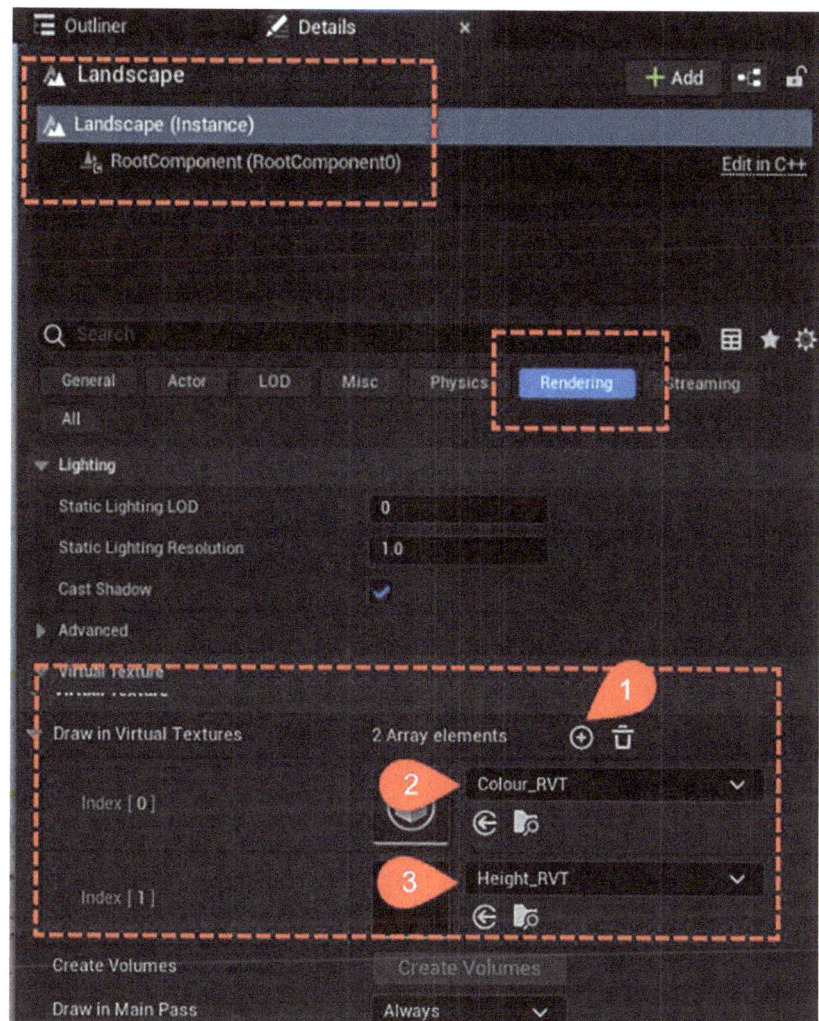

Figure 5-16. Attach both Colour_RVT and Height_RVT to all of the selected landscape partitions

RVT Output in Material Editor

The RVT Output node is a material expression node that enables us to write data to a Runtime Virtual Texture asset. It is utilized in materials that belong to the Virtual Texture domain, signifying that they are intended to render to a Runtime Virtual Texture instead of the screen.

CHAPTER 5 ENHANCING VISUAL REALISM WITH RUNTIME VIRTUAL TEXTURES AND MATERIAL BLENDING

This node allows us to designate the material attributes we wish to write to the Runtime Virtual Texture, including Base Color, Normal, Roughness, and more. Moreover, it grants us the capability to manage the blending mode and the virtual texture output level for each attribute. This serves as the primary focus of our usage within this chapter.

Set Up RVT Output Node

We need to decide which material serves as the base for the RVTOutput. In this case, it's the landscape AutoBlend_Height_MAT material (Figure 5-17).

1. Select the Materials folder.
2. Double-click to open up AutoBlend_Height_MAT material.

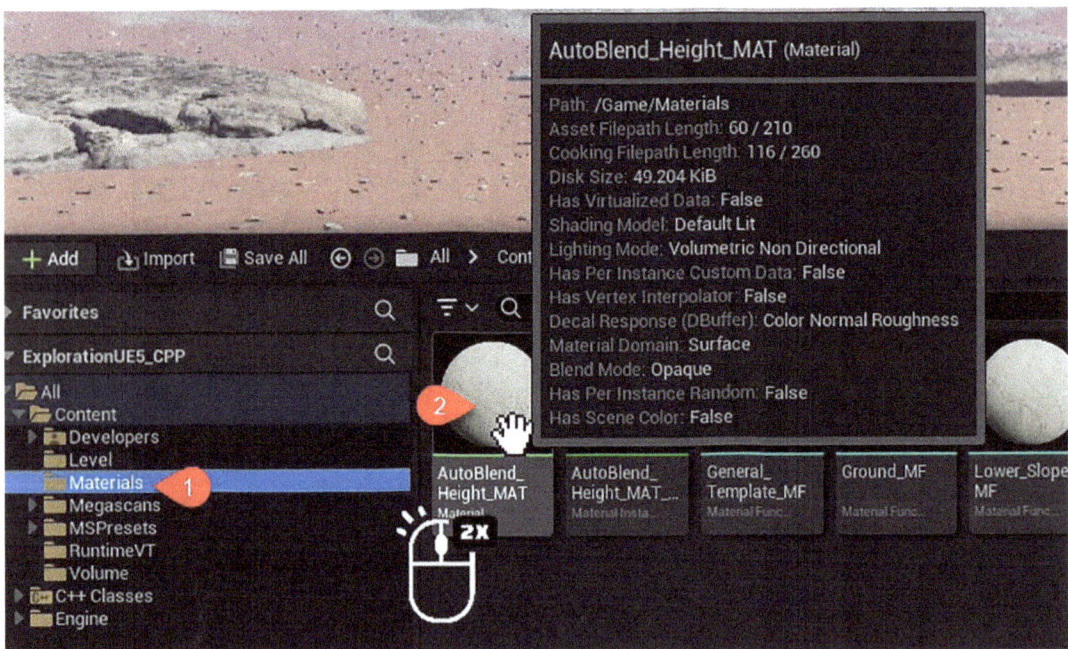

Figure 5-17. Open up the AutoBlend_Height_MAT landscape materials

We can add the RVTOutput node by following the steps shown in Figure 5-18:

1. Right-click the material editor to activate the node selection search box.
2. We can type runtime to filter out all the nodes that contain the word runtime.
3. We will select RuntimeVirtualTextureOutput for our RVT purpose.

CHAPTER 5   ENHANCING VISUAL REALISM WITH RUNTIME VIRTUAL TEXTURES AND MATERIAL BLENDING

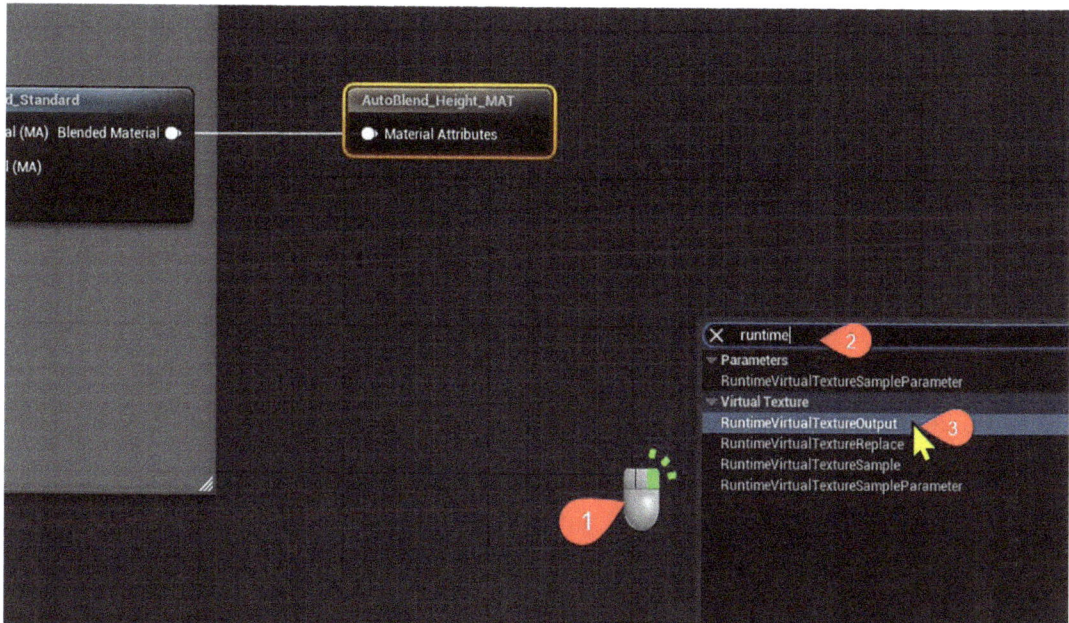

Figure 5-18. *Add the RuntimeVirtualTextureOutput node as one of end nodes to use*

Further, Figures 5-19 and 5-20 show how the material attributes are being processed and prepared for output to a Runtime Virtual Texture (RVT).

In Figure 5-19, we see a "MatLayerBlend_Standard" node where a base material and a top material are being blended together, with an alpha value controlling the blend ratio. The result of this blend is being fed into the "AutoBlend_Height_MAT" node, which contains the logic for blending materials based on the height of the landscape. Figure 5-20 shows the "GetMaterialAttributes" node that retrieves the material attributes from "AutoBlend_Height_MAT". The steps in Figure 5-20 can be described as follows:

1. In the "AutoBlend_Height_MAT" material graph, place a "GetMaterialAttributes" node from "Blended Material". This node will be used to extract material attributes like Base Color, Roughness, and Normal from a predefined set of attributes.

2. As shown in Step 1, connect the output pin of a "MatLayerBlend_Standard" node, which blends the base and top materials, to the "Material Attributes" input of the "GetMaterialAttributes" node.

CHAPTER 5   ENHANCING VISUAL REALISM WITH RUNTIME VIRTUAL TEXTURES AND MATERIAL BLENDING

3. From the "Details" panel on the left, as shown in Step 2, set up the "Attribute Get Types" by adding three elements to the array, configuring them to the corresponding material attributes: Base Color, Roughness, and Normal. Make sure that each attribute's index matches those that the "GetMaterialAttributes" node will extract.

4. Connect the outputs of the "GetMaterialAttributes" node, as shown in Steps 3 and 4, specifically Base Color and Roughness, to the respective inputs of a "Runtime Virtual Texture Output" node.

5. Regarding the Normal attribute, we need to add a "Transform" node set to convert from tangent to world space as "Need convert to World Space" before connecting it to the "Runtime Virtual Texture Output" node.

In short, we need to determine the necessary process to take the output from the Blended Material as input values into the Runtime Virtual Texture Output node (as shown in Figure 5-19). The red dotted box in the figure is where we need to add the necessary process to output the necessary channels into the necessary properties in the Runtime Virtual Texture Output node.

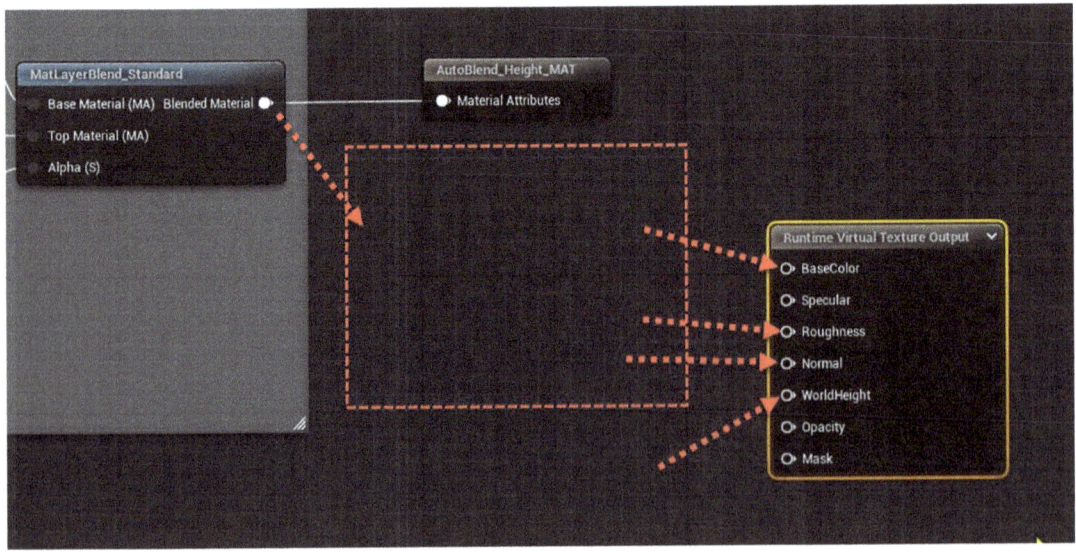

Figure 5-19. The pipeline that needs to create the necessary input value into the RVT Output

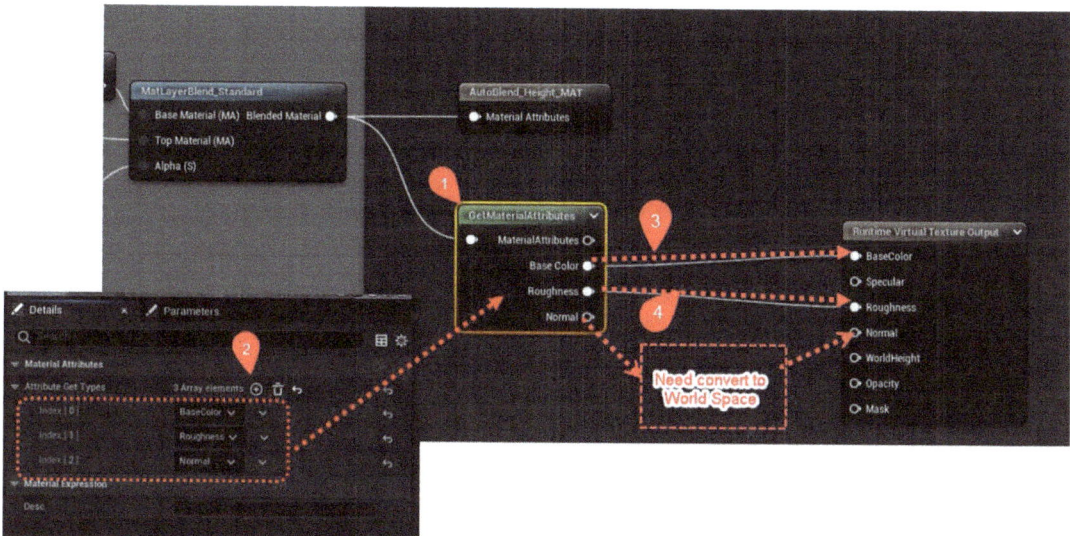

Figure 5-20. The Normal value needs to perform a conversion before inputting to RVT Output

---

The Specular attribute is a necessary component when setting up the RVT asset. However, there is no need to be concerned about the Specular attribute, as we are not utilizing it in any way. We can directly connect material attributes such as `Base_Color` and `Roughness` to the RVTOutput node.

Before these attributes can be output to the RVT, the `Normal` channel needs to be converted to `World Space`, as indicated by the note "Need convert to World Space" in Figure 5-20. This conversion is essential because the RVT requires the normals to be in world space to properly shade the terrain.

The whole process flow of creating a material that can blend different textures based on the landscape height, then preparing these blended material attributes for use in a Runtime Virtual Texture, which allows the material to be dynamically updated in the game world for various visual effects. The conversion to world space is a critical step in ensuring that the materials render correctly in the game environment.

The `Normal` value transforms the value from `Tangent Space` (a vector relative to the surface of an object) to `World Space` (the global coordinate system) before passing the Normal vector into the RVTOutput node (as shown in Figures 5-21 and 5-22) with the following steps:

1. Retrieve Material Attributes: First, the "GetMaterialAttributes" node is used to extract the "Normal" attribute, then we search for the word Transform, which represents the "TransformVector" node.

2. Apply Transformation: After adding the "TransformVector" node to the graph, it's connected between the "Normal" output of the "GetMaterialAttributes" node and the "Normal" input of the "Runtime Virtual Texture Output.

3. Set Transform Type: In the "TransformVector" node, set the transformation to convert from "Tangent Space" to "World Space". This ensures that the normal vectors are correctly oriented in relation to the world coordinates, which is necessary for accurate lighting and shading when the Runtime Virtual Texture is applied to the landscape.

This process is crucial because normal maps are often stored in tangent space to provide surface detail regardless of the object's orientation in the world.

However, for proper interaction with lighting in the world, they *need to be converted to world space*.

The "TransformVector" node performs this conversion, allowing the material to correctly interact with the game environment's global lighting.

CHAPTER 5 ENHANCING VISUAL REALISM WITH RUNTIME VIRTUAL TEXTURES AND MATERIAL BLENDING

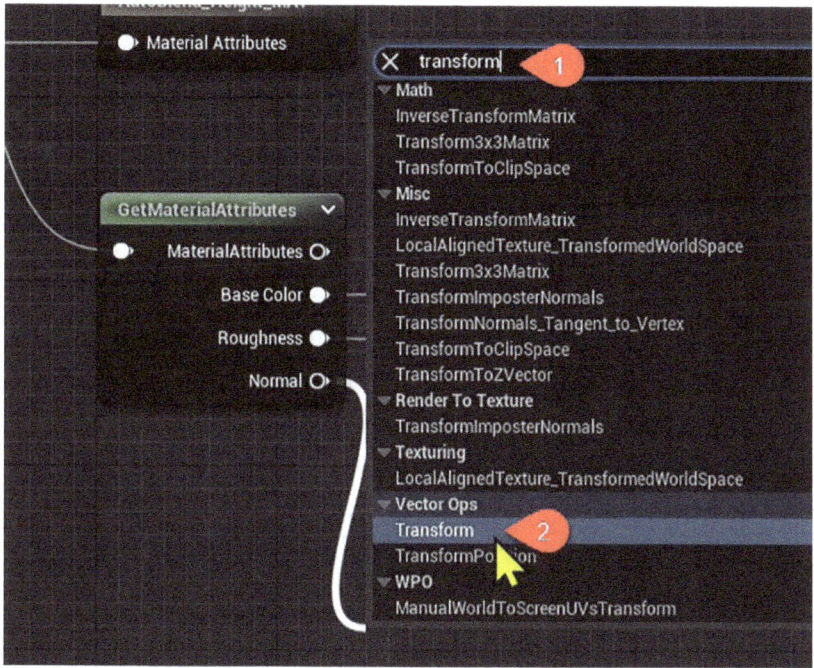

Figure 5-21. Add the Transform vector to convert the Normal value

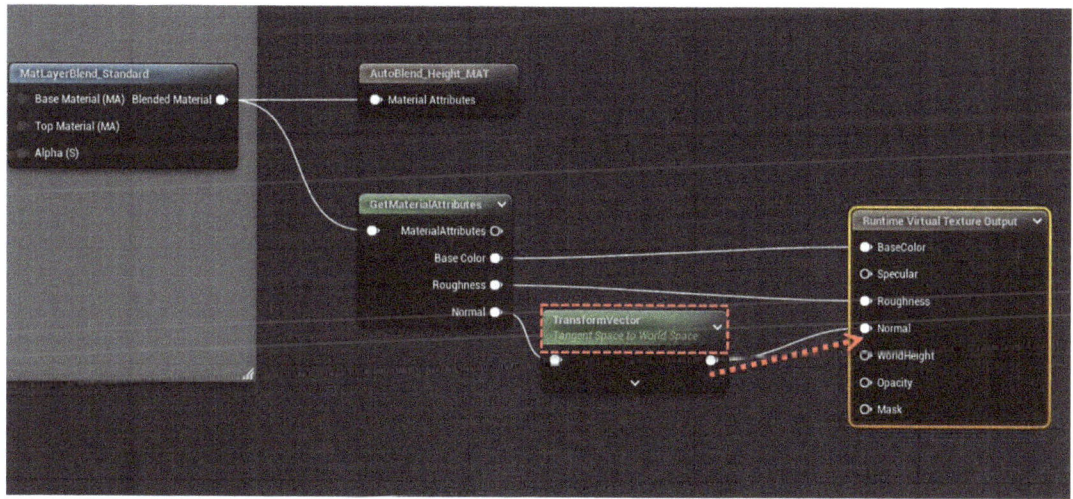

Figure 5-22. Transform Tangent Space to a World Space vector (it's set by default)

CHAPTER 5   ENHANCING VISUAL REALISM WITH RUNTIME VIRTUAL TEXTURES AND MATERIAL BLENDING

A Runtime Virtual Texture Output node in the Material Editor is a node that writes the material attributes of an object to a Runtime Virtual Texture Asset. A Runtime Virtual Texture Asset is a specialized type of texture generated on the fly by the GPU, capable of caching complex shading data over large areas. This can enhance the performance and quality of rendering landscapes and other objects that use layered or procedural materials.

When we provide the Normal input to the Runtime Virtual Texture Output node, we instruct the node to write the normal vector of the object to the Runtime Virtual Texture Asset. However, since the Runtime Virtual Texture Asset exists in world space, we must transform the normal vector from tangent space to world space before writing it. Otherwise, the object's shading won't align with the lighting and environment in world space.

Finally, we will need to input the value into the World Height attribute of the RVTOutput node. We can achieve this by extracting the Z value of the AbsoluteWorldPosition (as shown in Figure 5-23), as the Z-axis corresponds to the height position of the pixel in the level.

To utilize the Absolute World Position node in the UE material editor for generating world height information, you can follow these expanded steps:

1. Open the Material Editor: We begin by right-clicking within the material editor to access the node creation context menu.

2. Search for the Node: In the search bar that appears, type in world to filter through the available nodes. This will display nodes related to world coordinates and positioning.

3. Create the Node: From the filtered list, find and select the "WorldPosition" to create an Absolute World Position node in the material graph.

4. Connect the Node: Once the Absolute World Position node is in place, look for the "Z" output pin. This pin represents the Z component of the world position, which corresponds to the height in the game world.

CHAPTER 5   ENHANCING VISUAL REALISM WITH RUNTIME VIRTUAL TEXTURES AND MATERIAL BLENDING

5. Assign to WorldHeight: Drag a connection from the "Z" output pin of the Absolute World Position node directly to the "WorldHeight" input property on the Runtime Virtual Texture Output node. This connection maps the absolute height of each point in the material to the virtual texture, allowing for the creation of height-based effects within the game's terrain.

By following these steps, you effectively map the vertical position of textures within your game environment to a Runtime Virtual Texture, providing dynamic feedback and interactions based on the actual height of the landscape in your game world.

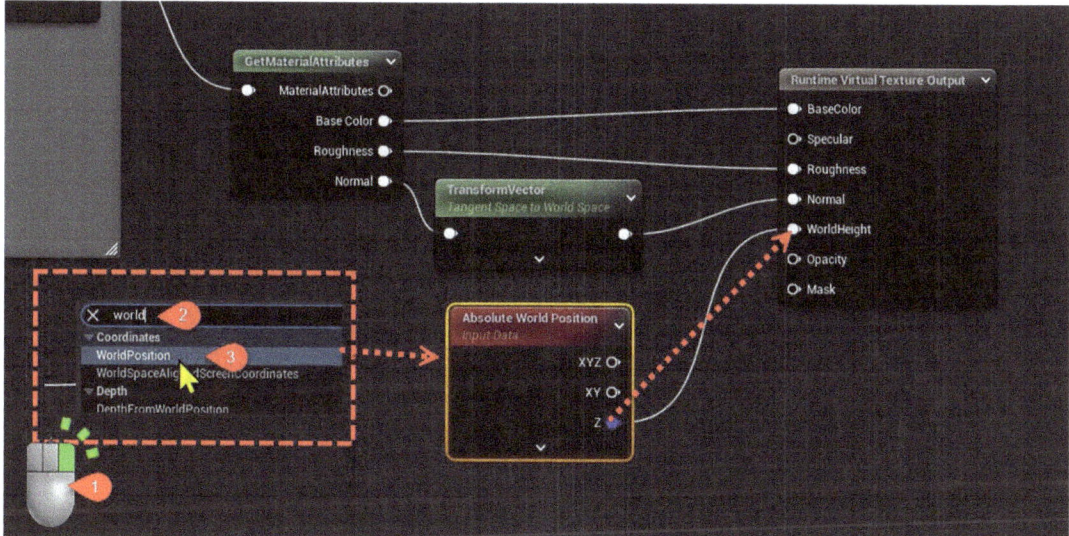

Figure 5-23. Connect the Z value (blue channel) from Absolute World Position to the WorldHeight property

See the Result on the RVT Assets

After configuring the RVTOutput as described in the previous section, it is important to ensure that the material logic is applied (as shown in Figure 5-24) before proceeding with the pipeline to generate the RVTOutput for the RVT assets.

1. Be sure that we are in the AutoBlend_Height_MAT editor.

2. Click the Apply button to activate the changes.

CHAPTER 5   ENHANCING VISUAL REALISM WITH RUNTIME VIRTUAL TEXTURES AND MATERIAL BLENDING

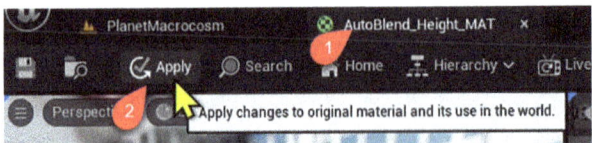

Figure 5-24. *Be sure to click Apply to activate the materials logic*

We can observe the entire RVT volume, the setup within the Landscape partitions, and the properly configured RVTOutput inside the material editor, all functioning correctly. This is evident as we are able to see the thumbnail displaying the color texture and the grayscale heightmap. Both of these components appear in the two RVT assets, as shown in Figure 5-25.

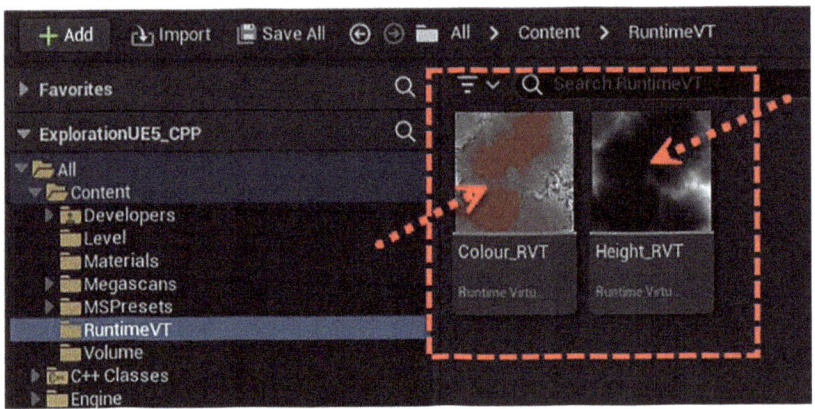

Figure 5-25. *The correct visual of the Colour_RVT and Height_RVT assets*

Material Blend with RVT Assets

Utilizing Runtime Virtual Texture (RVT) assets in UE serves a strategic purpose: it enables the flawless integration of 3D objects with the surrounding landscape material. By scrutinizing the material logic of these 3D assets, we ensure that they naturally coexist within the environment, sharing the same textural and lighting qualities as the terrain they stand on.

This meticulous process results in a cohesive visual field where the distinctions between modeled assets and the landscape are blur, enhancing the overall realism and immersion of the scene. Delving into this aspect of RVT assets not only elevates the aesthetic continuity of your project but also optimizes performance by streamlining the material complexity. The end goal is a more believable world that conserves resources, making it a crucial step for any project aiming for both visual excellence and technical efficiency.

As mentioned, the primary purpose for us to use RVT assets is to seamlessly blend the 3D asset with the landscape material. Therefore, we need to examine the material logic of the 3D assets. In this case, we should investigate the parent of the 3D asset (Forest_Rock_Formation) as shown in Figure 5-26:

1. Select the folder named Forest_Rock_Formation_wgquaam.

2. Double-click to open up the MI_Forest_Rock material instance.

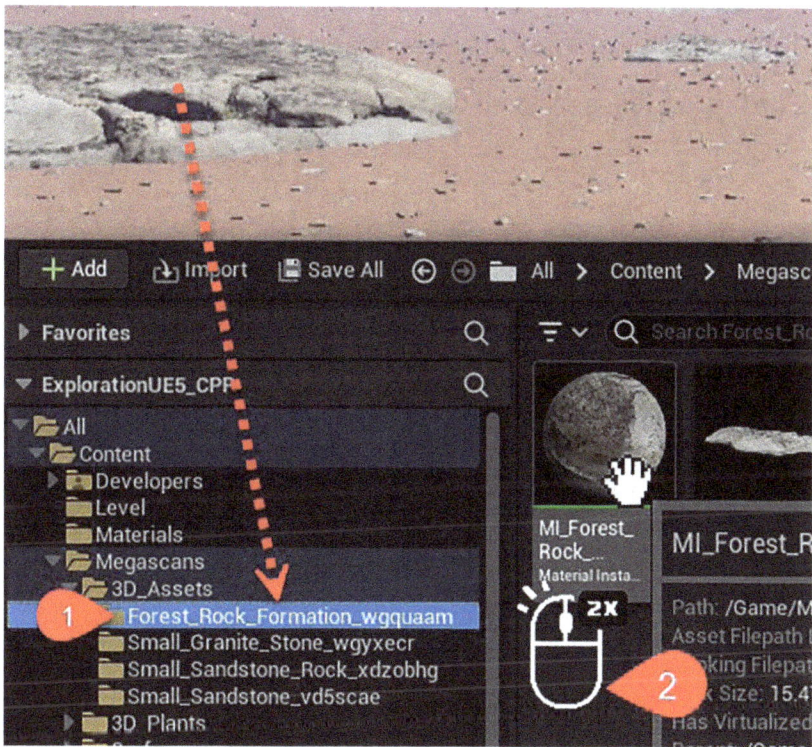

Figure 5-26. *Open up the materials for the 3D assets*

We need to integrate extra material logic into the parent material asset, M_MS_Default_Material_VT, in order to accurately render pixel values according to the supplied color and height data from the RVT assets. Nonetheless, we intend to avoid altering the original parent material asset due to its role as the foundation for all downloaded Megascan 3D assets.

CHAPTER 5   ENHANCING VISUAL REALISM WITH RUNTIME VIRTUAL TEXTURES AND MATERIAL BLENDING

Therefore, we will duplicate the material, assign this duplicated material to the Forest_Rock_Formation, and then proceed to make modifications on this duplicate material (instead of the original material M_MS_Default_Material_VT). This process is illustrated in the steps shown in Figures 5-27, 5-28, 5-29 (named RVT_Blend_M_MS_Default_Material_VT), and 5-30 (to assign to the rock formation 3D asset) or in the following steps.

Here are the steps to duplicate a material and use the duplicated material within the UE editor:

1. Duplicate the Material:

    a. As shown in Figure 5-27, locate the original material in the Content Browser, which is the M_MS.Default_Material_VT.

    b. Right-click the material to open the context menu as shown in Figure 5-28, Step 1.

    c. Select "Duplicate" from the context menu to create a copy of the material as shown in Step 2.

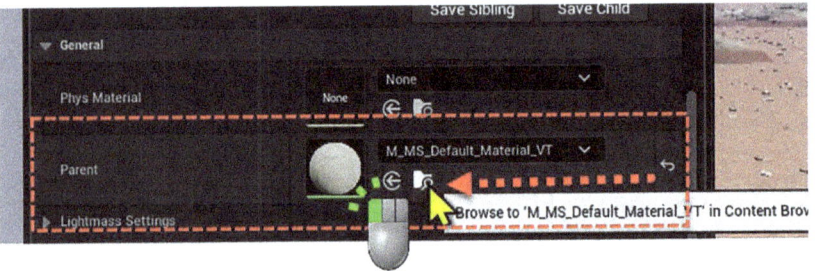

Figure 5-27. *Browse the original parent material*

CHAPTER 5 ENHANCING VISUAL REALISM WITH RUNTIME VIRTUAL TEXTURES AND MATERIAL BLENDING

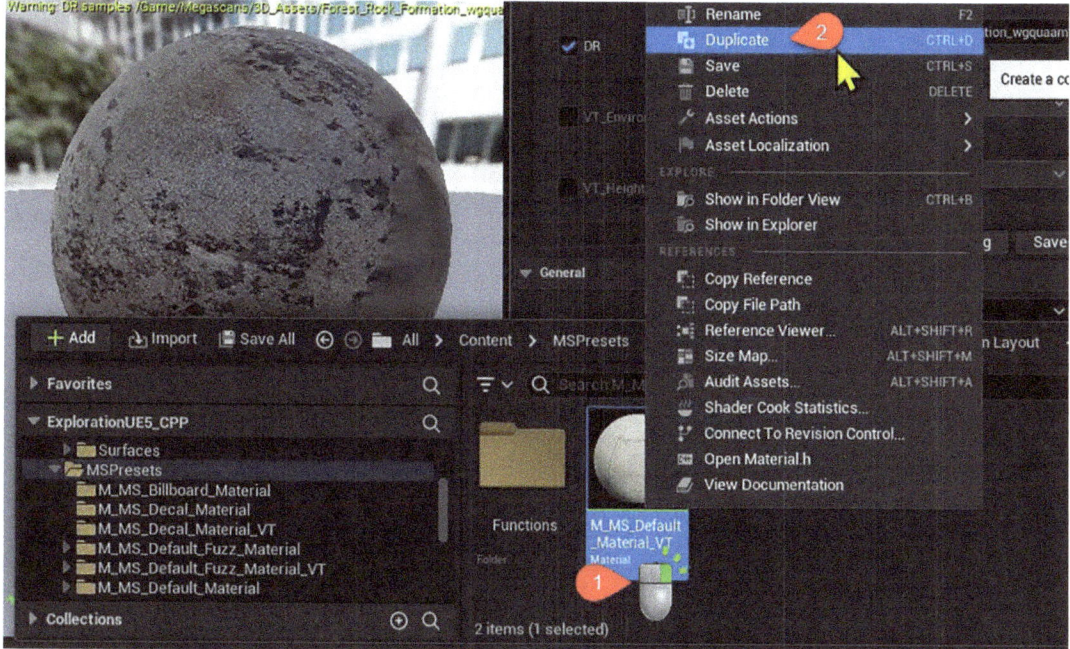

Figure 5-28. *Right-click to select Duplicate to make a duplicate of the parent material*

2. Rename the Duplicated Material:

   a. With the duplicated material selected, right-click again to open the context menu.

   b. Choose "Rename" from the context menu and give the duplicated material a new name, such as RVT_Blend_M_MS_Default_Material.

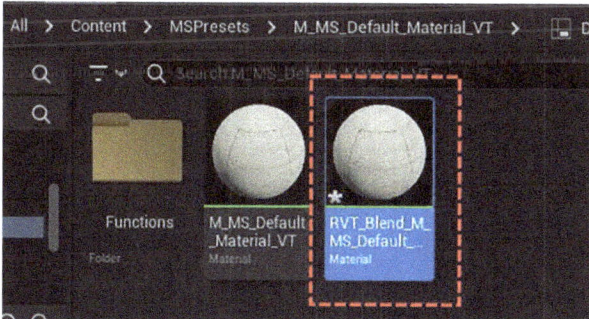

Figure 5-29. *The duplicated material renamed as* RVT_Blend_M_MS_Default_Material_VT

CHAPTER 5   ENHANCING VISUAL REALISM WITH RUNTIME VIRTUAL TEXTURES AND MATERIAL BLENDING

3. Assign the Duplicated Material As Shown in Figure 5-29:

   a. Open the details panel of the asset to which we want to apply the new material as shown in Step 1.

   b. Find the material slot where we want to assign the duplicated material. This might be labeled "Parent" under the material properties as shown in Step 2.

   c. Click the drop-down next to the material slot, enter the search term (rvt as shown in Step 3), and select the duplicated material, RVT_Blend_M_MS_Default_Material, from the list to apply it to the asset as shown in Step 4.

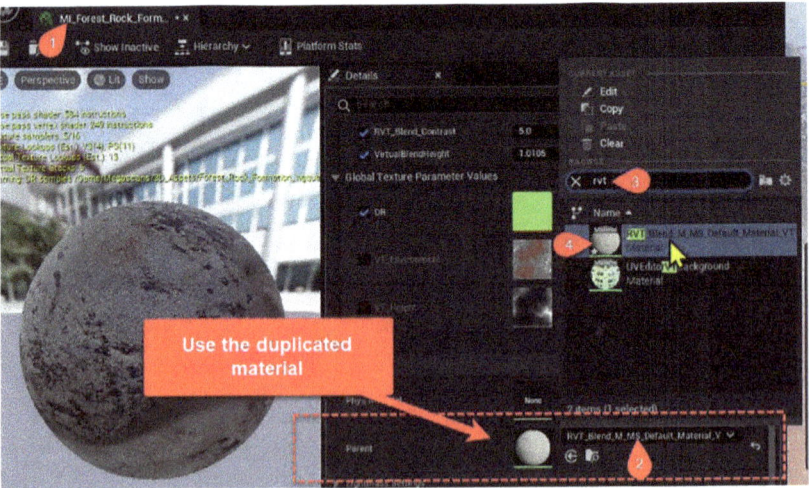

Figure 5-30. *Attach the duplicated material as the parent of the* MI_Forest_Rock_Formation

4. Verify the Material Assignment:

   a. Once assigned, we can verify the material is correctly applied by looking at the preview of the asset. The preview should now show the asset with the newly applied material.

   b. Optionally, we can save the asset to confirm the changes.

These steps facilitate the management of materials within our project, allowing us to create variations of materials without affecting the original. This is especially useful when working with Runtime Virtual Textures or when specific material variations are needed for different assets or level designs.

Additional Logic to This Duplicated Material (RVT)

In this section, we introduce three distinct material attributes – `Original_MAT`, `Landscape_Colour_MAT`, and `Height_MAT` – each serving a specific purpose in this `RVT_Blend_M_MS_Default_Material` material setup.

To clarify, these separate node trees we're constructing are essential for defining the parameters that will enable us to blend between the landscape material and the 3D asset's original material using the "`BlendMaterialAttributes`" node. It's important to understand that each node tree is a critical component in building the composite material, which might not have been evident initially. We're essentially layering these attributes to create a cohesive material that combines the original look of the 3D asset with the environment's texture details, achieved through the procedural steps detailed at the beginning of this section. This blending is key to integrating the asset into the landscape, ensuring visual continuity within the scene.

In summary, we are to incorporate the following additional logics into each of the corresponding named reroute declaration material nodes:

- Modify the asset's original material (`Original_MAT`)
- Create the landscape's Color RVT (`Landscape_Colour_MAT`)
- Create an alpha for the blending (`Height_MAT`)
- Final material output and result

Each of these logics is placed inside the attached duplicate parent material, which we can access by opening up the `RVT_Blend_M_MS_Default_Material` material as shown in Figure 5-31.

Figure 5-31. Double-click to open up the parent of the material instance (the duplicated material)

Asset's Original Materials

Initially, our objective is to blend the asset's own material, with the output derived from the "MF_ObjAdjustments" node. As indicated in Figure 5-32, the Normal output from this node is in Tangent Space.

To ensure compatibility with the RVT's Normal, which is in World Space, a conversion is required. This step is critical to align the Normal directions between the asset's material and the RVT, facilitating a coherent rendering of lighting and shading across the asset's surface.

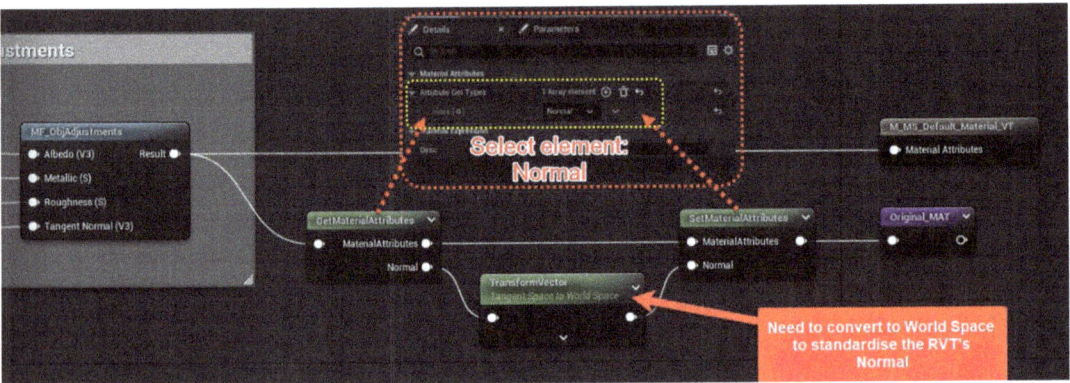

Figure 5-32. *Create a named reroute from the result material attributes, but with the conversion of the Normal*

Landscape's Color RVT

As we try to blend the materials betwee*n the landscape and the 3D assets, one of the source inputs will be the RVT color values, which can be retrieved from the RVT asset (as shown in Figure 5-33).

1. As indicated in Step 1, we are to right-click the material editor to activate the node search selection box.

2. Step 2 indicates that we type the search term **runtime** to filter out the node with the name that contains **runtime**.

3. Step 3 indicates that we select the node named `RuntimeVirtualTextureSampleParameter`.

CHAPTER 5   ENHANCING VISUAL REALISM WITH RUNTIME VIRTUAL TEXTURES AND MATERIAL BLENDING

4. Step 4 indicates that we connect the "Colour_RVT" node to the "VT_Environment" Runtime Virtual Texture sample parameter node, which will utilize the virtual texture data for sampling.

   The details of the "Colour_RVT" node, displayed in the Details panel, indicate it's configured to include Base Color, Normal, Roughness, and Specular data.

   Next, we will route the connection from "VT_Environment" node into the "SetMaterialAttributes" with the Colour_RVT output attributes (Base Color, Roughness, Normal).

   Note that we will have to set an array of material attributes as Base Color, Roughness, Normal in the "SetMaterialAttributes" node.

5. In Step 5, the single output from the "SetMaterialAttributes" node is directed through reroute nodes, which are named and organized for clarity, to the "Landscape_Colour_MAT". These reroute nodes act as intermediary pass-through points, keeping the material graph tidy. They will be utilized later in the material setup process.

Therefore, as shown in Figure 5-33, we are storing the result of this RVT asset as material attributes into a named rerouted node called Landscape_Colour_MAT.

Please note that the Normal retrieved from the RVT Sample Parameter node is the one that has already been converted to World Space.

We don't need to perform any conversions at this stage, until the final phase for the output material attribute in a later stage.

CHAPTER 5   ENHANCING VISUAL REALISM WITH RUNTIME VIRTUAL TEXTURES AND MATERIAL BLENDING

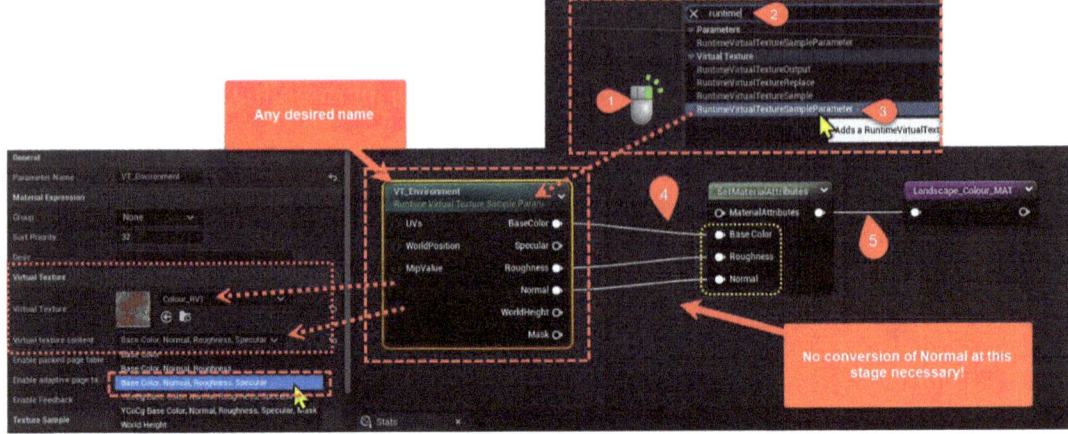

Figure 5-33. *Create the RVT Sample Parameter node*

Create an Alpha for the Blending

It is now time to determine the alpha value for the material blend between the color landscape (from RVT) and the material of the 3D asset itself. Based on the height-based RVT, we are able to blend the materials using RVT Sample Parameter nodes, considering the height of the 3D asset. We will use the `Object Bounds` node to specify the `BlendHeight` values (the first part of this setup is shown in Figure 5-34 and the second part in Figure 5-35).

1. As shown in Figure 5-34, we will right-click the material editor to activate the node search selection box.

2. We will type normal to filter out the node with the name that contains runtime.

3. Then, we select the node named `RuntimeVirtualTextureSampleParameter`.
   Be sure that we select the `Height_RVT` as the Virtual Texture property as shown in the figure.

4. The "`Object Bounds`" node, in conjunction with the "`VirtualTextureHeight`" parameter node, performs calculations taking into account the object's dimensions within the world and the height data from the virtual texture, both integral to the material's logic. Specifically, we focus on the "`Object Bounds`" height, which involves extracting the blue channel from the `Mask` node.

CHAPTER 5   ENHANCING VISUAL REALISM WITH RUNTIME VIRTUAL TEXTURES AND MATERIAL BLENDING

Note that we can select the blue channel by checking the "B" option in the Details of the `Mask` node.

5. Arithmetic nodes, such as "`Subtract`," "`Add`," and "`Multiply`," are also part of the network. These are used to adjust the sampled values and to blend the virtual texture data with other material properties or to factor in the landscape's elevation data when applying the texture.

6. The "`VirtualTextureHeight`" parameter node, with a default value of 1.05, serves to control the influence of the virtual texture's height data on the overall material appearance.

7. Finally, the overall setup implies that this material is designed to react to the landscape's elevation, blending the virtual texture with the material properties based on the object's position and dimensions, ensuring a cohesive and dynamic appearance in the 3D world space.

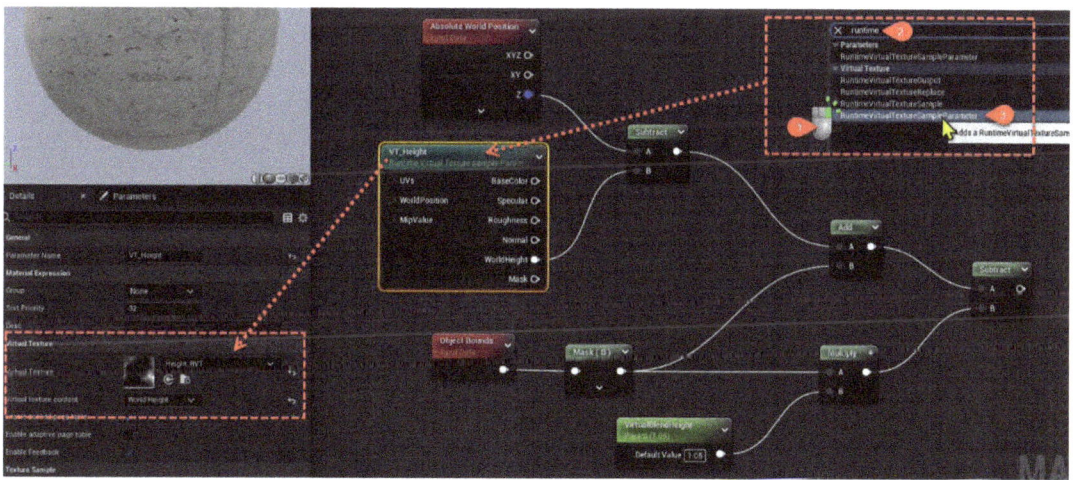

Figure 5-34. As part of the logic calculation with height, Height_RVT needed to be set up from RVTSampleParameter

CHAPTER 5   ENHANCING VISUAL REALISM WITH RUNTIME VIRTUAL TEXTURES AND MATERIAL BLENDING

As a follow-on to the logic of this material as shown in Figure 5-35, we will utilize the VertexNormalWS (in World Space) to indicate BlendContrast and multiply the FallOff values. Afterward, we will invert the total value using the One-Minus node before generating the Height_MAT. This will be achieved through the named reroute node as Height_MAT (as demonstrated in Figure 5-35).

1. As shown in Step 1 in Figure 5-35, we right-click the material editor to activate the node search selection box.

2. As indicated in Step 2, we type vertexnormal to filter out the node with the name that contains vertexnormal.

3. Then, we select the node named VertexNormalWS.

4. The network uses a "VertexNormalWS" node, which provides the normal of the vertices in world space. This is crucial for effects that depend on the direction each part of the mesh is facing relative to the world.

5. A "Mask" node is connected to the "VertexNormalWS" node, likely isolating a specific channel – possibly the blue (Z) channel, which would represent the up vector in world space. This can be used to apply effects only to certain orientations, such as the top faces of a landscape.

6. The output of the "Mask" node is then manipulated through a series of mathematical nodes – "Multiply," "Divide," "Subtract," and "Saturate" – to adjust the influence of the normal data on the material. These nodes are used to control how the texture blends across the surface based on the angle of the surface normal. For example, "Saturate" is used to clamp the values between 0 and 1, ensuring that they stay within valid ranges for use in shading.

7. The parameterized nodes named "Falloff" and "RVT_Blend_Contrast" are used to fine-tune the blending effect:

    a. "Falloff" adjusts how quickly the material effect fades from the peaks or ridges of the mesh.

    b. "RVT_Blend_Contrast" amplifies or reduces the contrast of the blending effect.

CHAPTER 5   ENHANCING VISUAL REALISM WITH RUNTIME VIRTUAL TEXTURES AND MATERIAL BLENDING

8. Finally, the output from these operations feeds into a reroute node named "Height_MAT", where the end result will influence a material parameter related to height, such as displacement or bump mapping, creating a material where the appearance changes depending on the mesh's curvature or angle relative to the world.

9. This node setup is typically used to create sophisticated material effects like snow accumulation on top surfaces, moss growing on shaded areas, or wind effects stronger on exposed edges. The exact purpose would depend on the rest of the material setup and the artistic goals of the project.

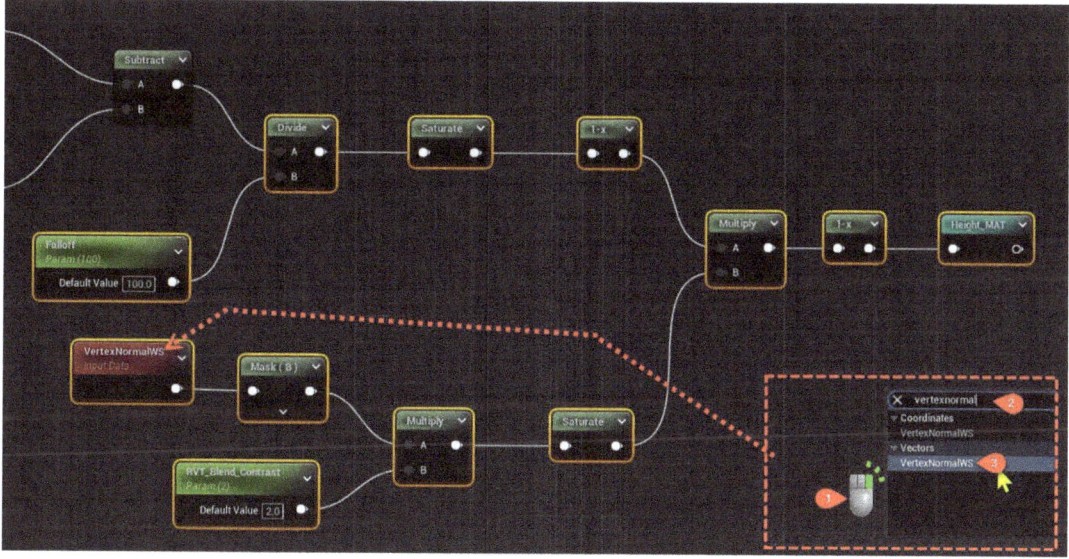

Figure 5-35. Connect the height (B value) from the VertexNormalWS node to blend the contrast

Final Material Output and Result

As mentioned at the start of this section, we have three different material attributes – Original_MAT (from the asset's original material), Landscape_Colour_MAT (from the landscape's Color RVT), and Height_MAT (from the alpha blending) – we can use the blending material attribute node to blend between the landscape and the 3D asset itself. This blending is determined by factors such as Falloff, Contrast, and the Height of the position.

Chapter 5   Enhancing Visual Realism with Runtime Virtual Textures and Material Blending

Firstly, we need to remove the original link to the resulting material attributes (as shown in Figure 5-36) before utilizing the `BlendMaterialAttribute` node shown in Figure 5-37.

1. Right-click the Material Attributes to activate the selection related to connecting line.

2. Select "`Break The Link`" to remove the original connecting line.

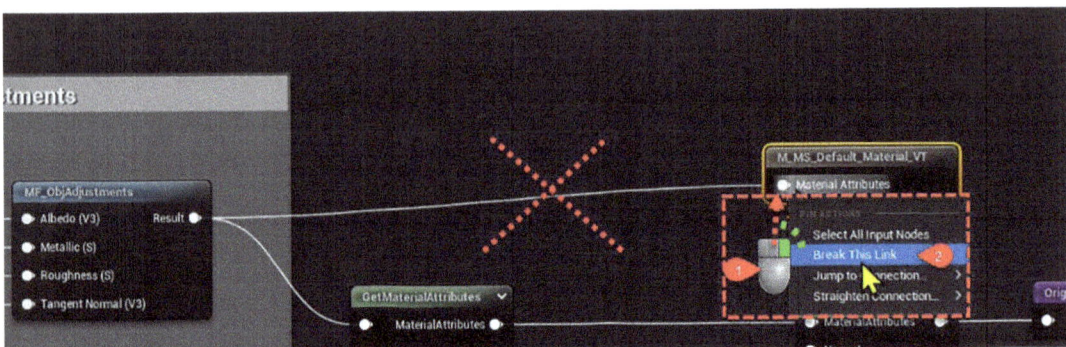

Figure 5-36. *Disconnect the original result to the M_MS_Default_Materials_VT as output*

Put Them Altogether

On the same material editor as before, we employ the `Height_MAT` (the named reroute node) as the alpha value, indicating the pixel color value (as shown in Figure 5-37).

- The logic in this material setup starts with combining the two existing named reroute declaration material nodes, "`Landscape_Colour_MAT`" and "`Original_MAT`", using a "`BlendMaterialAttributes`" node. The "`Height_MAT`" named reroute declaration node will be used as the "Alpha" input controls the blend ratio between these two formal mentioned materials.

- Next, we used "`GetMaterialAttributes`" node to extract the "Normal" attribute from the preceding blended material result, which holds information about surface angles and is crucial for realistic lighting and shading.

- The extracted "Normal" is then passed through a "TransformVector" node set to convert "World Space to Tangent Space." This is a crucial step because normal maps are typically used in tangent space, a coordinate system relative to the surface of the mesh that allows the normal map to remain consistent regardless of the mesh's orientation in the world.

- Next, the output of the "TransformVector" node is connected back to a "SetMaterialAttributes" node, where the converted normal map will apply toward the original output "BlendMaterialAttributes" material.

- Finally, the output from the "SetMaterialAttributes" node will set toward the final material, "M_MS_Default_Material_LVT", ensuring that lighting and shading perform correctly on the material's surface in the environment.

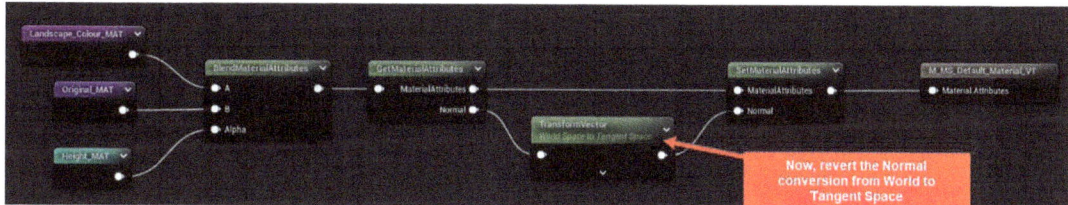

Figure 5-37. *Connect both the setup named reroute node to blend materials with the original based on the Height_MAT as alpha to generate the final material output*

As a result, as shown in Figure 5-38, we can adjust how much the landscape material should blend into the 3D asset using parameters like FallOff, BlendContrast, and VirtualBlendHeight. The illustration in Figure 5-38 serves as an example of the values and output of the blending materials at the junction between the edge of the 3D assets and the landscape.

CHAPTER 5   ENHANCING VISUAL REALISM WITH RUNTIME VIRTUAL TEXTURES AND MATERIAL BLENDING

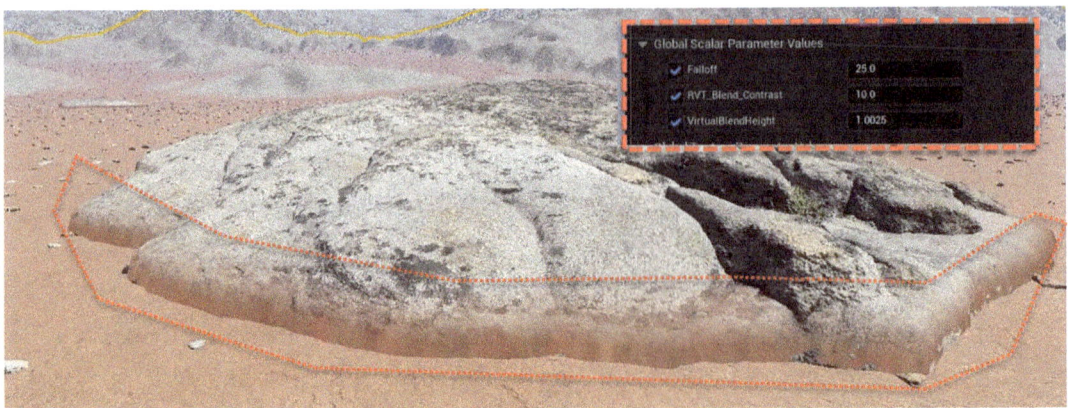

Figure 5-38. *Adjust the parameter values to suit the visual of the blend between the 3D assets and the landscape*

Summary

This chapter delved into the integration of RVT technology within UE5, aiming to enhance texture streaming, material blending, and overall graphics quality. RVT addresses the challenge of rendering intricate virtual environments by dynamically generating texture data on the GPU. Unlike traditional texture streaming, RVT directly streams textures from storage, mitigating memory overload.

The synergy between world height and RVT is pivotal for creating immersive virtual landscapes. World height data influences terrain details, and RVT efficiently streams textures based on player viewpoint. This combination improves visual quality and memory efficiency, particularly beneficial for expansive open-world games.

Material blending, a technique that smooths transitions between 3D objects and terrain, is advanced through RVT. RVT assets store shading information for seamless integration, enhancing realism and visual quality. RVT Output nodes within the Material Editor facilitate the writing of material attributes to virtual texture assets, enabling precise control over blending and output.

The practical implementation of these concepts involves setting up RVT assets, defining material blends based on RVT color and height, and achieving final material outputs. This process optimizes memory usage, elevates graphics quality, and empowers developers to create intricate, immersive, and visually captivating game worlds.

CHAPTER 5   ENHANCING VISUAL REALISM WITH RUNTIME VIRTUAL TEXTURES AND MATERIAL BLENDING

In the upcoming chapter, we will delve into one of UE5's groundbreaking features: Lumen. This revolutionary technology transforms virtual environments by enabling real-time global illumination and dynamic lighting interactions. By harnessing sophisticated methods such as ray tracing and voxelization, Lumen provides an unmatched level of visual fidelity. It facilitates seamless transitions between various lighting scenarios and times of the day. This innovation empowers developers to craft immersive and lifelike worlds, where light interacts authentically with surfaces and objects. As a result, it pushes the limits of realism and significantly elevates the overall visual quality of interactive experiences.

CHAPTER 6

Mastering Lumen Global Illumination in Unreal Engine 5

Lumen is a fully dynamic global illumination and reflection system that is designed for next-generation consoles and high-end PCs. It is the default lighting system in UE5, and it enables you to create realistic and immersive lighting scenarios for your games and applications. Lumen renders diffuse interreflection with infinite bounces and indirect specular reflections in large, detailed environments at scales ranging from millimeters to kilometers. It also supports features such as sky lighting, emissive materials, two-sided foliage, and virtual shadow maps. Lumen integrates well with other UE5 systems, such as Nanite, World Partition, and Chaos Physics. Numerous examples of Lumen lighting used for various purposes can be found at the following link:

https://docs.unrealengine.com/5.0/en-US/lumen-global-illumination-and-reflections-in-unreal-engine/

To use Lumen, we need to enable it in the **Project Settings** under **Rendering** as we will demonstrate this at the next section (shown in Figure 6-1). You can also adjust various parameters such as Lumen Scene and Lumen Reflections Quality, Indirect Lighting Cache, Screen Tracing Resolution, and more to fine-tune the performance and quality of Lumen. You can also use Post-Processing Volumes to control Lumen settings per camera or per area.

Lumen uses a combination of ray tracing, voxel cone tracing, and diffuse probes to compute the indirect lighting of the scene in real time. Lumen can handle dynamic changes in the scene such as moving lights, objects, and characters, as well as time of day and weather effects. Lumen can also produce accurate reflections, refractions, and shadows for both opaque and translucent materials. However, Lumen also has some limitations, such as requiring a minimum screen resolution of 1080p, not supporting multi-GPU configurations, and having some artifacts or noise in certain scenarios.

Lumen can work with any type of light source in UE, such as directional lights, point lights, spot lights, sky lights, and emissive materials. Lumen can also use the sky atmosphere and volumetric cloud components to create realistic sky and cloud lighting. To optimize your performance and quality, we can use the Stat Lumen console command to display the Lumen statistics and identify the bottlenecks. We can also use the Lumen GPU Lightmass feature to bake some of the indirect lighting and reduce the runtime cost of Lumen.

How Lumen Works

Lumen uses multiple ray tracing methods to solve global illumination and reflections. It first performs Screen Traces, which are fast and accurate for nearby surfaces, but have limited range and visibility. Then, it performs a more reliable method, which can be either Software Ray Tracing or Hardware Ray Tracing, depending on your platform and preference.

Software Ray Tracing uses Signed Distance Fields (SDFs) to approximate the scene geometry and perform ray marching. SDFs are volumetric representations of the scene that store the distance to the nearest surface at each voxel. SDFs are generated offline for each mesh and updated at runtime for dynamic objects. Software Ray Tracing is compatible with all platforms, but it has some limitations, such as lower resolution, aliasing, and artifacts.

Hardware Ray Tracing uses the dedicated ray tracing hardware on supported video cards to perform ray tracing. Hardware Ray Tracing can achieve higher quality and performance than Software Ray Tracing, but it requires a compatible video card and driver. Hardware Ray Tracing is currently experimental and not supported on all platforms.

Lumen uses a hybrid approach that combines Screen Traces, Software Ray Tracing, and Hardware Ray Tracing to achieve the best balance between quality and performance. Lumen also uses a Surface Cache, which is an automatic parameterization of the nearby scene's surface that stores the material properties and lighting information for each mesh. The Surface Cache is used to quickly look up lighting at ray hit points and to interpolate between different ray tracing methods.

Lumen also uses a Far Field, which is a simplified representation of the distant scene that stores the average color and normal for each voxel. The Far Field is used to provide lighting and reflections for large worlds and to fill in the gaps where Screen Traces and Software Ray Tracing fail.

Lumen updates its lighting and reflections at different rates depending on the scene complexity and the camera movement. Lumen uses Temporal Super Resolution (TSR) and Temporal Accumulation (TA) to improve the quality and stability of the lighting and reflections over time.

For more detailed information on how Lumen operates and its implementation in UE, you can explore the official documentation and resources provided by Epic Games at the following links:

Lumen in UE Documentation

`https://docs.unrealengine.com/5.0/en-US/lumen-global-illumination-and-reflections-in-unreal-engine/`

Introduction to Lumen Video Tutorial

`https://dev.epicgames.com/community/learning/tutorials/dL16/unreal-engine-introduction-to-lumen-essentials`

Enabling and Configuring Lumen

Lumen is enabled by default for new projects created in UE5:

1. The Lumen configurations can be accessed within the Project Settings, as shown in Figure 6-1.

2. Within the Rendering section of the Project Settings, we can find various Lumen-specific settings.

3. To quickly locate these settings, use the search function by typing "Lumen" into the filter bar.

4. Ensure that the "Dynamic Global Illumination" and "Reflection Method" options are both set to Lumen to utilize its full capabilities.

Chapter 6   Mastering Lumen Global Illumination in Unreal Engine 5

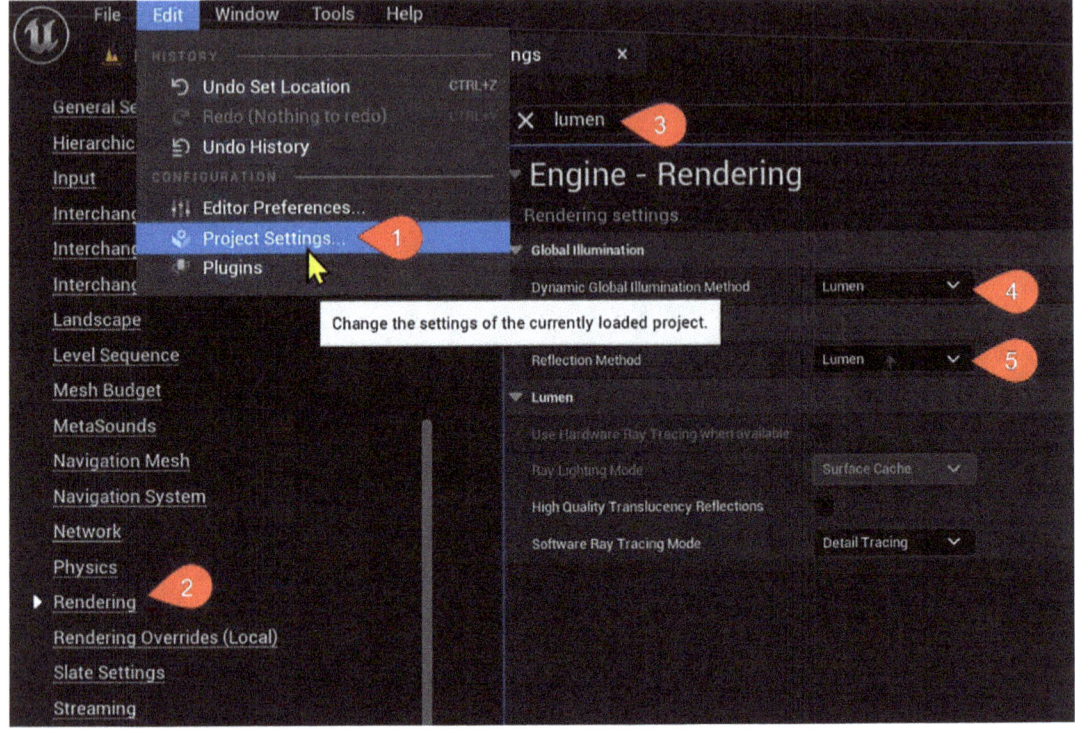

Figure 6-1. Lumen settings

It is also efficient to set the Software Ray Tracing Mode (as shown in Figure 6-2) to Global Tracing for Lumen to achieve faster tracing.

By setting the Software Ray Tracing Mode to Global Tracing for Lumen, the implication is that it enhances the speed or efficiency of the tracing process. Tracing, in this context, refers to the simulation of how light travels and interacts with the virtual scene. This can improve the speed of the tracing process, contributing to more efficient and faster rendering of realistic lighting effects in the virtual environment.

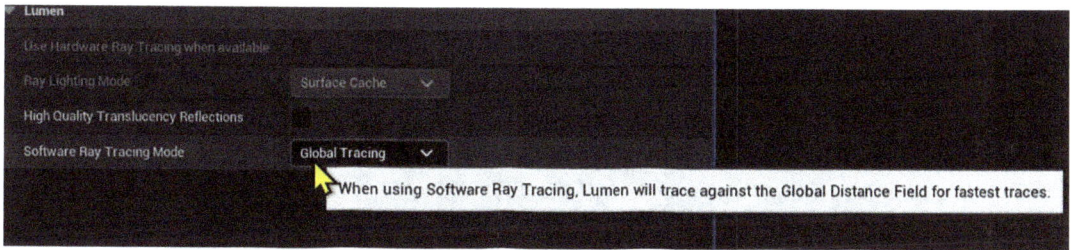

Figure 6-2. Software Ray Tracing method

Evaluating Lumen's Lighting Effects

We concluded Lumen in the previous section by noting that it is enabled by default in UE5 projects. Now, we will explore Lumen's impact by comparing scenes with Lumen both enabled and disabled. For this purpose, we will set up our level with a cave environment, as shown in Figure 6-3, to effectively demonstrate the shading and lighting effects.

> Instead of using a third-party 3D model for the cave, we will further develop our procedural generation techniques from the previous chapter, Chapter 4, to create a cave within our scene using Procedural Content Generation methods.

Furthermore, we will import two 3D models – a fuel cell and a macrovirus – as shown in Figures 6-4 and 6-5. These models will be outfitted with custom materials that feature animated emission properties. This setup will allow us to evaluate the impact of lighting effects on the models, thereby enabling us to appreciate the advantages of activating Lumen.

CHAPTER 6 MASTERING LUMEN GLOBAL ILLUMINATION IN UNREAL ENGINE 5

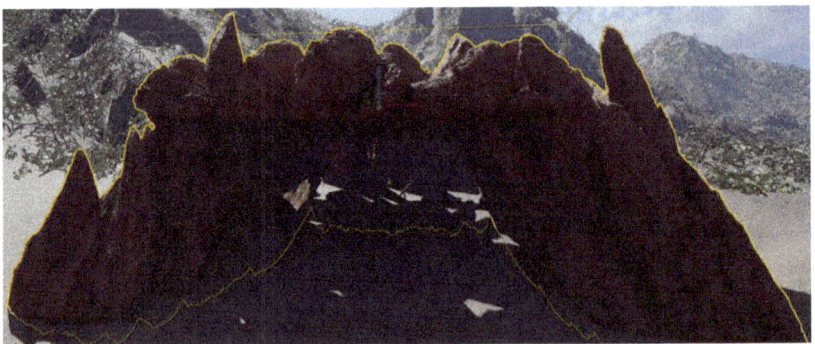

Figure 6-3. *Procedurally generated cave*

Figure 6-4. *Fuel cell*

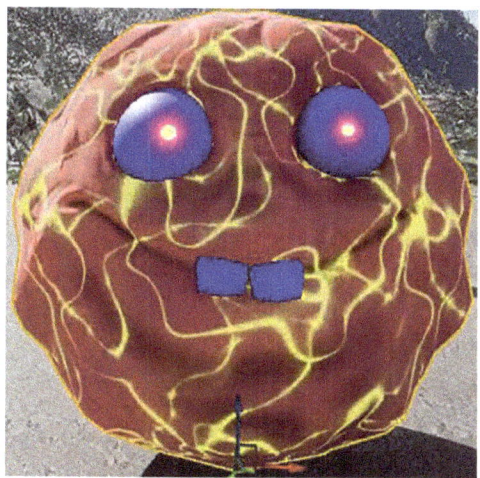

Figure 6-5. Macrovirus

Preparation Tasks

Firstly, let's move our TPP character to a location where we will construct our procedurally generated cave in front of this character. You are free to place the TPP character and construct the cave at any position, but the author has chosen the position for the TPP character as demonstrated in Figure 6-6:

- Position X: –68,707.0
- Position Y: –40,305.0
- Position Z: –69,410.0

CHAPTER 6   MASTERING LUMEN GLOBAL ILLUMINATION IN UNREAL ENGINE 5

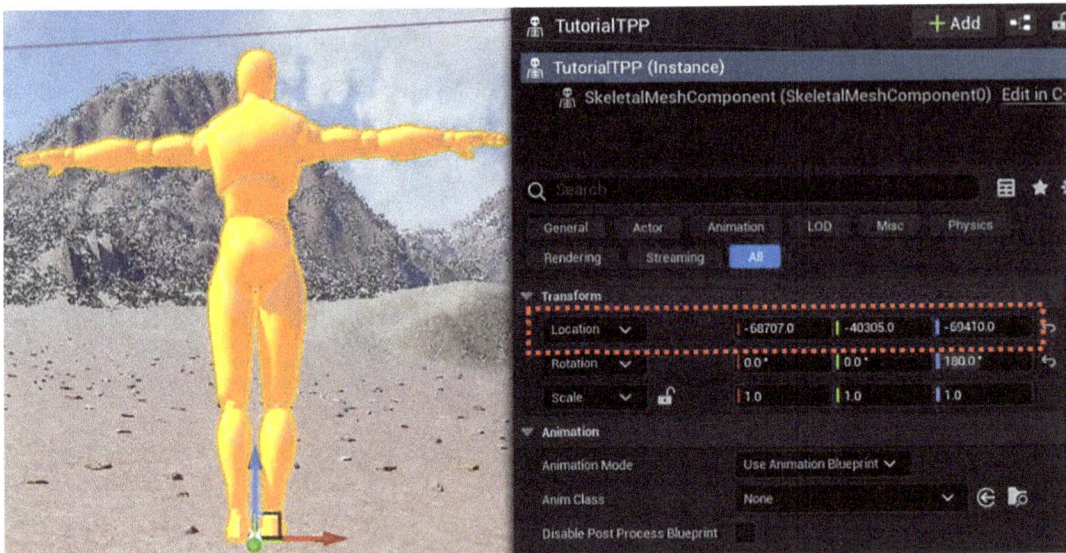

Figure 6-6. *Position the TPP character*

Procedural Content Generated Cave

Next, we will construct a cave using the procedural generation method discussed in Chapter 4. This time, we will use a mud rock 3D asset from Megascan as the foundation for generating the cave. While you are free to use any 3D rock assets, the author has chosen to utilize this specific mud rock (ID: **xd0mcf2**) from Megascan, as shown in Figure 6-7.

CHAPTER 6   MASTERING LUMEN GLOBAL ILLUMINATION IN UNREAL ENGINE 5

Figure 6-7. *Mud rock with Megascan ID xd0mcf2*

To generate our cave, we will create four separate planes to serve as foundations for each side (left, right, top, and rear) of the cave, as shown in Figure 6-8:

1. Click the Add New Actor shortcut.

2. Hover on top of Shapes to activate a list of shape actors.

3. Click to select the Plane actor.

CHAPTER 6   MASTERING LUMEN GLOBAL ILLUMINATION IN UNREAL ENGINE 5

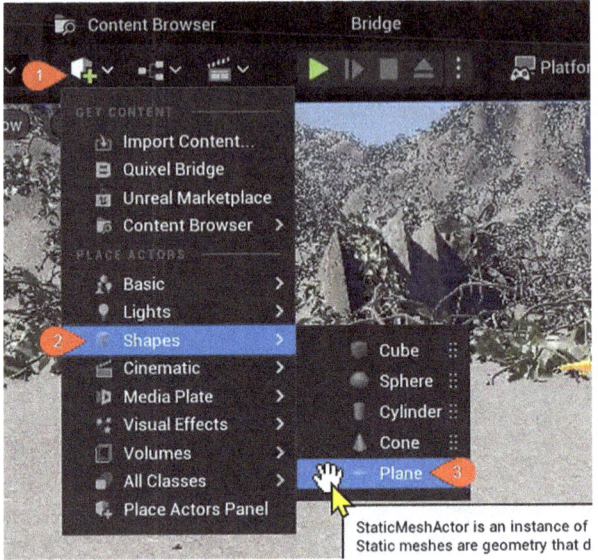

Figure 6-8. *Add a Plane actor*

It's crucial to set the Actor Tag – or at least to know the tag name assigned to each plane – since we will use the Actor Tag to identify the corresponding actor in the level that defines the position for each cave side. The author has assigned each plane an Actor Tag corresponding to the actor's name, as shown in Figures 6-9 and 6-10.

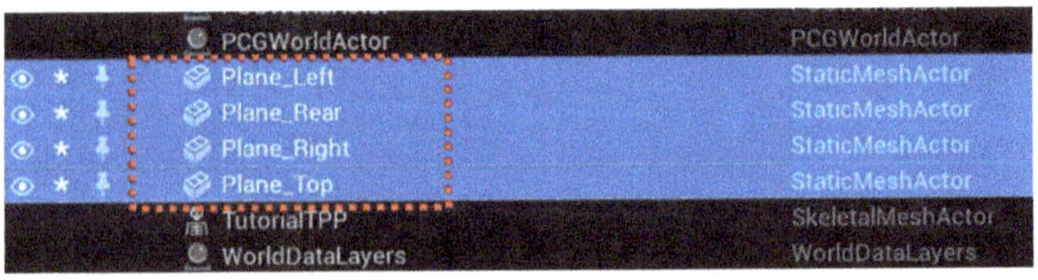

Figure 6-9. *Name the plane*

In order to add a tag to an actor in UE, we can follow these steps as shown in Figure 6-10:

1. Select the actor in the level editor. In this case, it's an instance named "Plane_Left".

2. Type in the word "tag" to see the Tag section in the Details panel.

3. Scroll down to the "Tags" section. Here, we can see a list of existing tags.

232

CHAPTER 6   MASTERING LUMEN GLOBAL ILLUMINATION IN UNREAL ENGINE 5

4. To add a new tag, click the "Add" button. This will create a new entry in the tags array.

5. Type in the desired tag name. For instance, here, we want to tag the actor as "Plane_Left"; we would enter this into the new tag field.

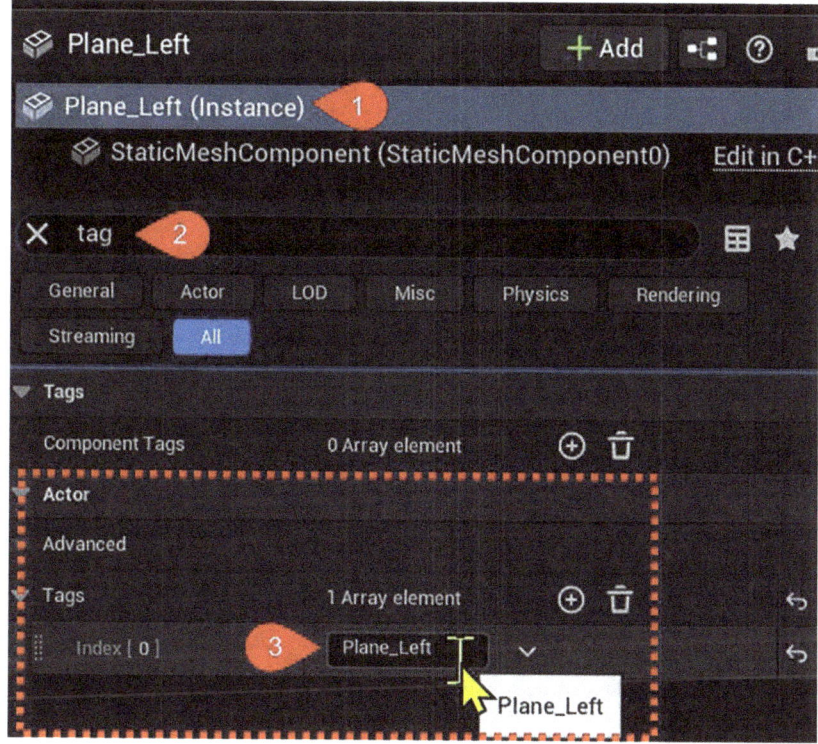

Figure 6-10. Set each of the Plane actors with the Tag accordingly by following these steps: Step 1, select the Plane actor; Step 2, type tag to filter out Tags; and Step 3, add a tag named accordingly

For using the planes as foundations for the cave's sides, the positions and scales for each of the four planes are detailed in Figures 6-11 and 6-12. These specifications will be applied to the planes in our procedural generation graph in the next step.

Plane_Left: Location X = –69,990; Y = –41,990; Z = –69,030

Scale X = 3.15; Y = 12.5; Z = 1.5

Plane_Rear: Position X = –69,110; Y = –42,520; Z = –69,060

Scale X = 12.5; Y = 3.15; Z = 1.5

Plane_Right: Position X = –68,120; Y = –42,240; Z = –69,130

Scale X = 3.0; Y = 12.5; Z = 1.5

Plane_Top: Position X = –69,130; Y = –42,140; Z = –68,650

Scale X = 12.5; Y = 12.5; Z = 1.5

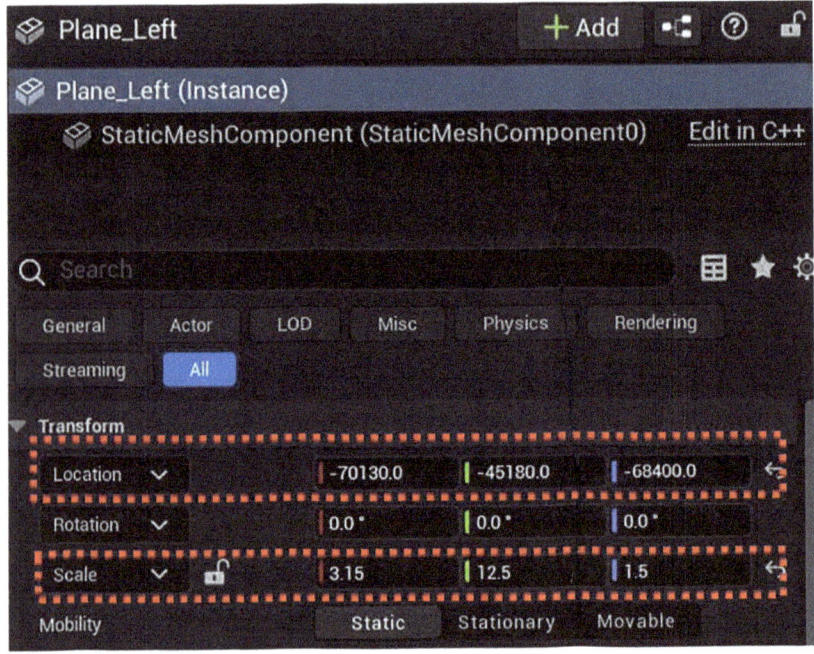

Figure 6-11. *Location and scale for Plane_Left*

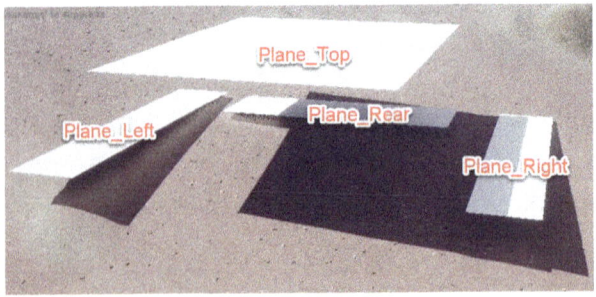

Figure 6-12. *Position of each plane*

Next, we will add a PCGVolume to the level by following these brief steps (as a reminder, you can refer to Chapter 4 if you need detailed steps).

To add a Procedural Content Generation Volume (PCGVolume) in UE, you can follow these brief steps:

1. Open the Place Actors panel in the UE editor.

2. Navigate to the Volumes category to find a list of available volume types.

3. Look for PCGVolume within the list or use the search bar to filter the options.

4. Click and drag the PCGVolume into the level to place it in the desired location.

5. With the PCGVolume selected, adjust its properties in the Details panel as needed for your Procedural Content Generation requirements.

6. Remember to save the level after placing the PCGVolume to ensure that all changes are retained.

The PCGVolume should cover the whole volume of the preceding four planes, as shown in Figure 6-13. The specifications are as follows:

Location X = –69,120; Y = –42,140; Z = –68,580

Scale X = 12.5; Y = 12.5; Z = 7.5

CHAPTER 6   MASTERING LUMEN GLOBAL ILLUMINATION IN UNREAL ENGINE 5

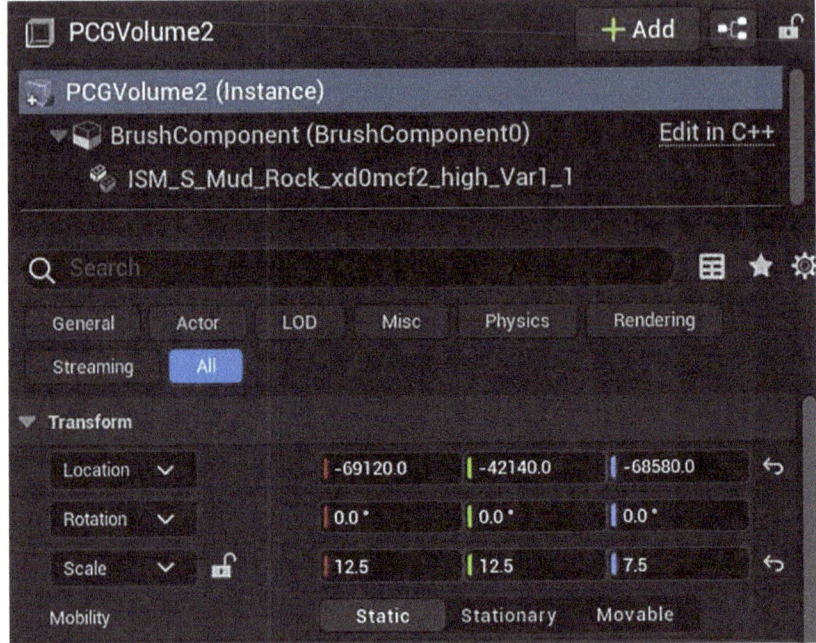

Figure 6-13. *Location and scale of the cave PCGVolume*

By creating a new PCG graph, as shown in Figure 6-14, we use the "Get Actor Data" node from the specified Actor Selection Tag (as shown in Figure 6-15) to retrieve the location.

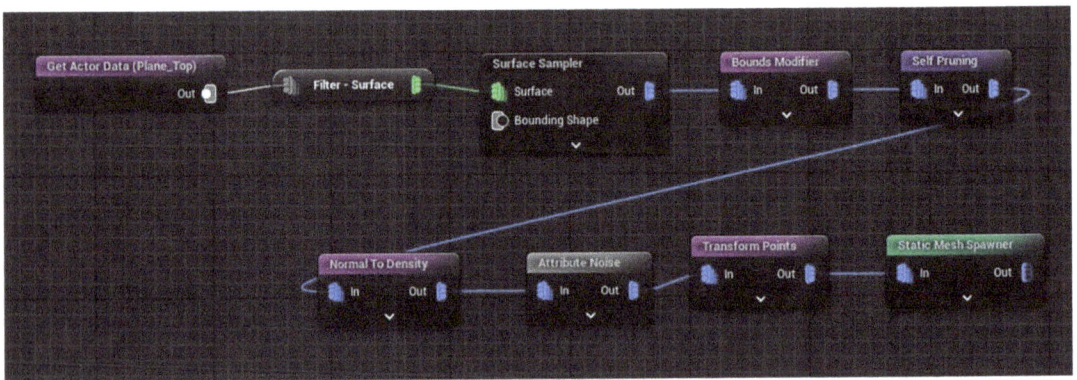

Figure 6-14. *The PCG nodes for caveLeft*

CHAPTER 6   MASTERING LUMEN GLOBAL ILLUMINATION IN UNREAL ENGINE 5

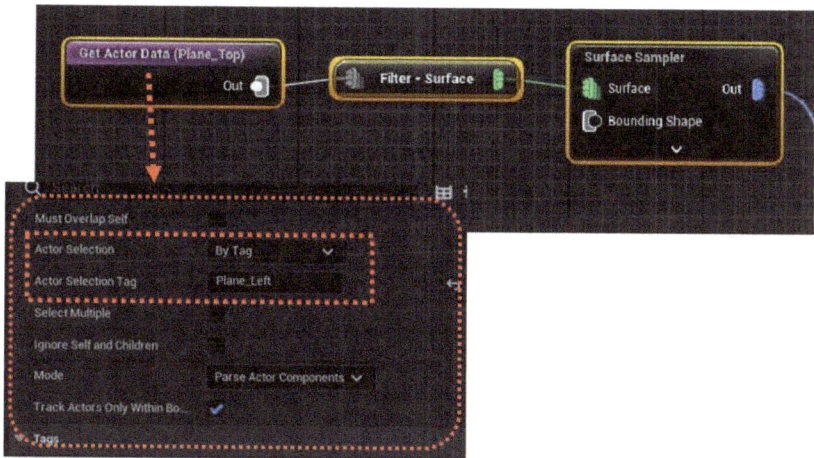

Figure 6-15. Set the Actor Tag to retrieve information about the actor

This is followed by nodes that randomly generate the mesh with varying rotations for each plane, as shown in Figure 6-16, and spawn the static mesh from the given mud rock, as shown in Figure 6-17. The result for the Plane_Left wall generated by the current PCG graph is shown in Figure 6-18.

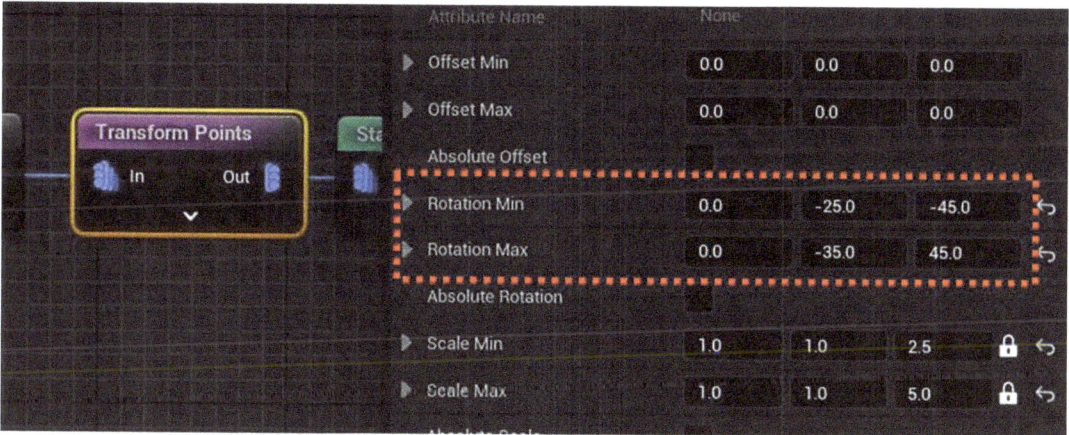

Figure 6-16. The random rotation value for the left side of the cave

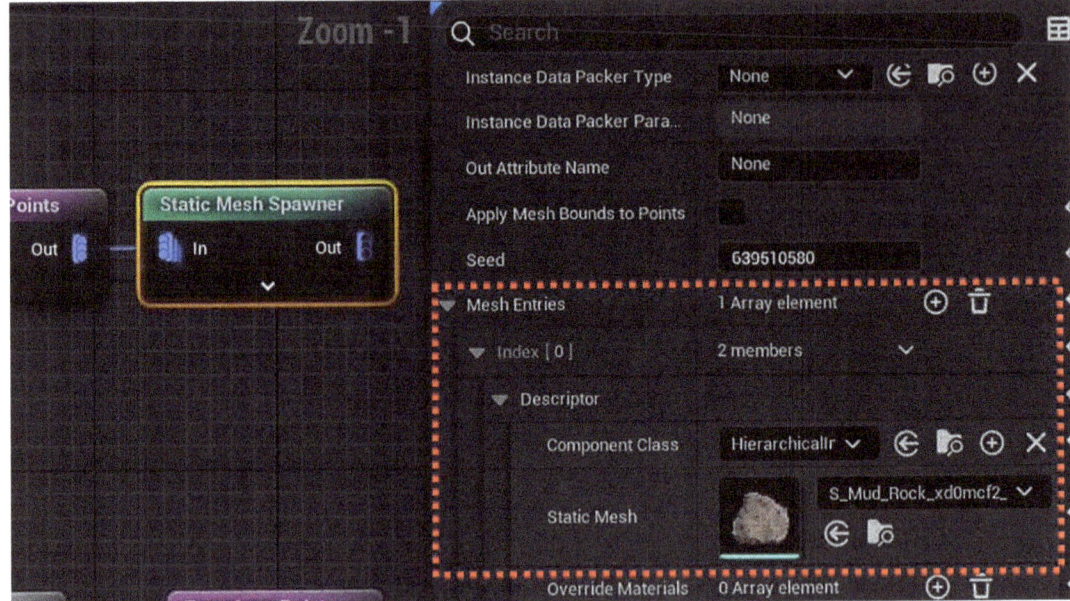

Figure 6-17. *Add a mud rock model as the static mesh spawner object*

The following are the summary steps to achieve this PCG volume, and by performing these steps and configuring the nodes correctly, we can leverage UE's PCG system to create dynamic and varied content within our level.

The author will not delve into the detailed construction of this set of PCG graphs, as the fundamental concepts are very similar to those discussed and presented in Chapter 4.

---

1. Begin by identifying the actor in our level that we want to use as a reference for the procedural generation. For instance, if we have an actor named "Plane_Left" that we want to use, locate it within our level's hierarchy.

2. With the actor selected, navigate to the details panel where we will find various properties of the actor. Look for the "Tags" section which allows us to assign a tag to your actor. Tags can be useful for identifying and organizing our actors within the level.

3. Add a new tag by clicking the plus icon or typing in the tag field if it's empty. Assign a meaningful tag to our actor, such as "Plane_Left" which corresponds to the actor's name or another identifier that makes sense for our project.

4. Once we have tagged our actor, we can use this tag to reference the actor within the PCG logic. For example, in the PCG graph, we can use a "Get Actor Data" node and set it to select actors by the tag we assigned.

5. Proceed to configure other nodes in the PCG graph such as "Surface Sampler," "Bounds Modifier," "Self Pruning," and "Static Mesh Spawner" based on the properties you want to generate. For instance, "Surface Sampler" can define where on the landscape the content will be generated, and "Bounds Modifier" can adjust the area of effect.

6. If we want to manipulate specific properties like rotation or scale of the generated items, use nodes like "Transform Points" where we can specify values for rotation and scale.

7. Finally, ensure we correctly set up the "Static Mesh Spawner" with the desired static mesh that will be spawned by the PCG system. This involves selecting the mesh and configuring attributes such as mobility, collision presets, and instance culling distances.

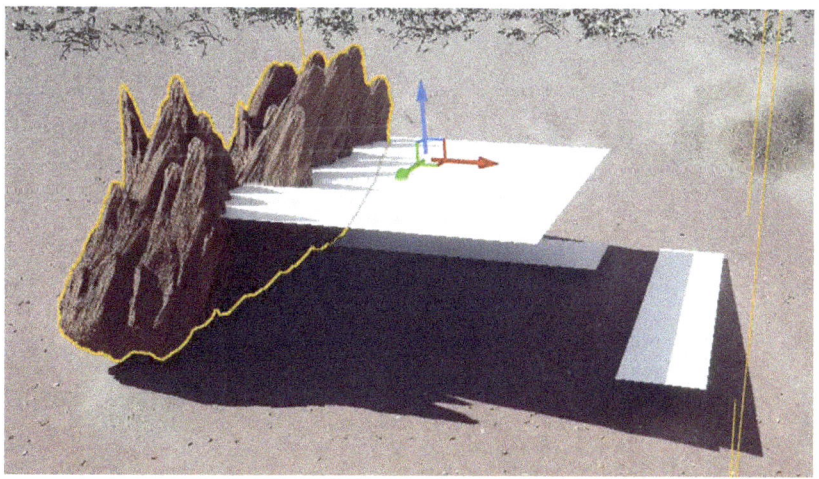

Figure 6-18. *The result from the PCG node of the Plane_Left*

EXERCISE: COMPLETE THE SETUP FOR PLANE_RIGHT, PLANE_REAR, AND PLANE_TOP

As an exercise, you will be expanding on the PCG setup by applying the techniques you have learned to additional planes in your environment. Here's how you can approach it:

- Step 1: Review the node setup we had used for Plane_Left. Understand the function of each node and how they contribute to the PCG outcome.

- Step 2: Decide on your approach. You can choose to use a single graph for all planes if their behavior will be identical or create multiple graphs if you plan to introduce variation between different planes. Splitting the graph for each plane might be beneficial if they require distinct settings or if you want to keep your work modular.

- Step 3: Duplicate the graph setup for Plane_Left and adjust it for Plane_Right, Plane_Rear, and Plane_Top. Pay close attention to the orientation and rotation settings for each plane to ensure they align correctly with your level's design.

- Step 4: Test each plane individually after applying the PCG nodes. Make necessary adjustments to ensure that the procedural content appears as intended without overlapping or conflicting with other elements.

This guided exercise will solidify your understanding of PCG in UE and provide a clearer path to managing multiple procedural elements in your projects.

The expected results are shown in Figures 6-19 and 6-20.

CHAPTER 6   MASTERING LUMEN GLOBAL ILLUMINATION IN UNREAL ENGINE 5

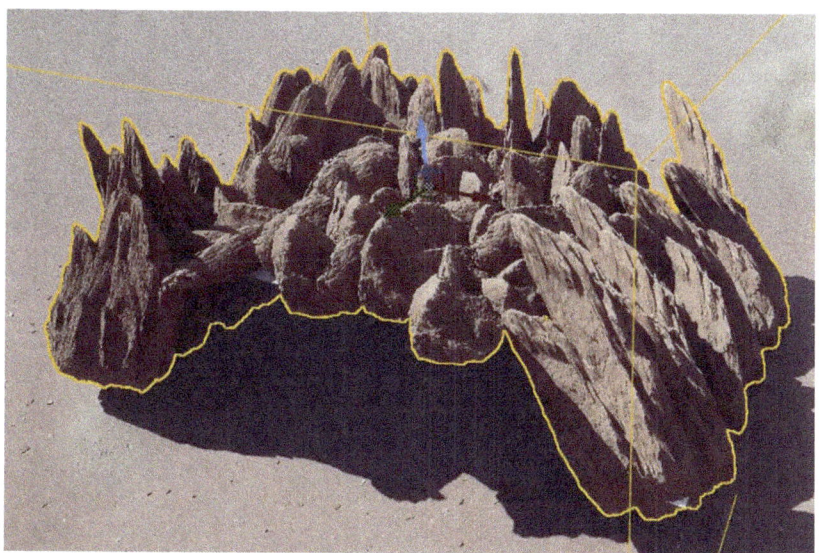

Figure 6-19. *Outcome of the cave from the top view perspective*

Figure 6-20. *Outcome of the cave from the front view perspective*

Finally, we set the visibility of all four planes to "False," as indicated in Figures 6-21 and 6-22, to hide the planes behind the generated rock formations that form the cave.

Figure 6-21. *Shift-click to select all four planes*

CHAPTER 6 MASTERING LUMEN GLOBAL ILLUMINATION IN UNREAL ENGINE 5

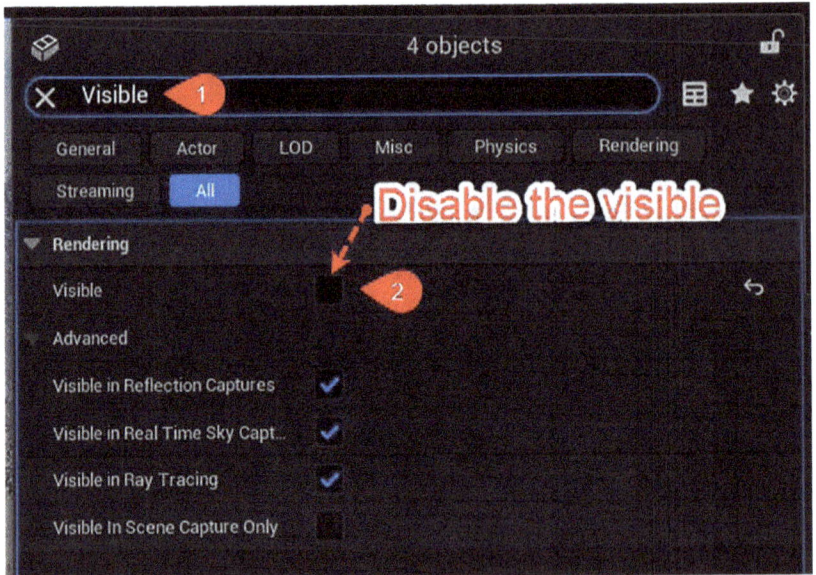

Figure 6-22. Steps to disable the visibility of the planes: Step 1, type Visible to filter the property that contains Visible; Step 2, uncheck the Visible property

Set Animate Emissive Materials

We will incorporate two sets of 3D models into the game that we are developing for this book: the fuel cell (Figure 6-4) and the macrovirus (Figure 6-5). Each 3D model can be freely downloaded from TurboSquid using the following links:

> **Note** While the original designs of these two models differ from our intended use, we are employing them as representations of the macrovirus and fuel cell for our own intentions in our game.

Fuel Cell: www.turbosquid.com/3d-models/3d-scifi-rainbow-barrel-model-1758254

Macrovirus: www.turbosquid.com/3d-models/free-character-virus-bacteria-herpes-3d-model/720389

Be sure to download the FBX format as shown in Figures 6-23 and 6-24.

CHAPTER 6   MASTERING LUMEN GLOBAL ILLUMINATION IN UNREAL ENGINE 5

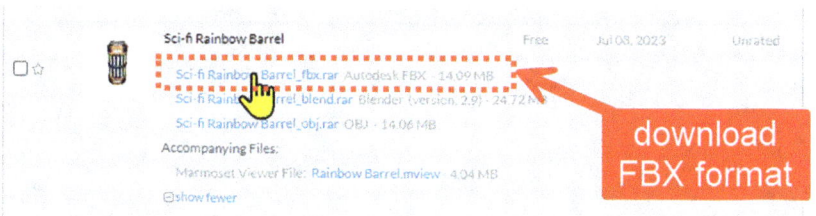

Figure 6-23.   *Download the fuel cell in FBX format*

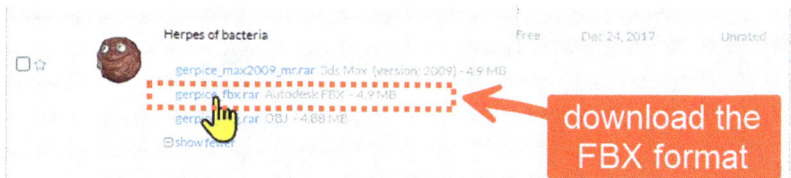

Figure 6-24.   *Download the macrovirus in FBX format*

The Workflow of Importing the Model in UE5

After extracting the downloaded model, we create a folder named Model within the UE5 Asset window for organization purposes, and we use the UE's Import feature to open up the FBX model; here, we are selecting Sci-fi_Barrel.fbx, as shown in Figure 6-25.

1. Create a folder named Fuel_Cell in the Content Browser by using +Add.

2. Click the Import button at the Content Browser.

3. Select the corresponding FBX file to be imported into the project.

243

CHAPTER 6 MASTERING LUMEN GLOBAL ILLUMINATION IN UNREAL ENGINE 5

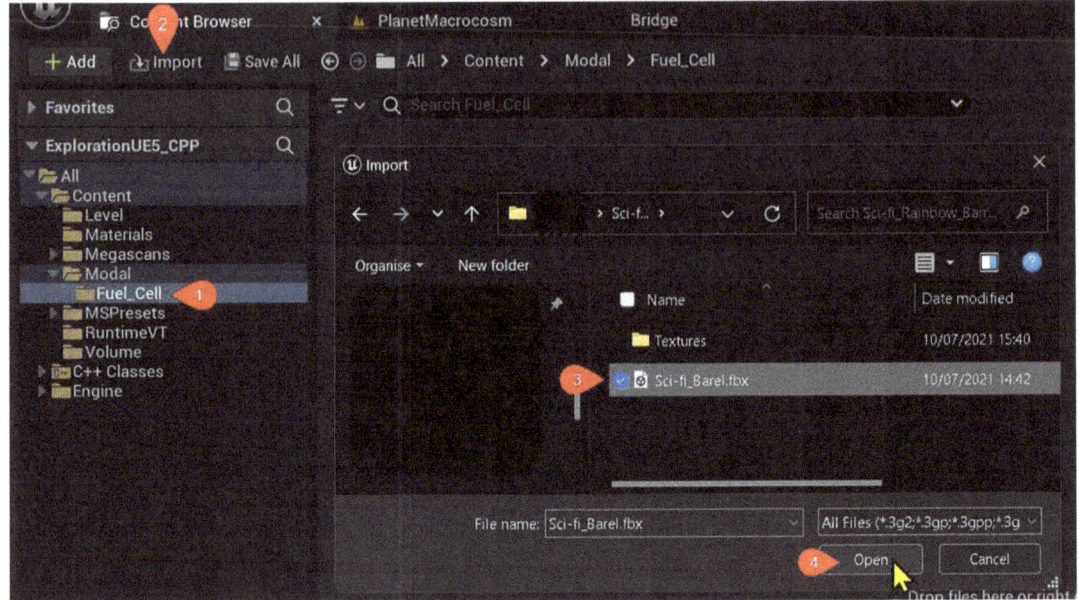

Figure 6-25. *Create a Fuel_Cell folder inside the Model folder and import the FBX model*

Within the FBX import options, we disabled the import of a Skeletal Mesh (since this model has no rigging and animation features within); we will need to ensure that "Combine Meshes" under the "Advanced" is enabled as shown in Figure 6-26.

1. Deselect the Skeletal Mesh.

2. Open up the Advanced option.

3. Be sure to select Combine Meshes.

4. Click the Import All/Import button.

CHAPTER 6   MASTERING LUMEN GLOBAL ILLUMINATION IN UNREAL ENGINE 5

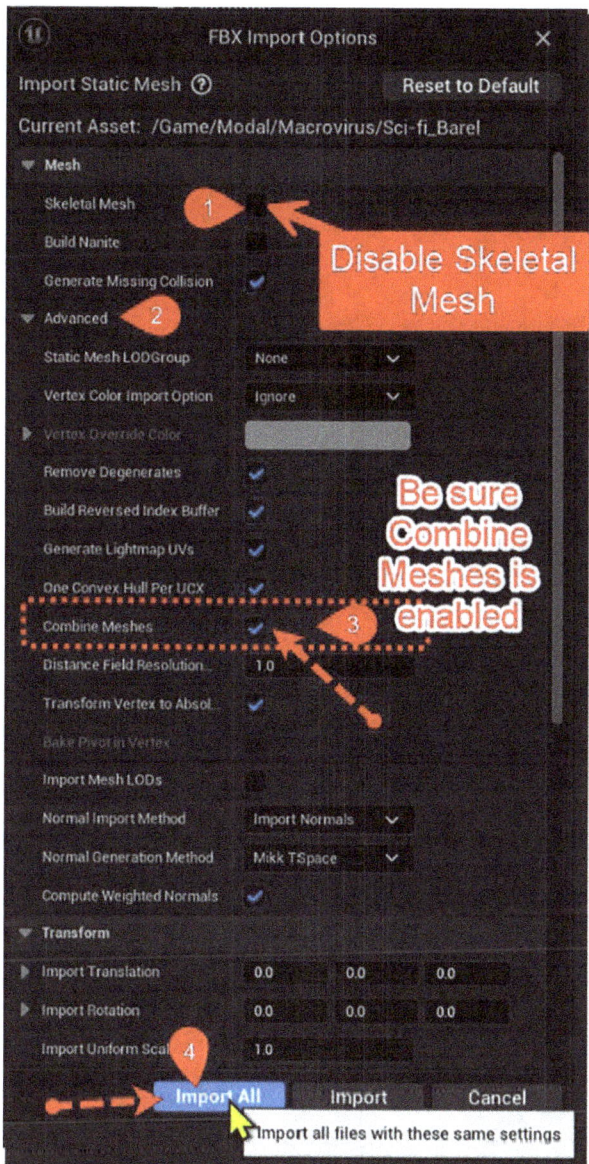

Figure 6-26. *Import settings*

Upon successfully importing the asset, we also notice that the normal map file of the model is also automatically imported along with the asset, which is often indicated by a thumbnail preview of the texture in full blue and purple colors (as shown in Figure 6-27).

CHAPTER 6   MASTERING LUMEN GLOBAL ILLUMINATION IN UNREAL ENGINE 5

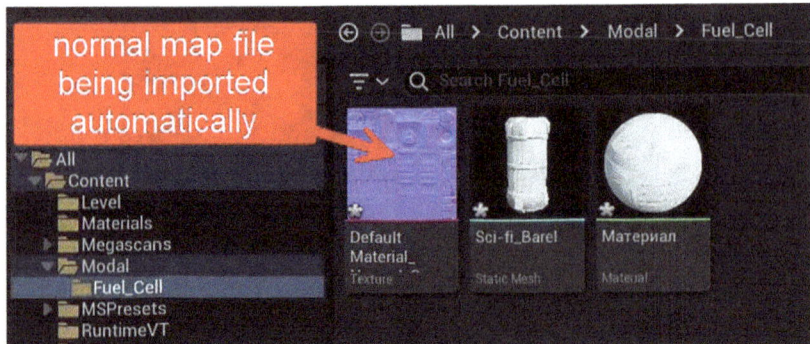

Figure 6-27. Imported Mesh and the Normal map file imported automatically

Besides the normal map, there are other textures that are needed to import to attach to the materials of this Fuel_Cell model. We can use the same Import feature to import the textures (ignoring the given normal map texture) as shown in Figure 6-28. The textures include maps for base color, emissive, metallic, ambient occlusion, opacity, and roughness.

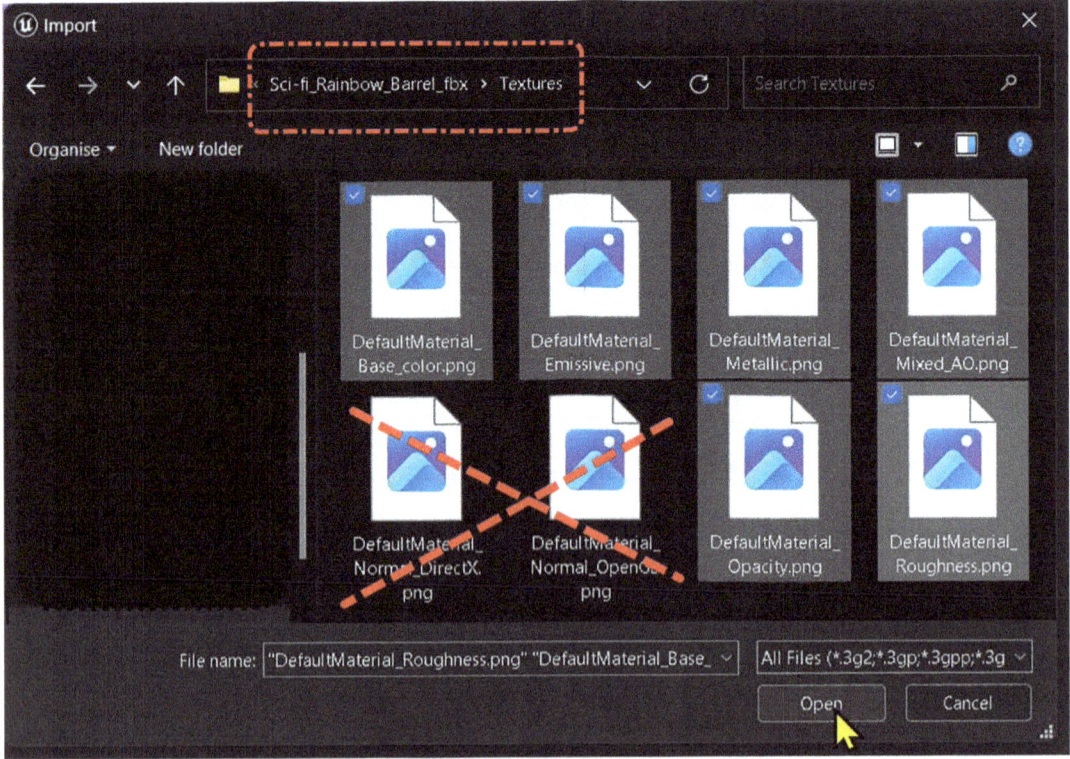

Figure 6-28. Import the Textures file (without needing the two normal map files)

After importing the 3D model and all the necessary texture files, we can now concentrate to work on the material asset named "Материал," which represents the final material applied to the 3D model, combining the imported textures to define the surface appearance of the model.

> You are welcome to rename the material asset by single-clicking the name text of the material itself, then typing in a new name.

To work on the material asset, we can double-click the material itself as shown in Figure 6-29, and this will open up the material editor (as we have encountered in the previous chapter).

We can now delete the default given color for the Base Color property for the material as shown in Figure 6-30.

Figure 6-29. Open up the Material file

CHAPTER 6 MASTERING LUMEN GLOBAL ILLUMINATION IN UNREAL ENGINE 5

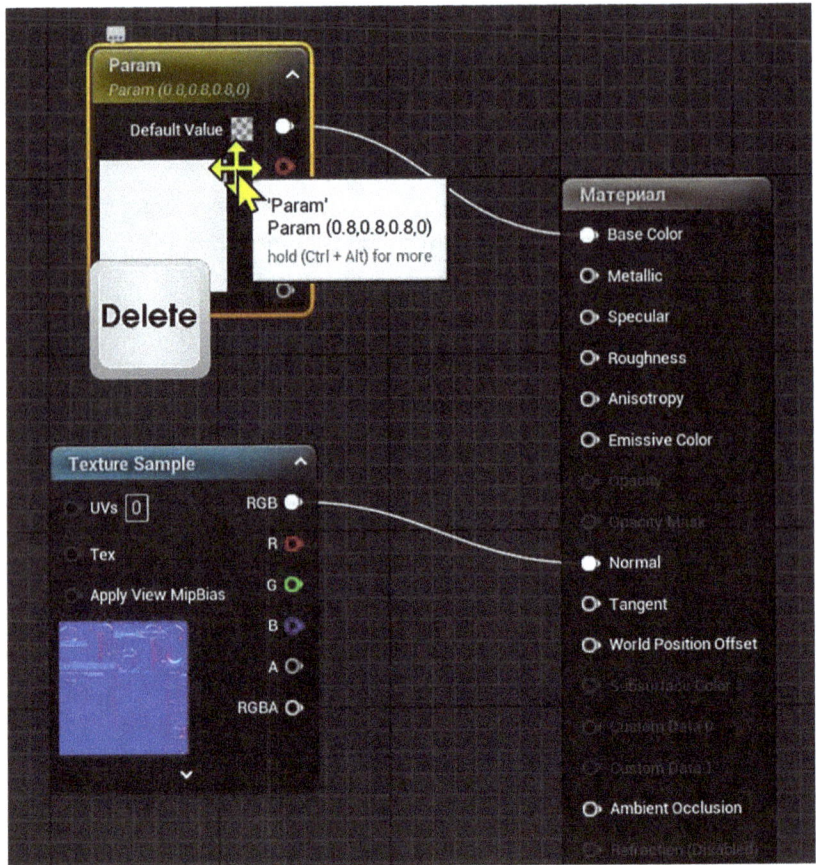

Figure 6-30. *Delete the default node for the Base Color property*

We can now start to bring the Base Color texture to the material editor by first single-clicking the thumbnail of the Base Color texture (as shown in Figure 6-31).

CHAPTER 6   MASTERING LUMEN GLOBAL ILLUMINATION IN UNREAL ENGINE 5

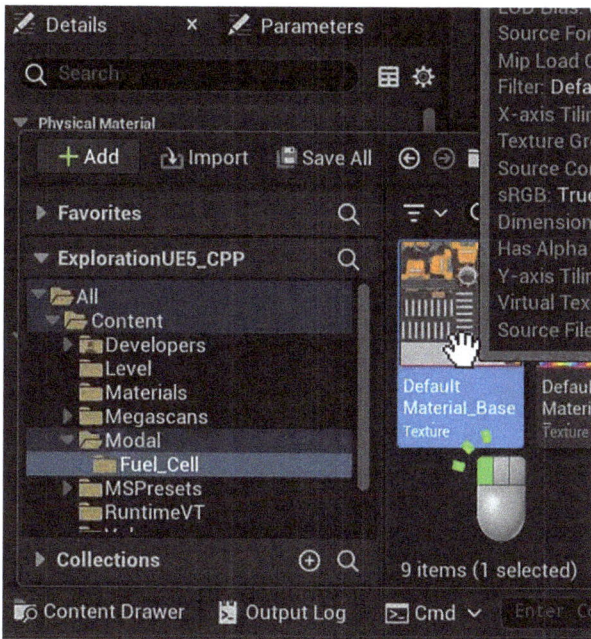

Figure 6-31. *Single-click to select the default Material_Base texture*

One method to import a selected texture into the material editor is by *holding down the "T" key and performing a single left-click* anywhere within the editor workspace. However, do note that this shortcut will only work if a material is already selected. So, ensure that you have the material selected in your content browser before using this method to bring in the texture.

The texture node with the selected texture will be created automatically, and we can connect the RGB channel from this texture node to the `Base Color` channel in the Material node itself (as shown in Figure 6-32).

1. Hold down the T key on the keyboard and left-click in the Material Editor to auto-create a texture node with the selected texture file automatically attached. (Ensure you select the desired texture in the Content Drawer before doing this.)

2. Connect the link from RGB to the `Base Color` property in the material node.

CHAPTER 6   MASTERING LUMEN GLOBAL ILLUMINATION IN UNREAL ENGINE 5

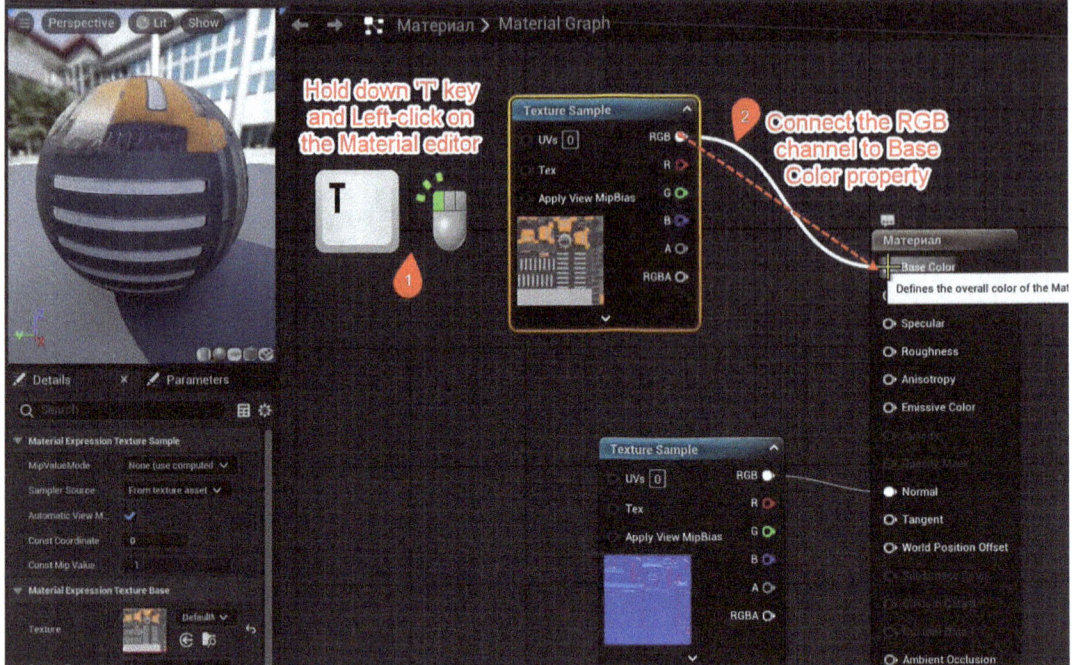

Figure 6-32. *Add the default Base Color texture to the Base Color property from the RGB channel*

To enable the `Opacity Mask` channel of the material, we will need to make sure the `Blend_Mode` is set to Masked, as shown in Figure 6-33:

1. Left-click the material box itself.
2. Select the Blend_Mode to be Masked.

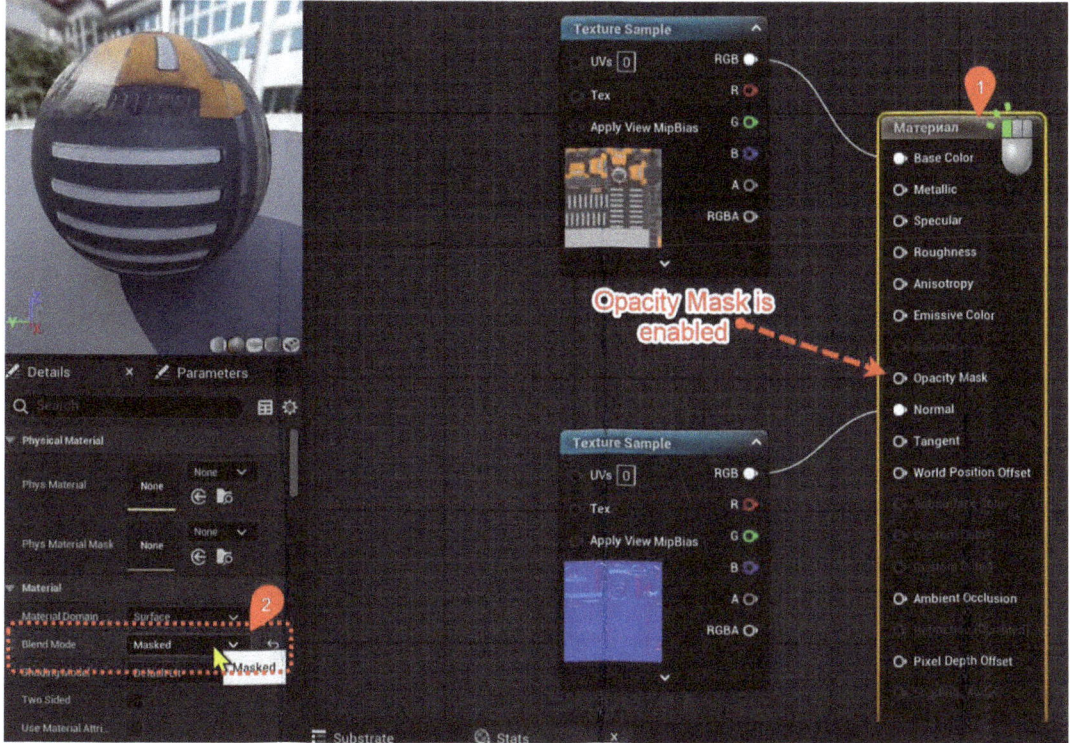

Figure 6-33. *To enable Opacity Mask, we need to set Blend Mode as Masked*

You should now be able to bring in other textures (Metallic, Opacity, Roughness, Ambient Occlusion, and Emissive) as shown in Figure 6-34 and part of Figure 6-35. Note that attaching the Opacity texture to the Opacity Mask channel will allow us to have the "see-through" gaps as shown in Figure 6-35.

CHAPTER 6   MASTERING LUMEN GLOBAL ILLUMINATION IN UNREAL ENGINE 5

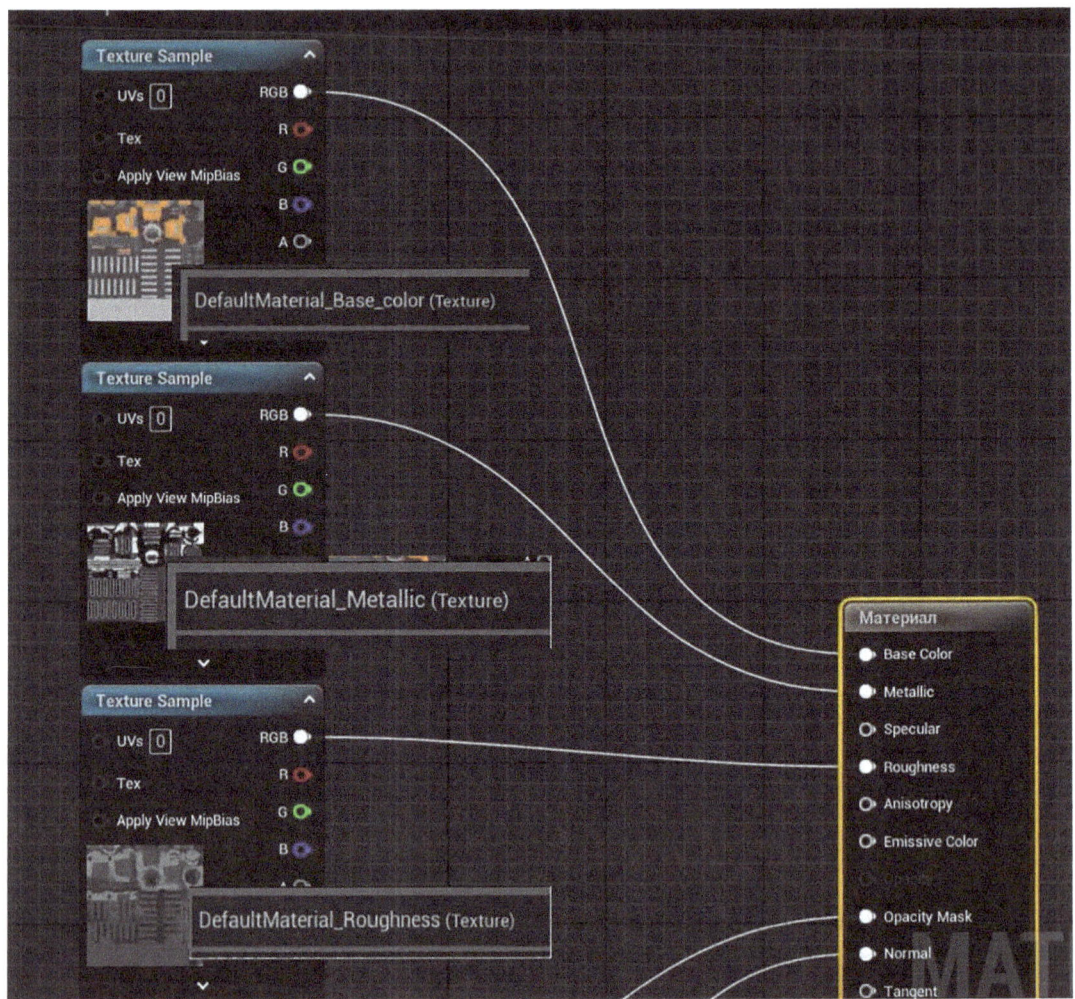

Figure 6-34. *Set up other properties (Metallic, Roughness)*

CHAPTER 6   MASTERING LUMEN GLOBAL ILLUMINATION IN UNREAL ENGINE 5

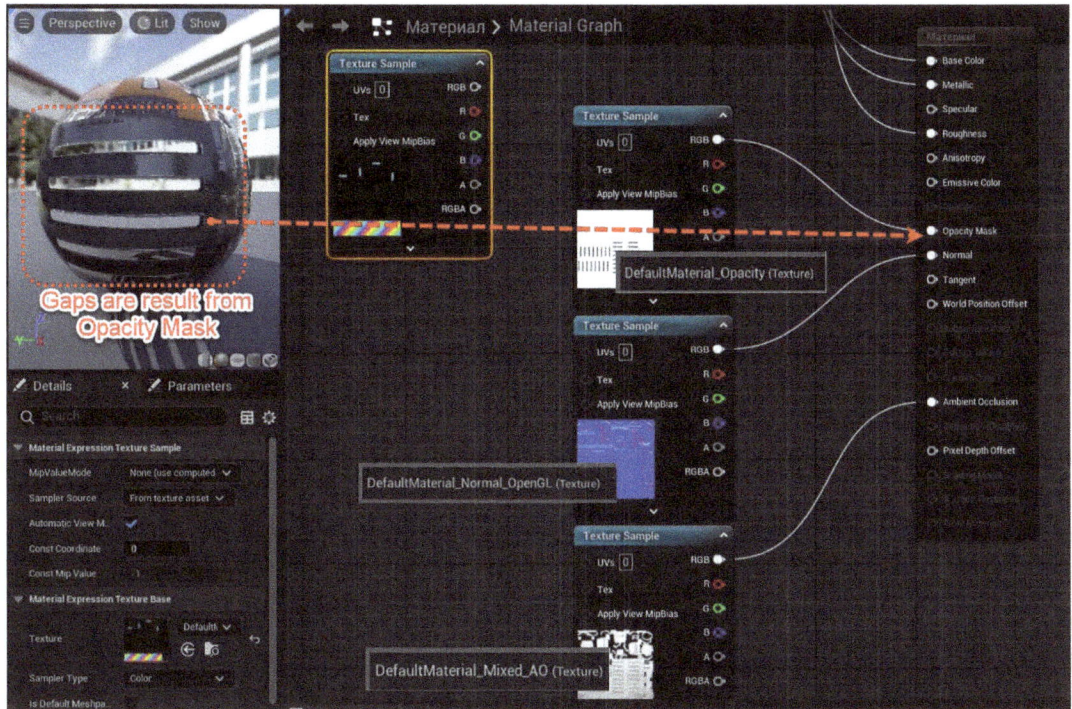

Figure 6-35. *Set up other properties, specifically the Opacity Mask*

Note that we have not yet connected the Emissive Texture to the material because we aim to animate the emission lighting. This can be achieved using a time-based node, specifically the Panner node. To add the effect of moving UV, we can follow these steps, which are also shown in Figure 6-36:

1. To create a panning effect on a texture, add a Panner node to the material graph. This can be done by right-clicking the graph space, searching for "Panner," and selecting it from the list.

2. Connect the Panner node to the UVs input of the Texture Sample node that contains the Emissive Texture.

3. Adjust the Speed parameter of the Panner node to control the direction and speed of the animation. In this example, we set the X Speed to 0.25 to move the texture along the X-axis.

4. To intensify the brightness of the texture, add a Multiply node to the graph. Connect the RGB output of the Texture Sample to the Multiply node.

253

CHAPTER 6   MASTERING LUMEN GLOBAL ILLUMINATION IN UNREAL ENGINE 5

5. Set one of the values of the `Multiply` node to 40, which will increase the emission's brightness by multiplying the texture's color values by this factor.

6. Finally, connect the output of the `Multiply` node to the `Emissive Color` input of the material to apply the animated and brightened texture to your material.

You are welcome to change the X and Y values of the Speed properties to adjust the speed and also the direction of the animation.

The author has decided to use Speed X as 0.25 and Speed Y as 0, so the animation only happens in the horizontal direction with the speed of 0.25.

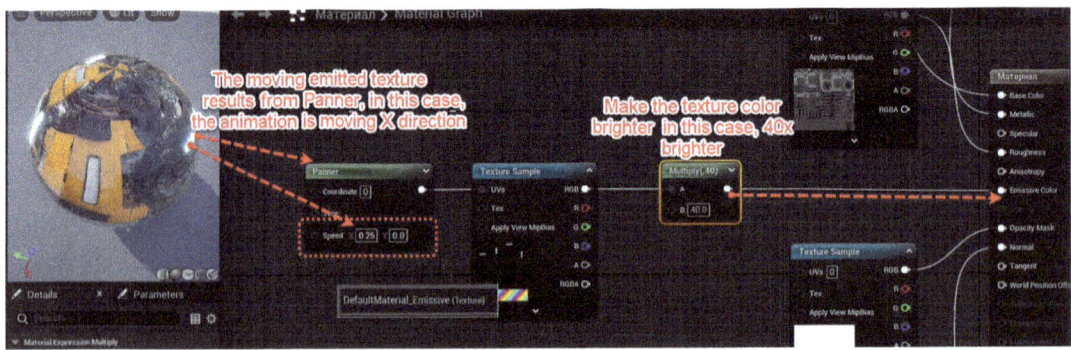

Figure 6-36. Add the important animated emissive texture color

Placing the Asset in the Level

One of the methods to place an asset in the level is to single-click to select the asset and right-click the level to select "`Place Actor`" to place the selected mesh asset as shown in Figures 6-37 and 6-38. We will place the asset itself inside the shaded cave using the following location values as shown in Figure 6-39:

Location X = –69,020; Y = –42,150; Z = –69,210

CHAPTER 6 MASTERING LUMEN GLOBAL ILLUMINATION IN UNREAL ENGINE 5

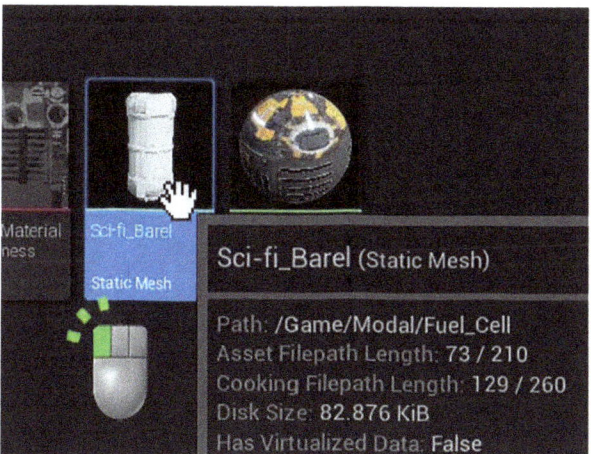

Figure 6-37. Select the static mesh

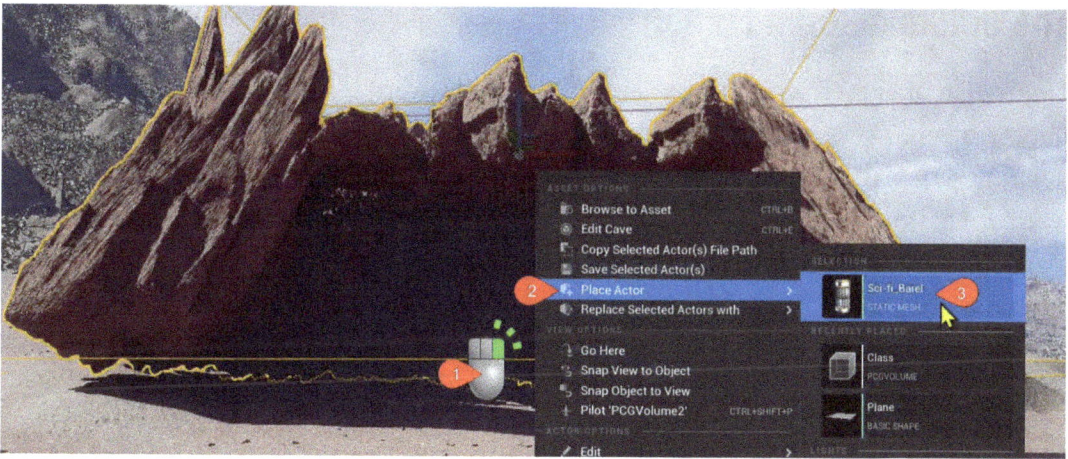

Figure 6-38. Place the selected static mesh to the level using these steps: Step 1, right-click the level to display the selection menu; Step 2, hover over the Place Actor; and Step 3, select the Sci-fi Barel as the Fuel_Cell object

CHAPTER 6 MASTERING LUMEN GLOBAL ILLUMINATION IN UNREAL ENGINE 5

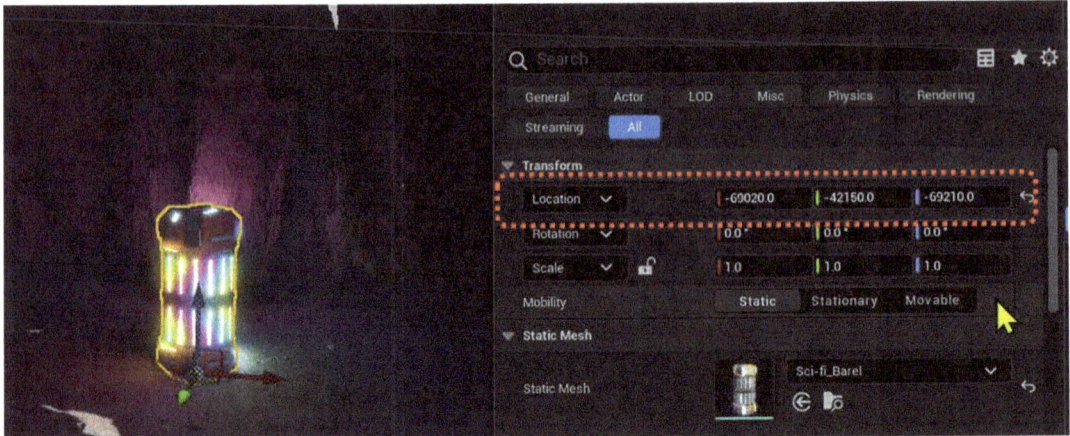

Figure 6-39. *Location of the fuel cell*

The Influence from Lumen

The advent of Lumen represents a significant leap forward in the pursuit of photorealism. This innovative lighting technology, as shown in Figure 6-40, offers a dynamic global illumination capability that transforms the visual landscape of digital worlds. As Lumen is enabled by default, this screenshot serves as a starting point, showcasing the Lumen system in its active state, endowing the environment with vibrant and nuanced lighting.

CHAPTER 6 MASTERING LUMEN GLOBAL ILLUMINATION IN UNREAL ENGINE 5

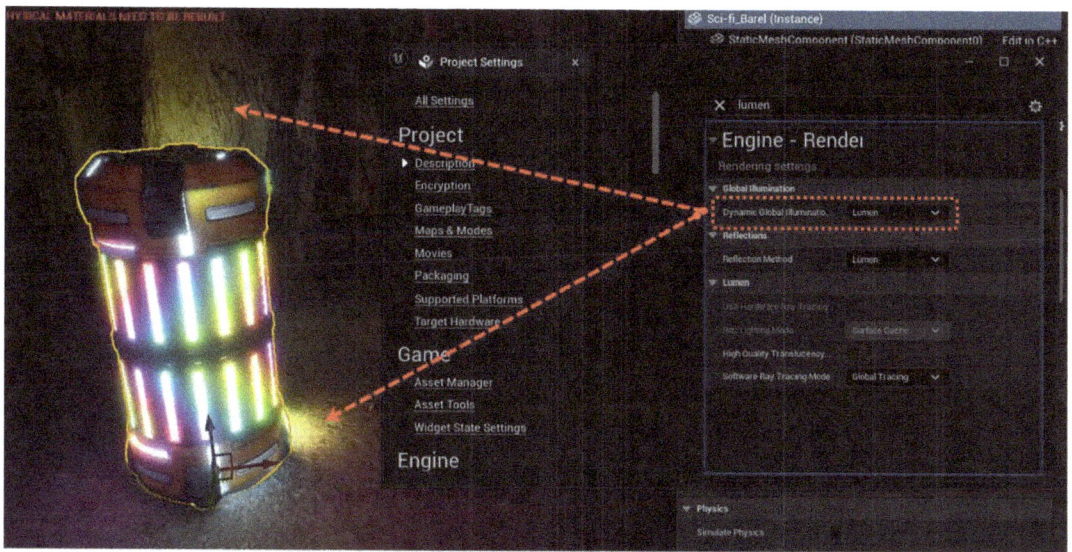

Figure 6-40. *Lumen enabled*

Proceeding to Figure 6-41, the narrative would contrast the absence of Lumen's dynamic illumination, drawing attention to the stark difference in environmental depth and realism. Here, the absence of Lumen's touch suggests a potential enhancement in performance, albeit at the expense of visual richness.

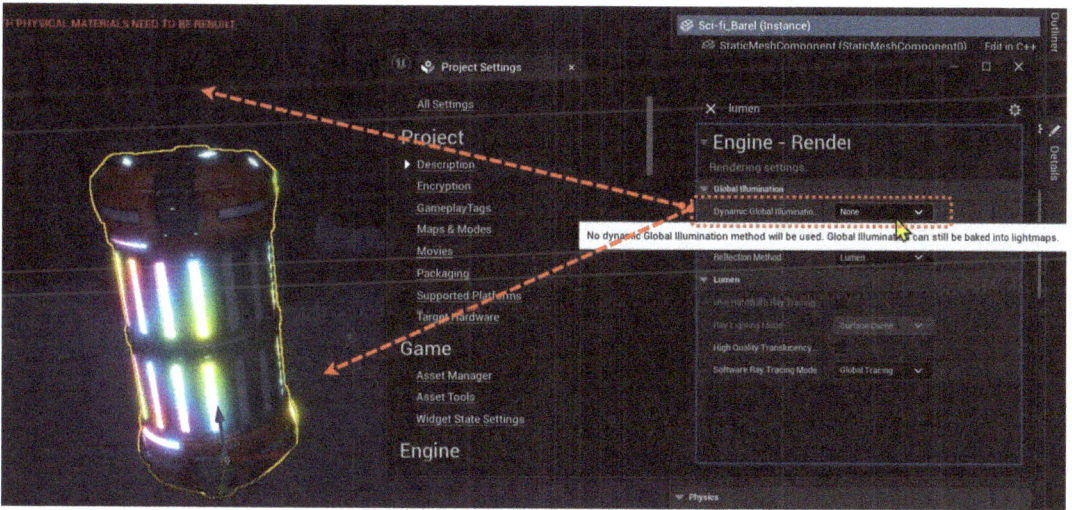

Figure 6-41. *Lumen is disabled*

CHAPTER 6 MASTERING LUMEN GLOBAL ILLUMINATION IN UNREAL ENGINE 5

Alternatively, Figure 6-42 could introduce a discussion on alternative illumination methodologies, such as the "Screen Space" technique, which, while innovative in its own right, is bounded by the limitations of on-screen data. This presents an opportunity to delve into the intricacies of lighting computations and their impact on the player's experience.

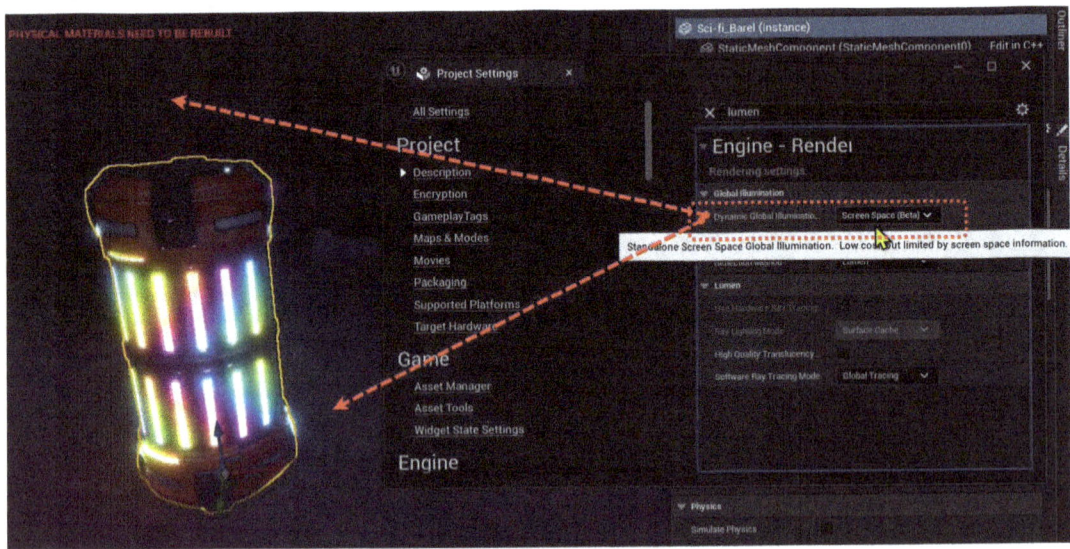

Figure 6-42. Using Screen Space illumination instead

Concluding with Figure 6-43, we can see the reflection on the comprehensive influence of Lumen when employed in full capacity; this emphasizes how it revolutionizes reflections and the overall luminosity of a scene.

CHAPTER 6 MASTERING LUMEN GLOBAL ILLUMINATION IN UNREAL ENGINE 5

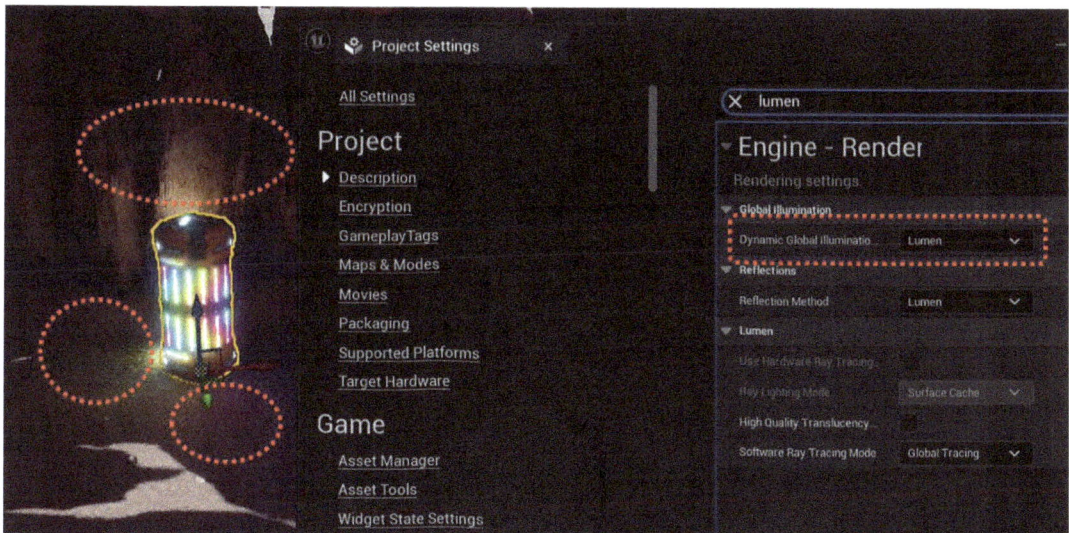

Figure 6-43. *See the result of Lumen from a distance*

We can encapsulate Lumen's role in crafting immersive digital environments, altering the very way light and shadow dance across the virtual landscape, thereby heightening the player's immersion and emotional response to the game world.

Import Macrovirus and Set Its Emissive Effects

As shown in section, we can import the FBX asset model, and the dialog box for FBX Import Options is shown in Figure 6-44. For this particular macrovirus model, it is essential to ensure that the "Combine Meshes" option is set to "True." This option is enabled because the 3D asset consists of multiple separated meshes. By activating this feature, we can merge these individual meshes into a single mesh.

> It's also important to note that combining meshes is not always desirable. For example, if we need to animate parts of the mesh separately, you should keep them separate. The decision to combine meshes depends on the needs of the project and the nature of the assets we are working with.

259

CHAPTER 6   MASTERING LUMEN GLOBAL ILLUMINATION IN UNREAL ENGINE 5

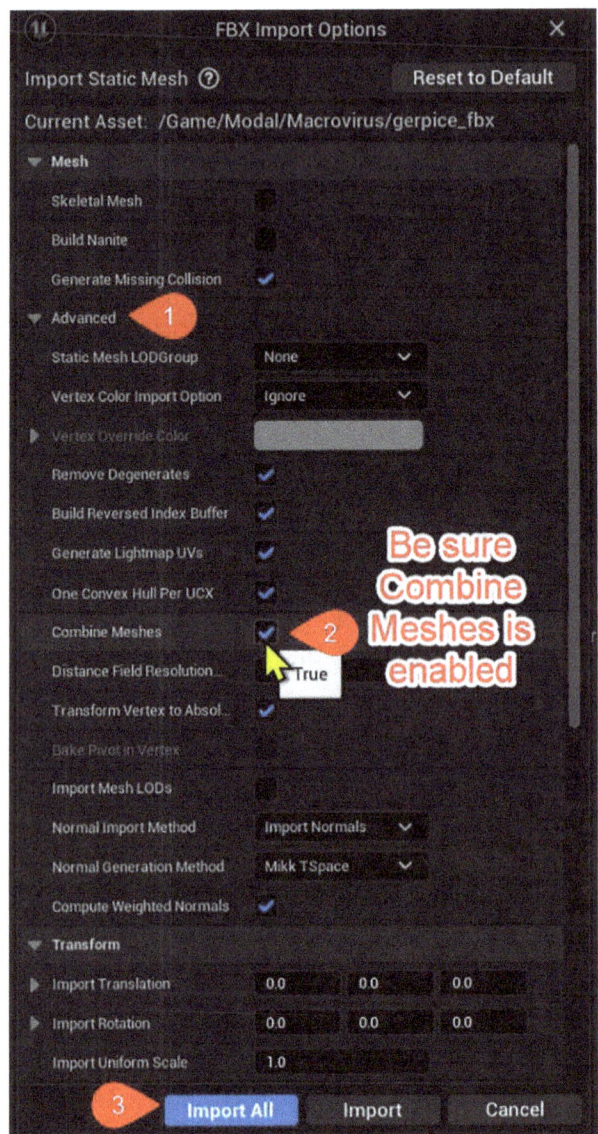

Figure 6-44.*   Be sure to select Combine Meshes*

SET ANIMATED EMISSIVE MATERIALS

As an exercise, you are to import the necessary one texture for this 3D model and set up the necessary materials for the model as indicated in Figure 6-45. Each of these materials will perform its purpose for each part of this 3D model.

CHAPTER 6   MASTERING LUMEN GLOBAL ILLUMINATION IN UNREAL ENGINE 5

You can follow Figure 6-46 to set up the materials for the material named black to represent the "Eye Pupils," Figure 6-47 for the material named Default to represent the "Eye," and Figure 6-48 for the material named Standard_7 to represent the "Body" of the 3D character. These screenshots mentioned illustrate the process of creating and applying materials for different parts of a character model in a 3D environment.

- In Figure 6-45, the materials are labeled for the character's body and eyes, indicating different textures and materials assigned to various parts such as the body and eye pupils.

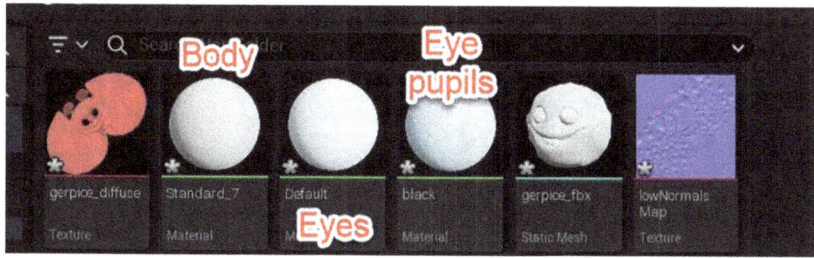

Figure 6-45. *Materials that assembled the visual of the macrovirus static mesh*

The following figures show how these materials are manipulated within the material editor to achieve the desired visual effects.

- For the eye pupils (Figure 6-46), a "black" material graph is shown with a Multiply node, intensifying the color to create a deeper and more pronounced pupil effect.

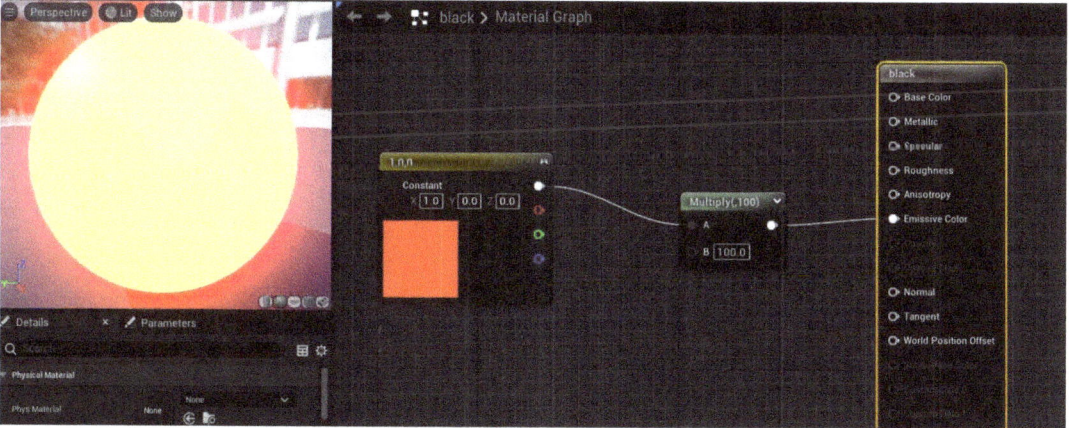

Figure 6-46. *Materials for the eye pupils*

261

- For the eyes (Figure 6-47), a "Default" material graph with a Constant node indicates a blue color applied to represent the iris or sclera part of the eyes.

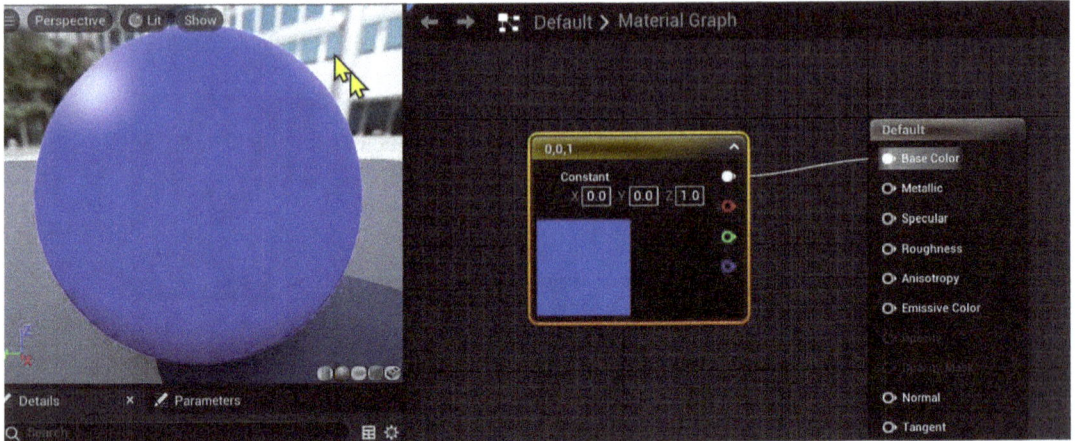

Figure 6-47. *Materials for the eyes*

- The body of the character (Figure 6-48) involves more complex material interactions.

  - A "Panner" node is used to create an animated effect on the texture, moving it along the X-axis, with the texture sampler set as /Engine/VREditor/Devices/Vive/UE4_Logo.

  - The texture's brightness is amplified using a Multiply node to enhance the emissive quality, giving the character a glowing, animated appearance.

  - The body of the character will have dynamic, moving textures that contribute to its overall animation.

CHAPTER 6   MASTERING LUMEN GLOBAL ILLUMINATION IN UNREAL ENGINE 5

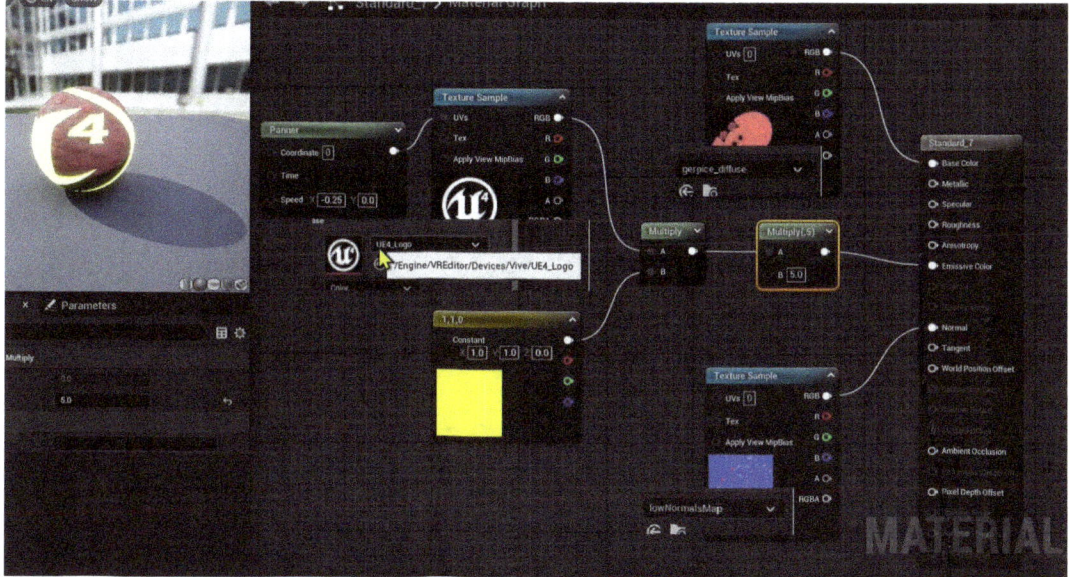

Figure 6-48. *Materials for the body of the macrovirus*

- Lastly, the assembly of these materials is shown on the character model (Figure 6-49). Each material slot is filled with the created materials for the body, eyes, and pupils, which is then rendered in the environment to bring the character to life with its glowing and vivid textures.

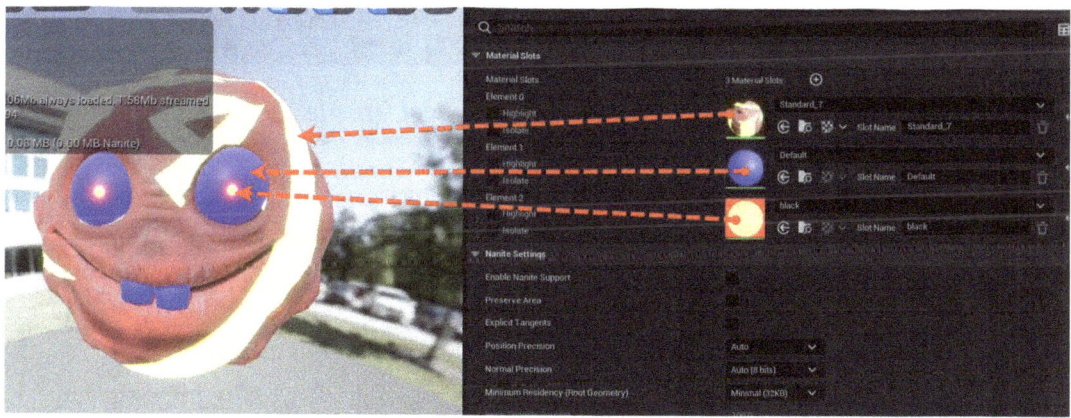

Figure 6-49. *The outcome of the macrovirus*

263

CHAPTER 6　MASTERING LUMEN GLOBAL ILLUMINATION IN UNREAL ENGINE 5

- This final figure (Figure 6-50) showcases the character with all the materials applied, glowing vibrantly in a dimly lit environment, demonstrating the successful application of the materials and textures to achieve the intended visual design.

Figure 6-50. *Place the macrovirus mesh in the level (inside the cave)*

Since this model has no transparent requirement, we can leave the Blend node as its default setting: Opaque.

As the material graph showcased in Standard_7 material (Figure 6-48), the node known as "Texture Sample" is attached from a "Panner" node, infusing the texture with the illusion of motion over the inexorable march of time by the given parameters labeled "Time" and "Speed."

The author has decided to use the value indicated in the screenshot with the emissive color and intensity presented before connecting to Emissive properties. The outcome of the macrovirus from the preceding materials can be seen in Figure 6-49.

We will place this macrovirus asset inside the shaded cave using the following location values as shown in Figure 6-50:

Location X = –69,200; Y = –41,610; Z = –69,260

Again, we can see the Lumen effect on this macrovirus model; the emissive lighting of the model does more than merely reveal; it transforms the space, bringing a radiance that seems almost otherworldly. Here, in the shaded cave, the lighting and shadow becomes a character of its own, playing across the contours of the cave with a life that is as dynamic as the virtual world it inhabits.

Lumen in Post-Processing Effects

Incorporating Post-Processing Effects in discussions about UE's Lumen lighting system is crucial, particularly because these effects can be finely tuned within a Post-Processing Volume, independent of the general project settings. This localized control allows for granular manipulation of the Lumen effects, enabling designers to dictate how light behaves and interacts within specific areas of a scene.

Whether intensifying the ambient glow in a dimly lit corridor or softening the harshness of daylight in an open landscape, the Post-Processing Volume becomes an essential tool for artists to express nuanced lighting narratives.

This tailored approach within the volume ensures that the Lumen's global illumination and reflections are not just global in scope but also delicately directed, contributing to the creation of an atmosphere that is both cohesive and contextually rich.

We can add a set of Post-Processing Volume as shown in Figures 6-51 and 6-52 to set the value of the location and scale, with the following steps on how to add Post-Processing Volume in the level:

1. In the "Place Actors" panel located on the left-hand side, which is categorized for ease of navigation, select the "Visual Effects" category.

2. Once the "Visual Effects" category is expanded, a list of available effects-related actors and volumes is displayed. Look for the specific volume we wish to add to our scene.

3. From the list, click the desired volume type to select it. In this case, we wish to add a volume that affects post-processing; we would select "`PostProcessVolume`."

4. After adding the volume, we can set the location and scale accordingly (the values had been added as follows) as shown in Figure 6-52:

    - Location X = –69,250; Y = –41,700; Z = –68,000
    - Scale X = 20; Y = 20; Z = 20

CHAPTER 6　MASTERING LUMEN GLOBAL ILLUMINATION IN UNREAL ENGINE 5

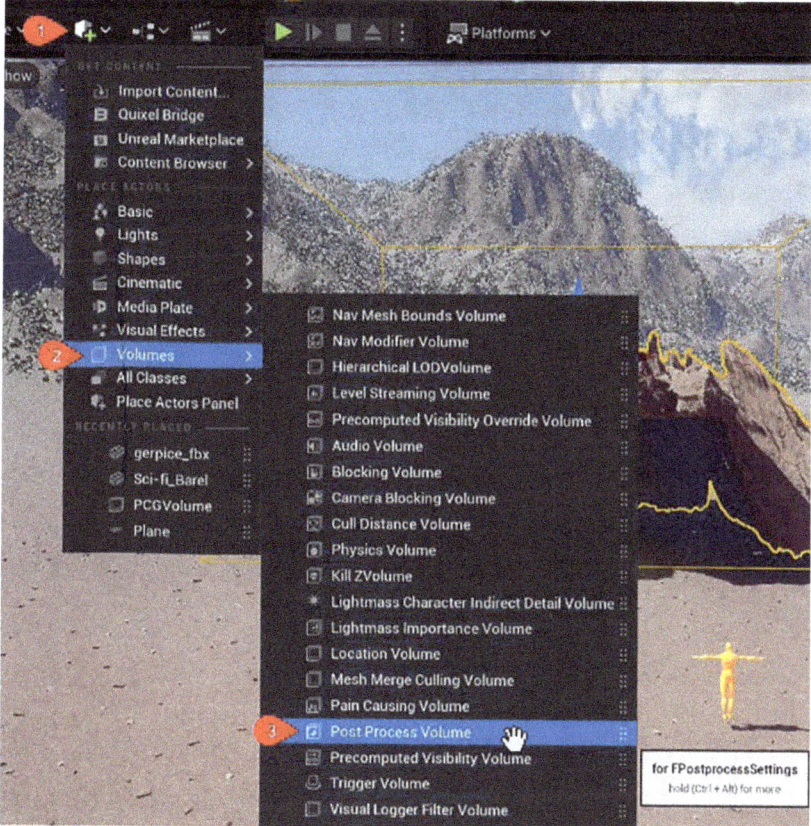

Figure 6-51. *Add the Post-Processing Volume using these steps: Step 1, add a new actor; Step 2, hover over the Volume to display a selection of volumes; and Step 3, select the Post-Processing Volume*

CHAPTER 6   MASTERING LUMEN GLOBAL ILLUMINATION IN UNREAL ENGINE 5

Figure 6-52. Position and scale of the Post-Processing Volume

Settings of the Post-Processing Volume

There are many settings within the Post-Processing Volume in UE, which can directly influence the appearance of scenes illuminated by Lumen, the engine's global illumination system.

As shown in Figure 6-53, the author had set the Exposure settings, with parameters like "Metering Mode" set to "Auto Exposure Histogram" and values for "Min EV100" and "Max EV100" to control the dynamic range of lighting, allowing for the simulation of a camera's response to different light levels within the environment.

This can affect how Lumen's lighting is perceived, ensuring that bright areas are not overexposed while darker areas retain detail.

CHAPTER 6   MASTERING LUMEN GLOBAL ILLUMINATION IN UNREAL ENGINE 5

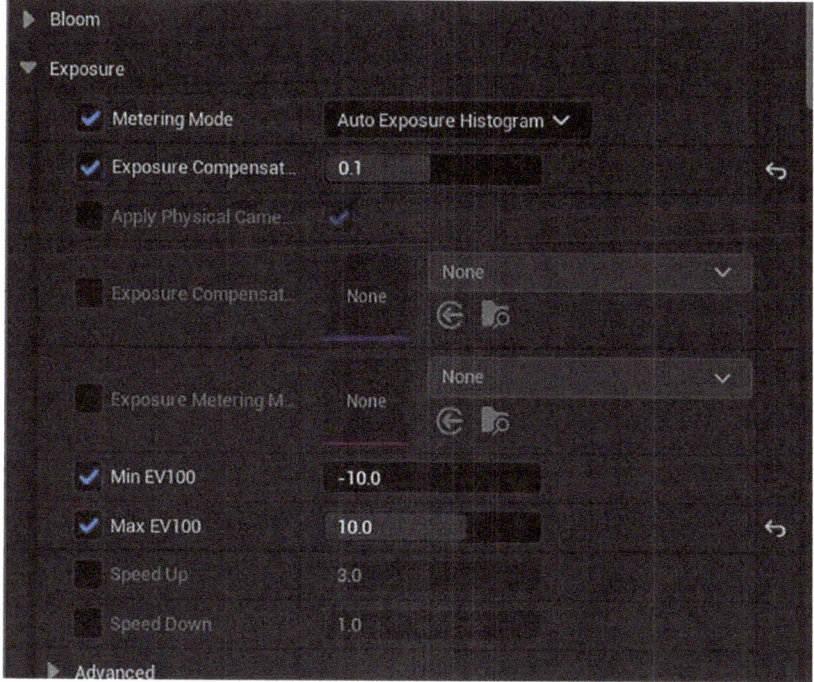

Figure 6-53. *Set the exposure value*

Additionally, as shown in Figure 6-54, the author decided to set the "Vignette Intensity" to offer artistic control over the edges of the screen, subtly darkening corners to draw the viewer's focus to the center or to create a certain mood or period look that complements the Lumen lighting.

1. Type "vignette" in the filter to display properties containing the term "vignette." Ensure the Post-Processing Volume is selected.

2. Adjust the Vignette intensity to 0.6 (as the author decided).

269

CHAPTER 6   MASTERING LUMEN GLOBAL ILLUMINATION IN UNREAL ENGINE 5

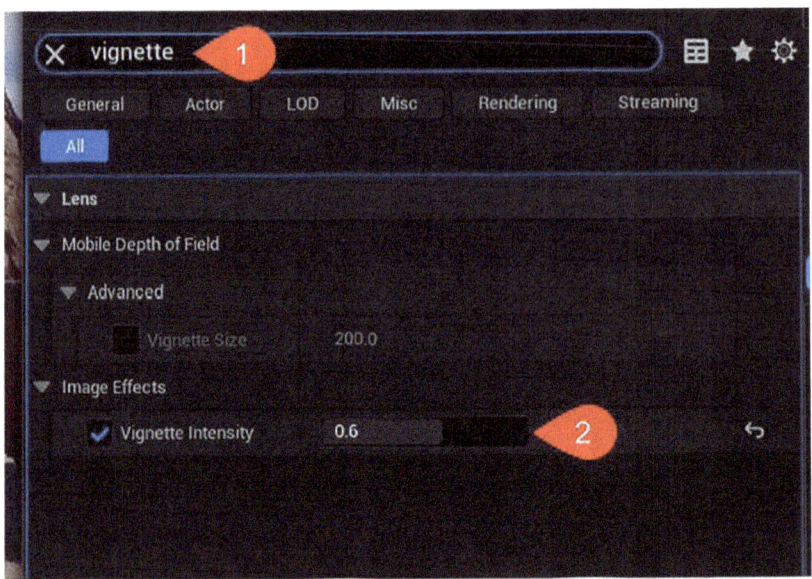

Figure 6-54. *Set the intensity of the vignette*

Finally, as shown in Figures 6-55 and 6-56, the Film settings show control over the tonal range with options like "Slope," "Toe," "Shoulder," "Black Clip," and "White Clip," which adjust the overall filmic look of the scene. These settings can enhance the naturalistic lighting provided by Lumen, ensuring that the visuals not only mimic real-world lighting conditions but also carry the desired cinematic quality.

Together, these Post-Processing settings allow artists to fine-tune how Lumen's lighting integrates with the scene, ensuring that the final output matches the creative intent, whether it's aiming for photorealism or a more stylized look.

CHAPTER 6 MASTERING LUMEN GLOBAL ILLUMINATION IN UNREAL ENGINE 5

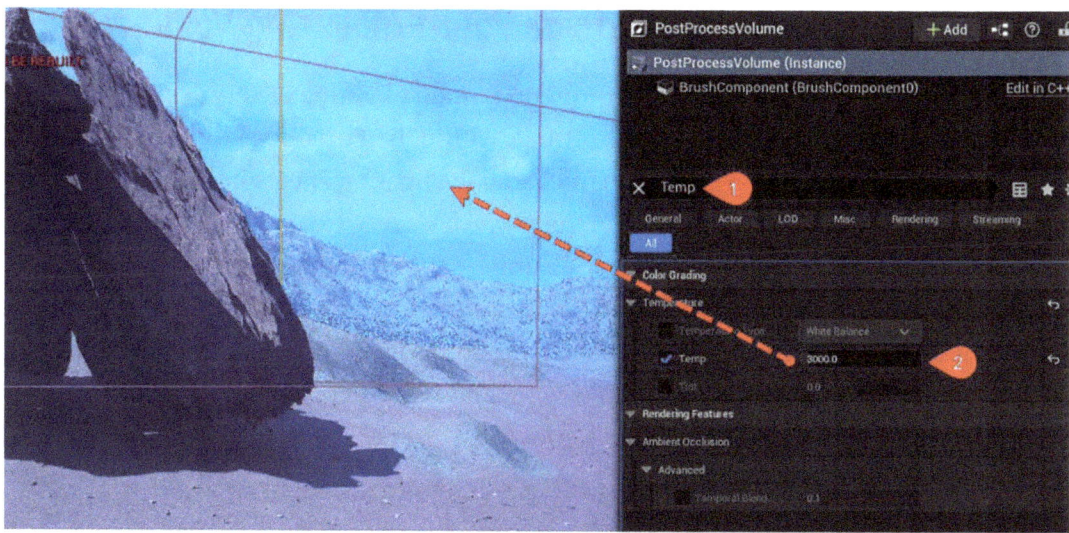

Figure 6-55. *Set the Temperature value to "cold" using these steps: Step 1, type in Temp to filter out the property that contains the word Temp; Step 2, type in value 3000 (a lower value to represent blue cold temperature)*

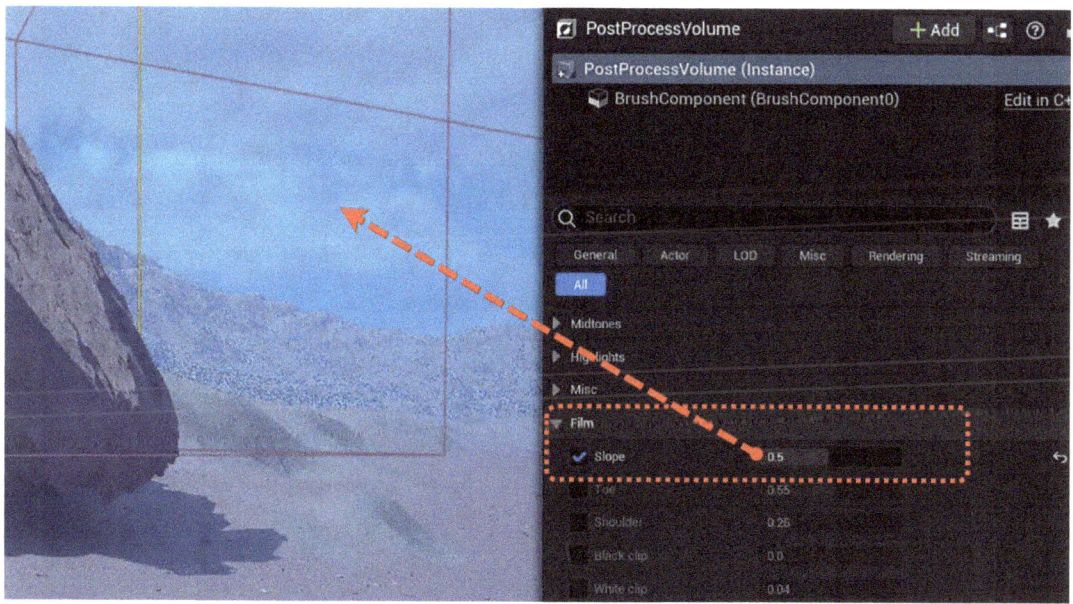

Figure 6-56. *Set the Slope value in the Film effect*

271

CHAPTER 6 MASTERING LUMEN GLOBAL ILLUMINATION IN UNREAL ENGINE 5

Lumen Effects Inside and Outside Post-Processing Volume

Within the confines of the digital environment, Lumen's lighting effects display a remarkable versatility, governed by the settings within the Post-Processing Volume. This pivotal feature allows the developers to fine-tune the lighting effects in specific areas, independent of the general ambiance of the project.

As shown in Figures 6-57 and 6-58, the Lumen settings within the volume can be meticulously adjusted, providing the ability to craft distinct atmospheres in localized spaces, offering a stark contrast to the broader lighting effects at play in the rest of the scene. To enable Lumen for global illumination and reflections in the Post-Processing Volume within UE, we can follow these steps:

1. In the Post-Processing Volume settings, use the search bar to filter properties by typing "Lumen" to quickly find the relevant settings.

2. Under the "Global Illumination" section, set the "Method" drop-down to "Lumen" to enable Lumen for dynamic global illumination.

3. Adjust the "Final Gather Quality" slider to fine-tune the accuracy of the global illumination. This setting determines the quality of the indirect lighting.

4. Scroll to the "Reflections" section, and again set the "Method" to "Lumen" to use it for dynamic reflections.

5. Lastly, adjust the "Quality" slider under the "Reflections" section to set the desired quality level for reflective surfaces. Higher values will result in more accurate and detailed reflections but may impact performance.

These steps help to optimize the visual fidelity of the scene by utilizing Lumen, UE's fully dynamic global illumination and reflection system, ensuring a more realistic and immersive environment.

CHAPTER 6 MASTERING LUMEN GLOBAL ILLUMINATION IN UNREAL ENGINE 5

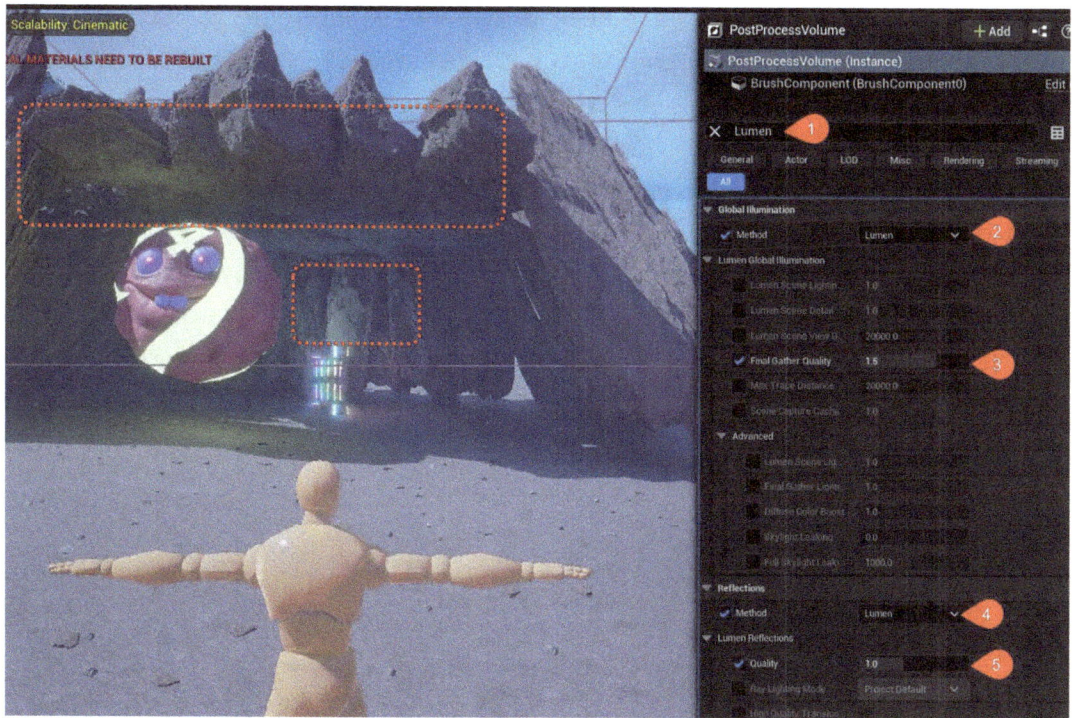

Figure 6-57. Enable Lumen in this Post-Processing Volume

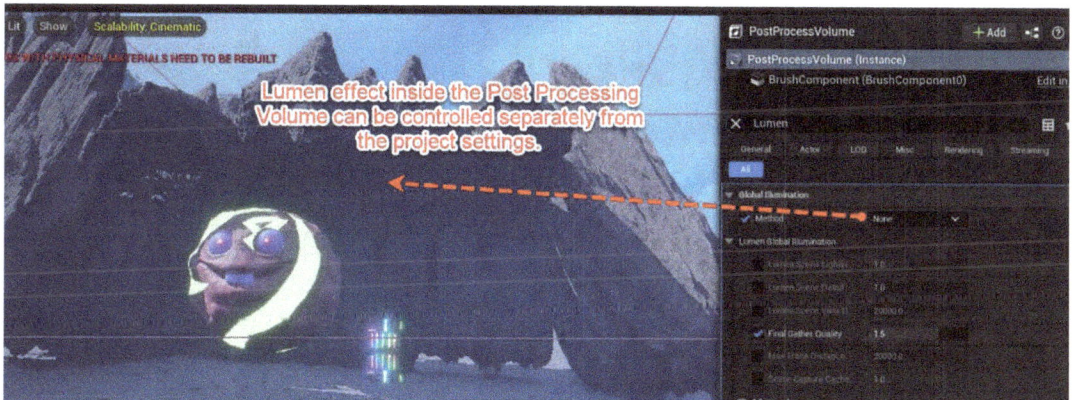

Figure 6-58. Disable Lumen in this Post-Processing Volume

The influence of the Post-Processing Volume on lighting effects is crucial for defining the atmosphere within its boundaries, as demonstrated in the image provided. Inside the volume, specific Lumen settings are applied to create a distinct lighting environment.

273

CHAPTER 6 MASTERING LUMEN GLOBAL ILLUMINATION IN UNREAL ENGINE 5

However, once a character steps outside the volume, as shown in Figure 6-59, the lighting settings from the global Project Settings take precedence, which may result in a different visual experience. This illustrates how the engine transitions from localized lighting effects to broader, scene-wide illumination.

When considering character movement, it's important to understand that as the character moves from an area outside the Post-Processing Volume into its interior, the lighting and atmospheric effects will gradually shift to those defined by the volume's settings. Conversely, as the character exits the volume, the global lighting settings defined in the Project Settings will again become apparent. This dynamic capability of UE allows for a fluid and immersive transition between different lighting environments, providing a nuanced and sophisticated approach to scene lighting.

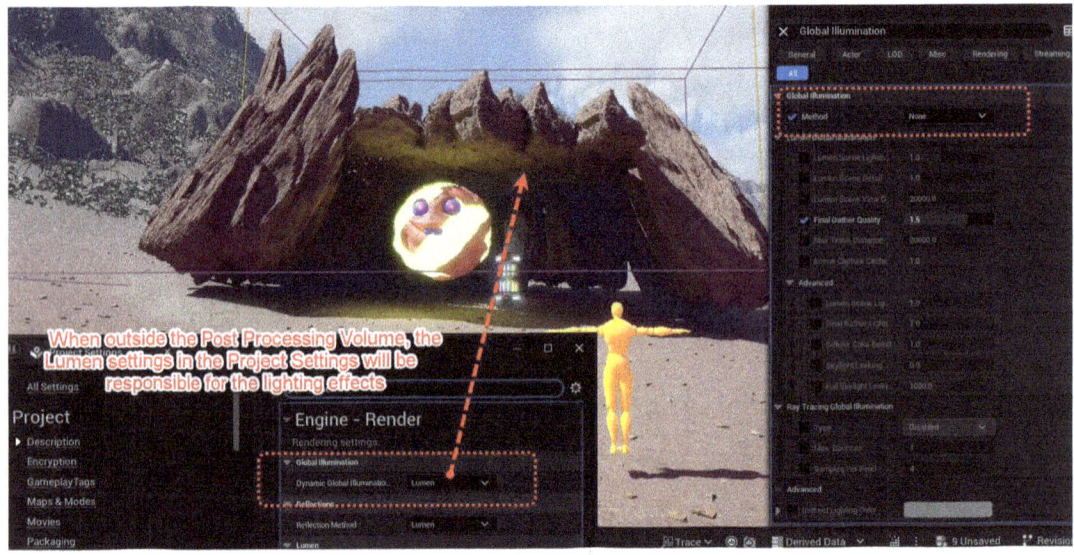

Figure 6-59. *Demonstrate that Lumen effects act differently inside and outside the Post-Processing Volume*

Here, the screenshot in Figure 6-60 illustrates how to re-enable the Lumen settings within the Post-Processing Volume configurations.

CHAPTER 6   MASTERING LUMEN GLOBAL ILLUMINATION IN UNREAL ENGINE 5

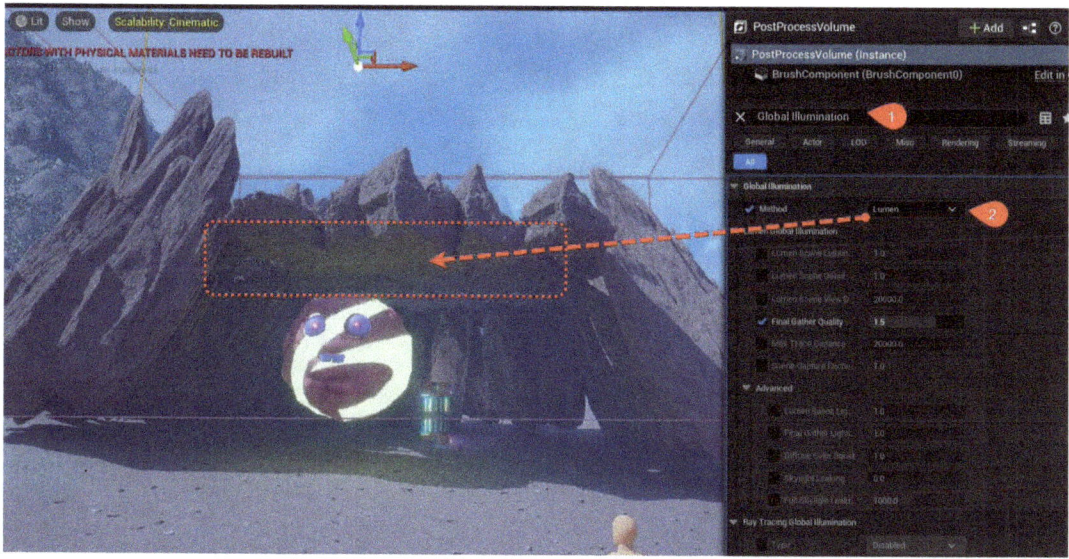

Figure 6-60. *Re-enable the Lumen inside the Post-Processing Volume by typing global illumination as the text to filter out, before we can select Lumen as the method*

Summary

This chapter delved into the capabilities of Lumen within UE5, a cutting-edge lighting system engineered for the creation of strikingly realistic scenes on next-gen consoles and high-end PCs. Lumen stands out with its dynamic global illumination and reflections, handling complex environments and a wide range of scales with ease.

Setting up Lumen requires enabling specific settings in the project configuration, allowing for customization of quality and performance metrics. The system achieves real-time lighting through a blend of ray tracing, voxel cone tracing, and diffuse probes, which can adapt to dynamic scene changes – be it in lighting, objects, or environmental conditions like weather or time of day.

By default, Lumen is enabled in new projects, but its settings can be finely tuned for specific cameras or areas using Post-Processing Volumes. For developers seeking to optimize the balance between performance and visual fidelity, Lumen provides tools to assess and adjust its impact on scenes, illustrating its transformative effect on in-game lighting and atmosphere.

CHAPTER 7

Harnessing the Power of Niagara: Practical Examples in Unreal Engine 5

In the dynamic realm of real-time graphics, the introduction of Niagara in UE5 marked a significant advancement in visual effects and particle simulation systems. This chapter delves into the heart of Niagara, exploring its capabilities within the powerful UE5 framework. It offers readers a comprehensive understanding of how Niagara revolutionizes the creation and management of complex particle systems, which are crucial for adding realism and dynamism to virtual environments.

We begin by familiarizing with the basic concepts and components of Niagara, setting the stage for a deeper exploration. As we progress, the chapter presents a series of practical examples designed to showcase the versatility and control that Niagara offers. These examples serve not just as instructional guides but as a springboard for inspiration, encouraging readers to experiment and innovate in their visual effects work.

From creating simple particle emissions to orchestrating intricate interactive effects that respond to environmental variables and user input, this chapter aims to equip us with the knowledge and confidence to harness the full potential of Niagara in our UE5 projects. Whether you're a game developer, a visual effects artist, or simply an enthusiast of cutting-edge graphical technology, the insights offered here will enhance your toolkit and elevate our creative endeavors to new heights.

So buckle up, there's a lot to get through.

CHAPTER 7   HARNESSING THE POWER OF NIAGARA: PRACTICAL EXAMPLES IN UNREAL ENGINE 5

Getting Started with Niagara

As we embark on this journey to unlock the full potential of Niagara, this section will serve as our foundational guide, tailored to help us navigate the intricacies of particle systems and the visual wizardry they can unleash in our projects.

The Niagara system in UE5 boasts a multitude of features. We will focus on the most important and frequently used features, such as deploying emitters to create particle systems. Subsequently, we will explore three sets of examples that demonstrate how to generate smoke, create spiral effects, and simulate dust storms.

Creating First Particle Effect

Now, let's create a simple particle effect to get a feel for the process as shown in steps in Figure 7-1:

1. We will create a new folder named VFX, where the visual effect files will be stored.

2. Right-click to activate the pop-up of the "Create Basic Asset" dialog box.

3. We can create a new Niagara system by selecting the existing shortcut as shown in Figure 7-1.

CHAPTER 7   HARNESSING THE POWER OF NIAGARA: PRACTICAL EXAMPLES IN UNREAL ENGINE 5

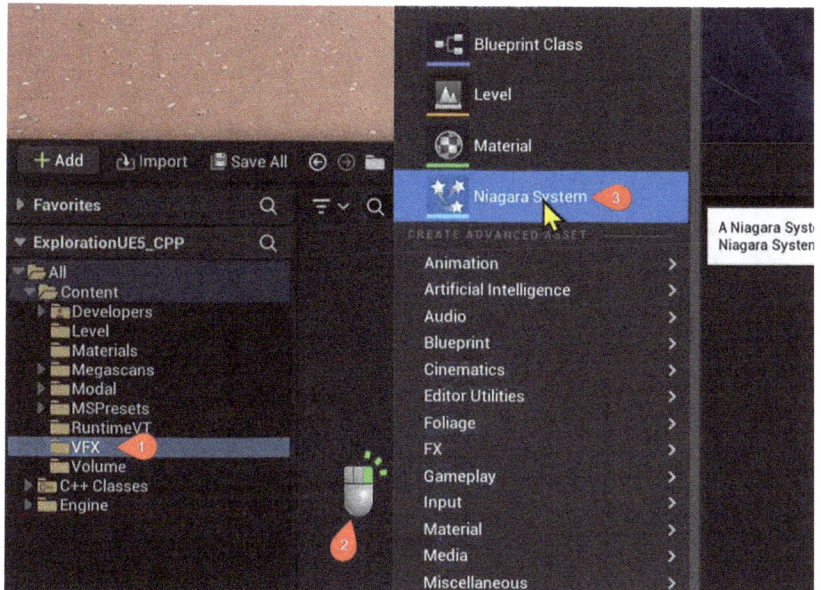

Figure 7-1. *Create a new Niagara system*

As shown in Figure 7-2, we can see the first part of the Niagara wizard dialog box, and the steps can be described as follows:

1. Options for Starting a New Niagara System

   We will choose "New system from selected emitter(s)" since this option allows the user to create a new system by selecting from a mix of existing emitters. The description indicates that the user can choose between inherited emitters and emitter templates or behavior examples which do not carry inheritance.

2. We can click the "Next" button to proceed to the next step in the creation process after selecting one of the preceding options.

CHAPTER 7 HARNESSING THE POWER OF NIAGARA: PRACTICAL EXAMPLES IN UNREAL ENGINE 5

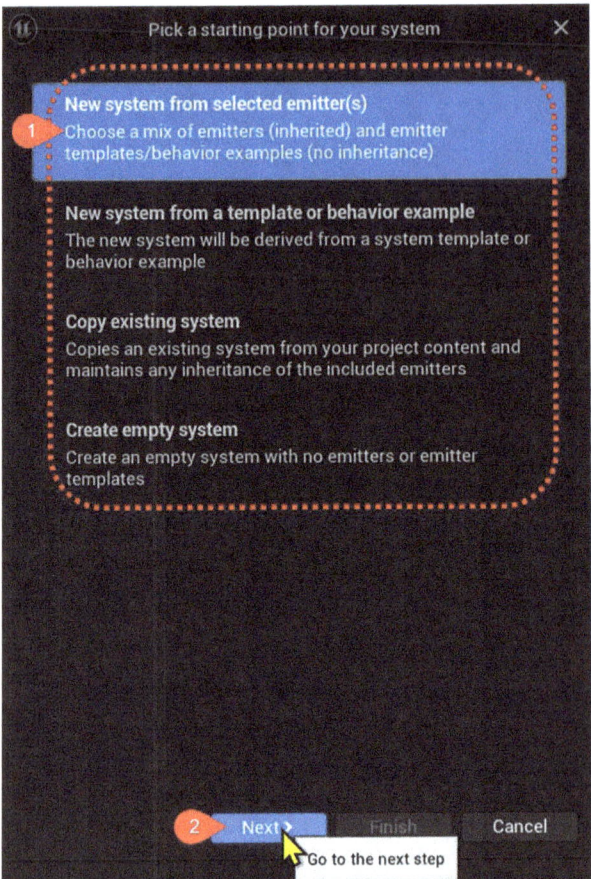

Figure 7-2. *Select a starting emitter as a starting point*

Following on to Figure 7-3, we are creating a new Niagara system based on the existing emitter template, as discussed in the following steps:

1. Emitter Selection: In this step, we select the "Fountain" emitter, described as "a looping fountain spray". This emitter template can be chosen from the provided list to create particle effects that resemble a water fountain.

2. Adding Emitters: In this step, Step 2, we click to add this "Emitters to Add:", with "Fountain" ready to be added to the system.

3. We can finalize the selection as shown in Step 3.

CHAPTER 7   HARNESSING THE POWER OF NIAGARA: PRACTICAL EXAMPLES IN UNREAL ENGINE 5

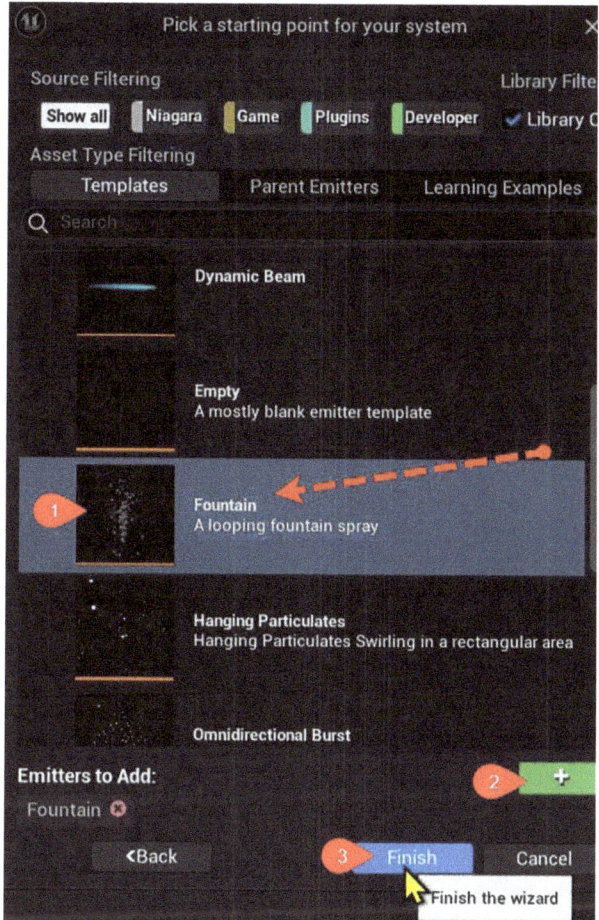

Figure 7-3. *Add existing emitters (Fountain)*

Before we can carry on to look into the Niagara system itself, as shown in Figure 7-4, we will make sure to rename this system as "Test_NS" (as Step 1), then we can double-click the asset (as Step 2) to open up the Niagara system editor box.

CHAPTER 7 HARNESSING THE POWER OF NIAGARA: PRACTICAL EXAMPLES IN UNREAL ENGINE 5

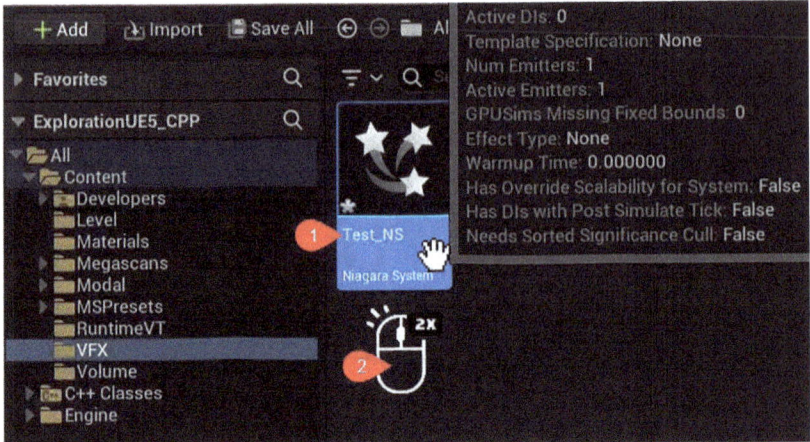

Figure 7-4. *Rename and double-click to open the Niagara system*

Learning the Basics of Particle Systems

Understanding the basics of particle systems is crucial as we start with the Niagara system. A particle system consists of emitters that produce particles, which can be anything from simple sprites to complex meshes. Each particle has properties such as size, color, and velocity, which can be randomized or controlled to create diverse effects.

As a start of the process, we can see the breakdown of the steps and elements as indicated by the numbered labels as shown in Figure 7-5:

1. System Overview Panel (labeled as *Test_NS*): This is the main control panel for the Niagara system named "Test_NS". It gives us an overview of the system and allows us to adjust properties, spawn rates, and other system-related settings.

2. Emitter Properties (labeled as *Fountain*): We can have multiple emitters to build up a whole Niagara system, but in this case, we have one emitter named *Fountain*. This section detailed an emitter suggesting it creates a fountain-like particle effect. We can adjust properties like spawn rate, particle life, etc., here.

3. Details Panel (from the *Test_NS* System): This panel provides detailed properties for the selected object, which in this case is the "Test_NS" system. We can usually tweak advanced settings such as rendering options, system parameters, and performance settings here.

4. Preview Window: This window shows a real-time preview of the particle system. It lets us see the effects of our adjustments without having to run the game or application.

5. Timeline and Controls: This part of the UI includes a timeline that allows us to scrub through the particle system's animation and effects. It's useful for setting up sequences and understanding how our particle system evolves over time.

Each of these panels and windows works together to allow a developer or technical artist to create complex particle systems within the UE5 environment. The specific steps to create or modify a particle system would depend on the desired outcome and could involve setting up emitters, defining particle properties, and using the timeline to synchronize effects with game events or animations.

Figure 7-5. The breakdown of the steps and elements in the Niagara system editor

Rename Emitter

If we want to rename an emitter within a particle system, we can do so by following the labels indicated in Figure 7-6:

1. Selected Emitter (in this case, *Fountain*): Step 1 indicates the currently selected emitter in the Niagara system editor.

2. Context Menu Gear Icon: Step 2 indicates the gear icon that represents the context menu for the selected emitter. Clicking it would bring up various actions related to the emitter, such as creating a new asset from it, setting a new parent emitter, or other emitter-related actions.

3. Rename Option with Arrow and Hotkey: Step 3 indicates the yellow arrow pointing to the "Rename" option within the context menu. This option is used to rename the selected emitter. The shortcut key "F2" is listed beside the "Rename" option, indicating that pressing F2 on the keyboard would also activate the renaming function for the selected emitter.

4. As indicated in Figure 7-7, by selecting the "Rename" option or pressing F2, we can change the name of the emitter from "Fountain" to a name of our choice. This could be part of organizing and managing various emitters within a particle system to make it easier to identify and modify them as needed during development.

CHAPTER 7 HARNESSING THE POWER OF NIAGARA: PRACTICAL EXAMPLES IN UNREAL ENGINE 5

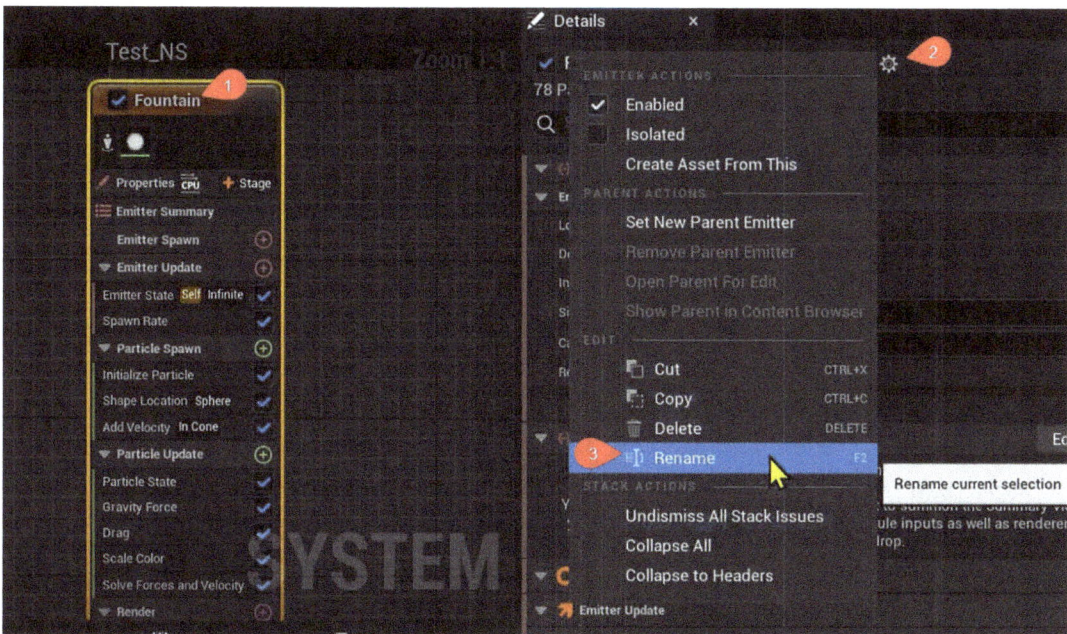

Figure 7-6. To rename the label of an emitter

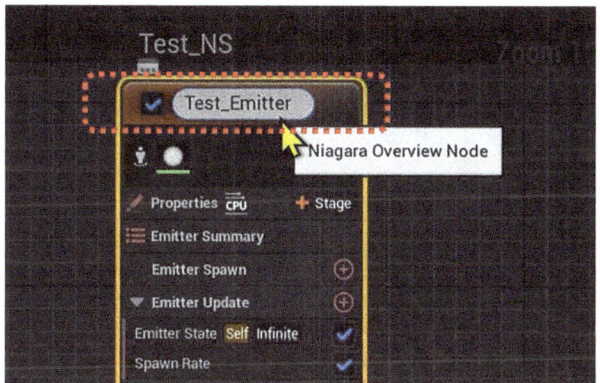

Figure 7-7. Type in the desired new label

Enable an Emitter's Local Space

The "Local Space" option within the "Emitter" category is a key property to understand, as indicated in Figure 7-8. When this option is activated, it ensures that any transformations applied to the particles are relative to the emitter's own origin point. Conversely, when deactivated, the particles are positioned and moved within the global coordinates of the entire scene. This toggle essentially dictates whether the particles' behavior is linked to the emitter or independent of it. Understanding the "Local Space" setting is crucial as it determines the behavior of particles in relation to the emitter. If "Local Space" is enabled, the particles will maintain their position and movement in sync with the emitter's orientation and position. This means that as the emitter moves or rotates within the game world, the particles will follow suit as if they were a part of the emitter itself.

On the other hand, if "Local Space" is disabled and the particles are in "World Space," they will not move or rotate with the emitter. Instead, they will behave as if they are independent entities within the global space of the game world. Their position and motion will be consistent with the world's coordinate system, remaining unaffected by changes to the emitter's location or orientation. This distinction is important to consider when designing particle effects, as the choice between "Local Space" and "World Space" will significantly influence the visual outcome of the effect.

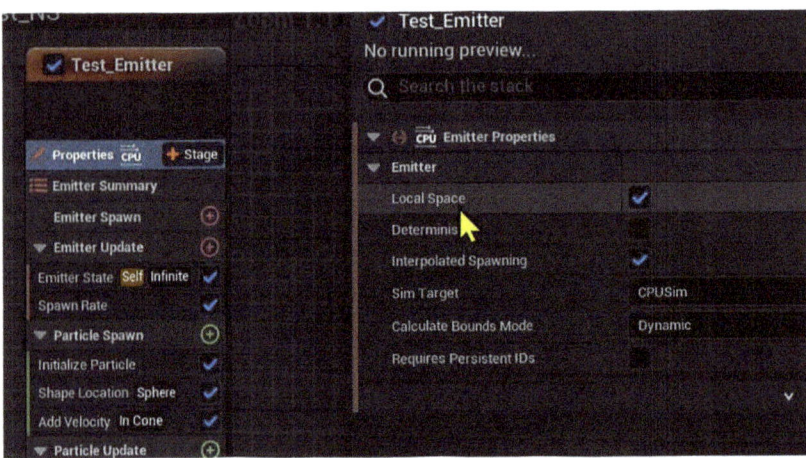

Figure 7-8. *Local Space within the Emitter property*

As shown in Figure 7-9, the number "2" emphasized the function of this particular setting. Understanding and manipulating this property is crucial for achieving the desired behavior of the particle effects in relation to the game environment and the movement of the emitter itself.

CHAPTER 7   HARNESSING THE POWER OF NIAGARA: PRACTICAL EXAMPLES IN UNREAL ENGINE 5

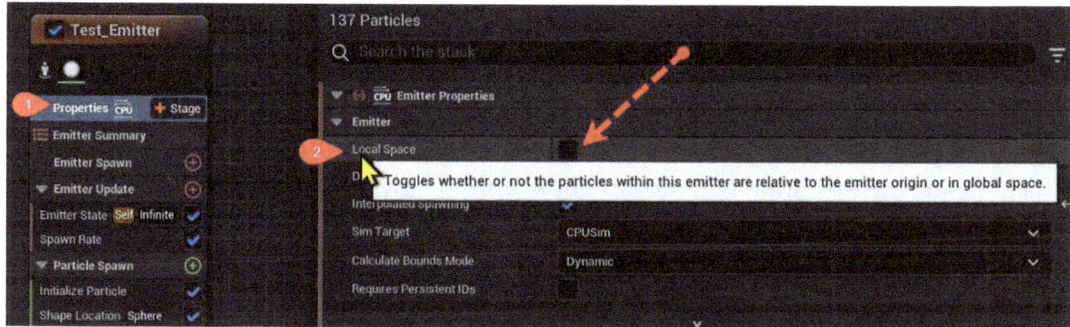

Figure 7-9. *Enable or disable local space*

Result of Enabled/Disabled Local Space

Using the same method for positioning assets as previously illustrated when placing 3D assets, we can apply the same technique to position the sample Niagara system asset, as shown in Figure 7-10.

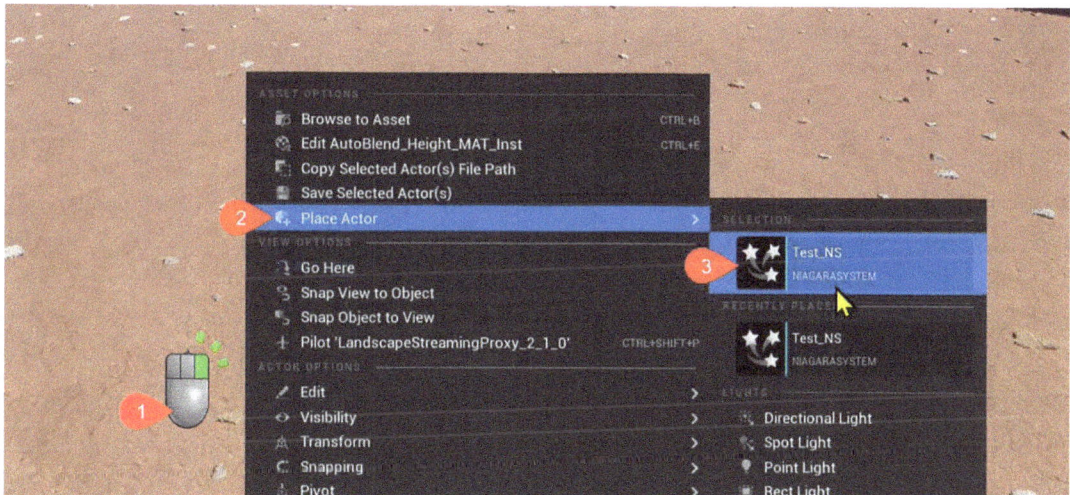

Figure 7-10. *Place the Niagara system asset in the level*

We can observe the outcome of the particle system in Figure 7-11, which shows the result with local space disabled. When the local space is disabled, the particles are distributed across the screen in a widespread pattern, not moving or bound to any specific object in the scene. This suggests that they are being emitted in global space, which means their position and movement are not relative to a moving object or an emitter that could be in motion.

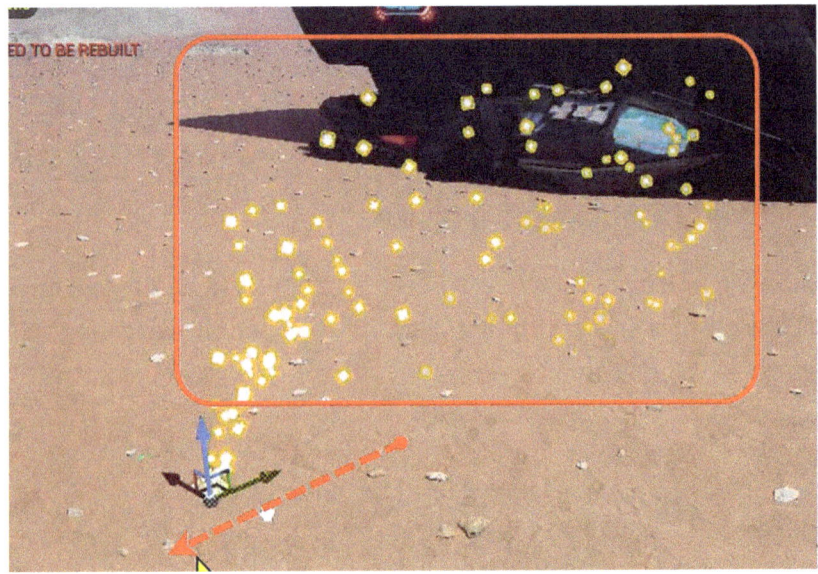

Figure 7-11. *Disabled local space (try to move the particles around to see the effects)*

Figure 7-12 displays the result when local space is enabled, which shows the particles that appear to be concentrated and moving along with the origin of the emitter, indicating that they are relative to the object's local coordinate system.

When "Local Space" is enabled, the particles are bound to the object, inheriting its movement and orientation – imagine them as an extension of the object itself. Thus, if the object is animated to move within the scene, the particles faithfully follow, maintaining their relative positions and paths as if they are an integral part of the object.

CHAPTER 7   HARNESSING THE POWER OF NIAGARA: PRACTICAL EXAMPLES IN UNREAL ENGINE 5

Figure 7-12. *Enabled local space (try to move the particles around to see the effects)*

To clearly observe this, you can experiment by moving the particle system within your scene. With the "Local Space" option enabled, you'll notice that the particles adhere to the emitter's movement. Conversely, with "Local Space" disabled, the particles will act independently, illustrating a starkly different effect. This visual comparison is crucial for understanding how the "Local Space" setting influences the particle system's behavior.

In summary, the key difference here is that with local space enabled, the particles maintain a fixed relationship with the moving object, which can be useful for effects like exhaust or thrusters on a vehicle, where you want the particles to appear as if they are being emitted from the object no matter where it goes or how it rotates. Conversely, with local space disabled, the particles do not follow the object and instead behave independently within the world space.

Sim Target

In UE5's Niagara system, a simulation target refers to the computational context in which particle simulations are processed. The system provides two types of simulation targets: the default "CPUCompute Sim," which runs particle simulations on the computer's central processing unit (CPU), and "GPUCompute Sim," which leverages the graphics processing unit (GPU) for the same purpose.

The choice between these targets determines where the calculations for particle behavior – such as movement, collision, and other effects – take place, which can affect performance and capabilities depending on the complexity of the particle system and the hardware specifications of the computer.

A GPUCompute Sim leverages the GPU for particle simulation calculations:

- The GPU is highly efficient at handling thousands of parallel operations, which makes it well suited for complex and large-scale particle systems that require significant parallel processing power.

- GPU simulations can handle a larger number of particles and more complex simulations without a significant impact on frame rates, as GPUs are designed for high-throughput calculations.

- However, GPU simulations can be more complex to set up and may have specific limitations or requirements, such as the need to ensure that all data needed for the simulation is available on the GPU.

A CPUCompute Sim, on the other hand, relies on the CPU to process particle simulations:

- The CPU is generally better at tasks that require sequential processing and can handle more complex logic per particle than the GPU.

- CPU simulations may be easier to set up and debug and can offer more precise control over individual particles.

- However, the CPU is less capable of handling the parallel processing of large numbers of particles, which can result in lower performance or reduced complexity in the particle system compared to a GPU-based simulation.

CHAPTER 7   HARNESSING THE POWER OF NIAGARA: PRACTICAL EXAMPLES IN UNREAL ENGINE 5

Choosing between GPUCompute Sim and CPUCompute Sim depends on the specific requirements and constraints of the project, including performance considerations, the complexity of particle interactions, and the total number of particles needed.

Figure 7-14 demonstrates the process of setting up and modifying the bounds calculation mode for a GPU particle emitter. As we can notice in the figure, it shows a warning message indicating that the emitter is using a dynamic bounds mode on the GPU and that there are missing fixed bounds. This suggests that the system expects a fixed bounds setting for optimal performance or correctness but is currently set to dynamically calculate the bounds. The "Calculate Bounds Mode" is set to "Dynamic," which means the system calculates the bounds of the particles in real time. This can be performance intensive for the GPU. There is an option to change the bounds mode to "Fixed," which would require the user to manually specify the bounds, or "Programmable," which implies the bounds can be set through a script or a programmatic method.

As shown in Figure 7-13, altering the simulation target results in a warning. To address this issue, we switch the "Calculate Bounds Mode" to "Fixed," as shown in Figure 7-14, which successfully resolves the warning. This adjustment ensures that the system uses a predetermined boundary for the particle simulation, thus eliminating any alerts related to the dynamic calculation of bounds. With "Fixed" bounds, we have specified the minimum and maximum bounds for the particles manually, as seen with the example value we entered here for Min (−100.0) and Max (100.0). This sets a defined space within which the particles are allowed to exist.

This configuration is important for performance optimization because it prevents the GPU from having to calculate the bounds of the particle system dynamically every frame, which can be computationally expensive.

When a particle emitter's simulation target is set to use Fixed Bounds, it means that the simulation space for the particles is constrained to a defined volume within the 3D world. Here are a few reasons why we might use Fixed Bounds:

Performance Optimization: By limiting the simulation to a fixed volume, the engine doesn't need to calculate particles that would exist outside of these bounds, which can improve performance, especially for particles that wouldn't be visible to the player or contribute to the scene.

CHAPTER 7 HARNESSING THE POWER OF NIAGARA: PRACTICAL EXAMPLES IN UNREAL ENGINE 5

Culling: Fixed Bounds can help with culling strategies, where particles outside the camera's view are not processed or rendered, further enhancing performance.

Stability: Fixed Bounds can ensure that the particle simulation behaves consistently, preventing particles from accidentally being spawned outside of the intended area, which could cause visual anomalies or performance issues.

Collision and Interaction: When particles are meant to interact with the environment, such as smoke colliding with walls or the ground, Fixed Bounds can help keep the simulation contained so that it interacts only with the intended surfaces within the defined volume.

Predictability: For simulations that require precise control, such as a smoke effect that should not exceed a certain area, Fixed Bounds make the behavior of the particles predictable and easier to manage.

In essence, Fixed Bounds are used to define the "play area" for the particle simulation, ensuring that all particles are emitted, exist, and interact within a controlled and optimized volume in the game world.

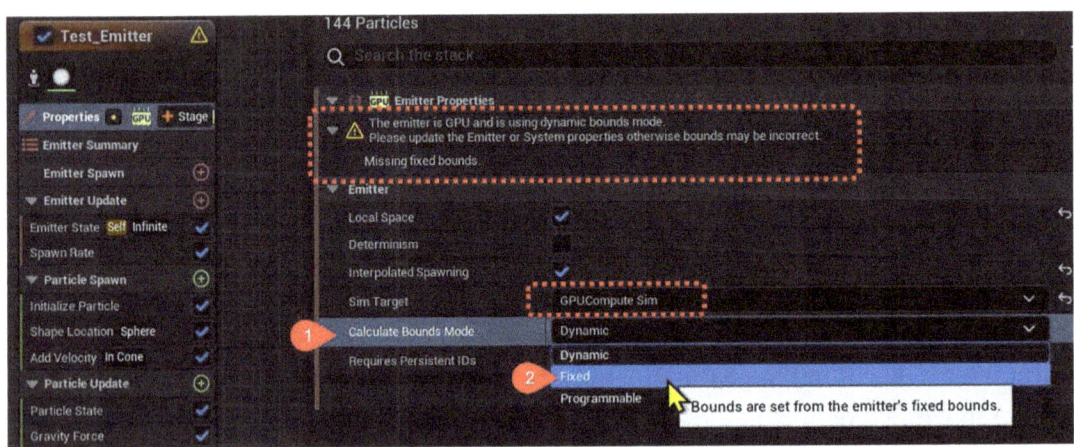

Figure 7-13. Set to GPUCompute Sim

CHAPTER 7   HARNESSING THE POWER OF NIAGARA: PRACTICAL EXAMPLES IN UNREAL ENGINE 5

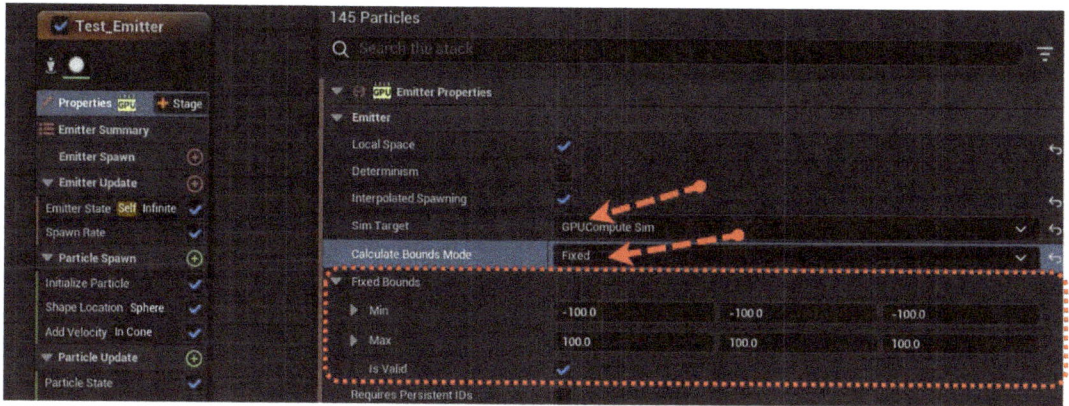

Figure 7-14. *Be sure to have fixed-size boundary for GPUCompute Sim*

Emitter Update: Loop Duration and Spawn Rate

Figure 7-15 shows the "Emitter Update" section of the emitter properties. It is highlighting the "Loop Duration" setting within the "Emitter State" subsection. The tooltip associated with the "Loop Duration" explains that this setting determines the duration of the emitter's life cycle. It is set to a value of 2.0, indicating the duration or time for each loop of the particle emission cycle. The tooltip notes that this value can be initialized with a random number, but by default, it is cached off and persists as the duration for all subsequent loop iterations. There is also a mention that to allow the emitter to choose a new duration each time the loop iterates, one should check the option for "Recalculate Duration Each Loop." A value of 0 is not allowed in this field, and if entered, it will be interpreted as approximately 0.01667, which likely corresponds to a frame duration at 60 frames per second.

CHAPTER 7   HARNESSING THE POWER OF NIAGARA: PRACTICAL EXAMPLES IN UNREAL ENGINE 5

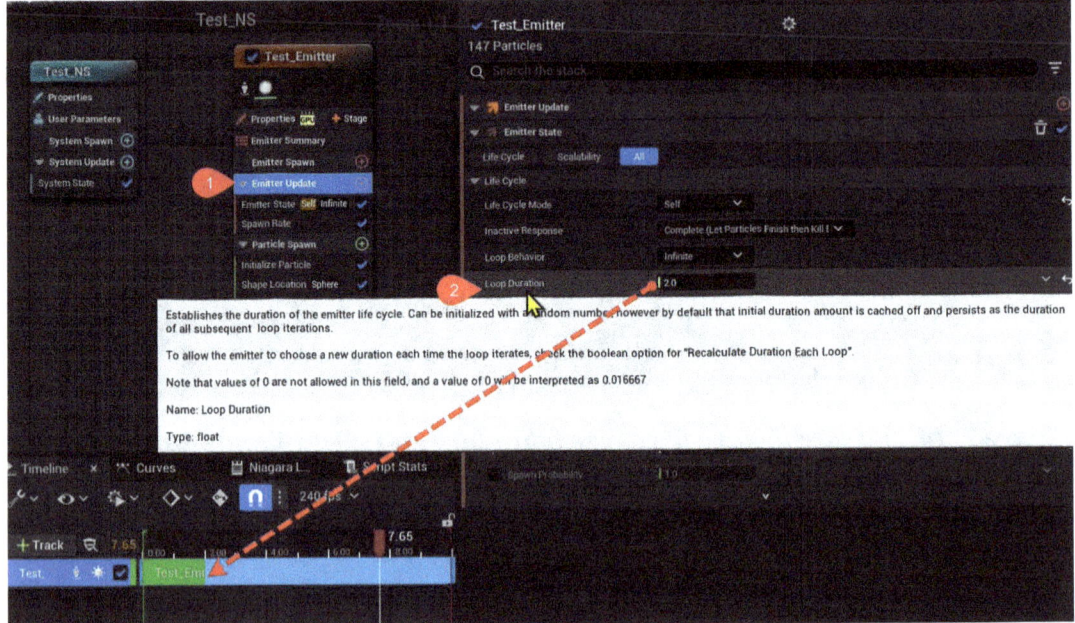

Figure 7-15. Loop duration for Emitter Update

Figure 7-16 focuses on the "Spawn Rate" section of the emitter properties. The "Spawn Rate" setting, set to 30.0, dictates the number of particles spawned per second. This parameter controls the density of the particle emission, with higher numbers resulting in more particles being generated in a given timeframe. The tooltip provides the name of the setting ("SpawnRate") and its type ("float"), indicating that it accepts decimal values for fine-tuned control over the particle emission rate.

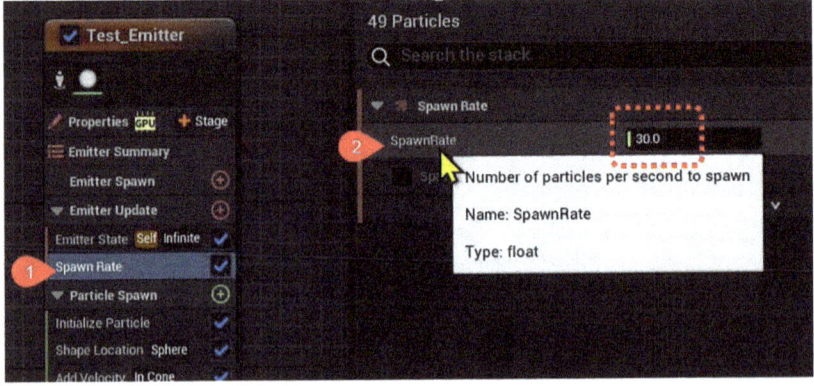

Figure 7-16. Spawn Rate

CHAPTER 7   HARNESSING THE POWER OF NIAGARA: PRACTICAL EXAMPLES IN UNREAL ENGINE 5

The basic purpose of these two properties illustrates how specific parameters within the Niagara system editor are used to define the behavior of particle systems in UE5, affecting how particle effects are generated and displayed and how they behave over time.

Modifying Particle Emission: From Continuous Flow to Burst Effect

The following steps illustrate a fundamental process of modifying a particle system within the UE5 Niagara editor. Specifically, these will show the deletion of an existing module and the addition of a new "burst" module to alter the particle emission behavior, along with the resulting visual effect.

Figure 7-17 shows the "Spawn Rate" module checked in the Niagara system editor. The tooltip indicates that this module causes particles to spawn at a continuous rate. There's also a "Delete" option selected, which suggests that this module can be removed from the particle system.

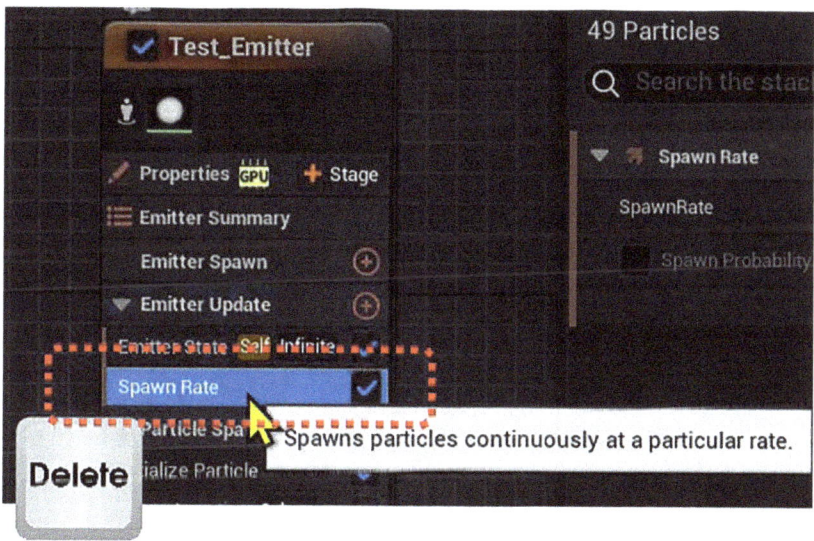

Figure 7-17. Module deletion

Figure 7-18 shows that we need to select the "Emitter Update" section (as shown in Step 1) of the Niagara system editor with a "+" sign, indicating the addition of a new module (as shown in Step 2). The tooltip prompts to "Add a new Module to this group."

295

CHAPTER 7   HARNESSING THE POWER OF NIAGARA: PRACTICAL EXAMPLES IN UNREAL ENGINE 5

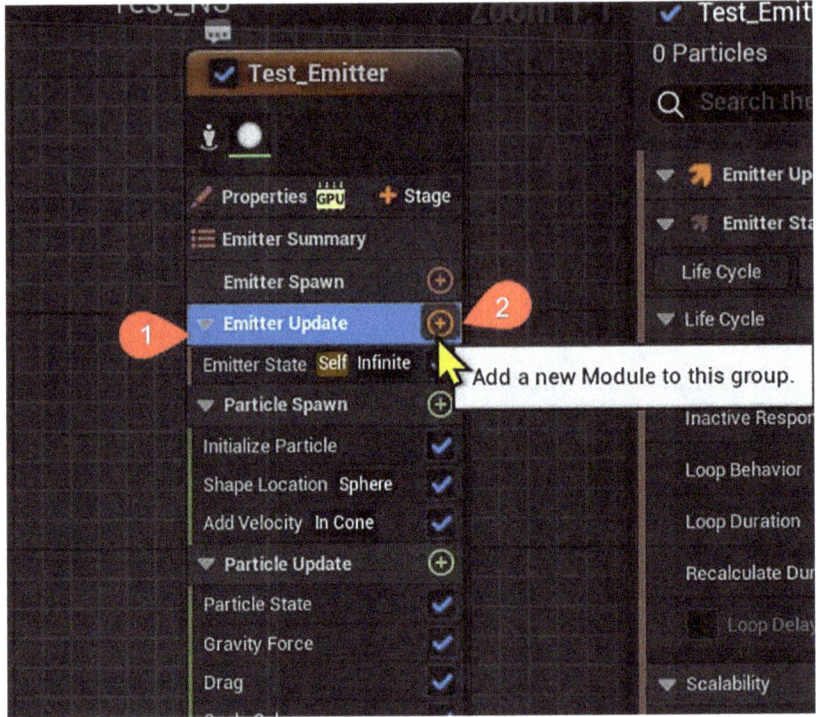

Figure 7-18. *Adding a new module*

Figure 7-19 shows a search for "spawn"-related modules is conducted in the "Add new Module" section. As shown in Step 1, we can type in the keyword to indicate the search results with various spawning modules. Step 2: Select "Spawn Burst Instantaneous", which will cause particles to spawn in a sudden burst rather than continuously over time.

CHAPTER 7   HARNESSING THE POWER OF NIAGARA: PRACTICAL EXAMPLES IN UNREAL ENGINE 5

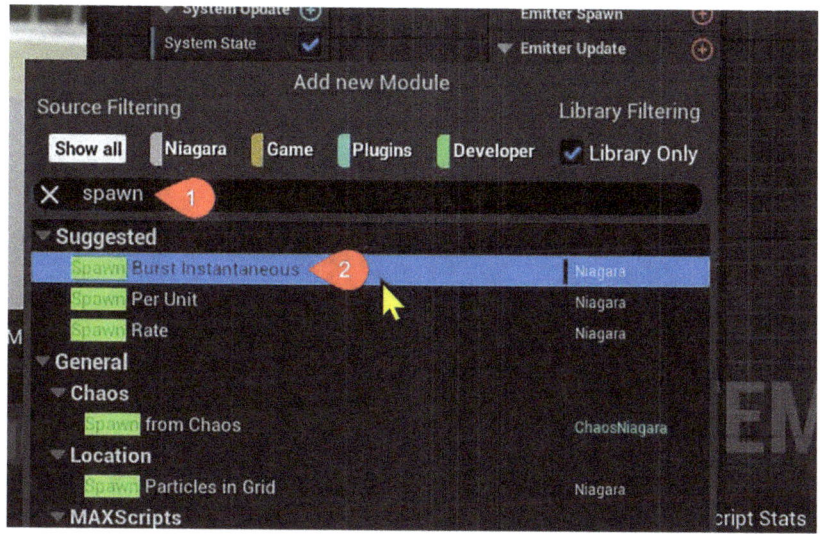

Figure 7-19. *Module selection*

Figure 7-20 shows the "Spawn Burst Instantaneous" module's properties after it has been added to the particle system. The "Spawn Count" is set to 300, which specifies the number of particles to be emitted in a single burst.

We can see the result in Figure 7-21 with the visual result in the editor's preview window. A burst of particles is visible, demonstrating the effect of the "Spawn Burst Instantaneous" module after the continuous spawn rate module was removed.

As the previous steps illustrated, the process of deleting the continuous "Spawn Rate" module and adding a "Spawn Burst Instantaneous" module changes the particle emission from a steady flow to a one-time burst, which can be used for effects like explosions, fireworks, or splashes.

CHAPTER 7   HARNESSING THE POWER OF NIAGARA: PRACTICAL EXAMPLES IN UNREAL ENGINE 5

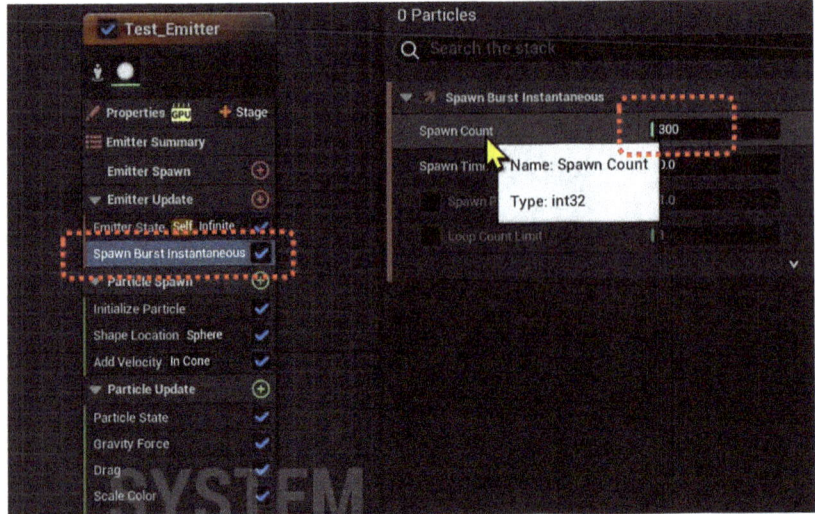

Figure 7-20. *Configuring the burst module*

Figure 7-21. *Resulting effect*

Adjusting Particle Burst Frequency

The following figures will illustrate how we can use the Niagara particle system editor within UE to observe the effect of changing the loop duration on a particle system.

Figure 7-22 shows the "Emitter Update" section with "Loop Duration" set to 5.0 seconds. This means that the burst of particles will occur every 5 seconds. "Spawn

Burst Instantaneous" is configured with a "Spawn Count" of 300 particles, meaning 300 particles will be emitted in each burst. The timeline at the bottom indicates that the playback head is at 2.59 seconds, suggesting the system is partway through a loop.

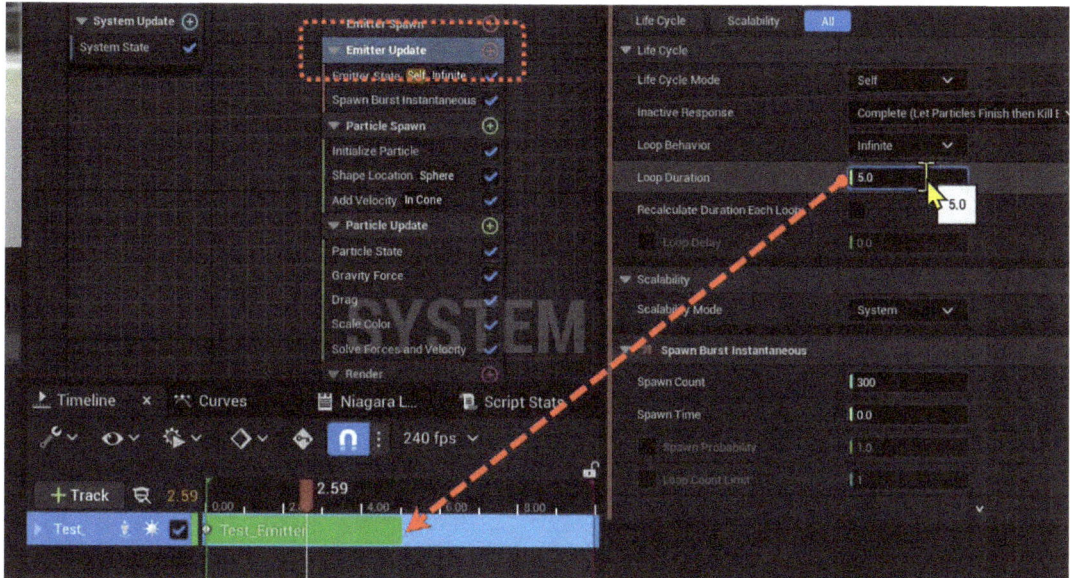

Figure 7-22. *Loop duration of 5 seconds*

Figure 7-23 illustrates the "Loop Duration" has been changed to 0.5 seconds, which will result in the particles bursting more frequently, every half a second. The "Emitter Update" settings are otherwise unchanged, with "Spawn Burst Instantaneous" still set to emit 300 particles each time.

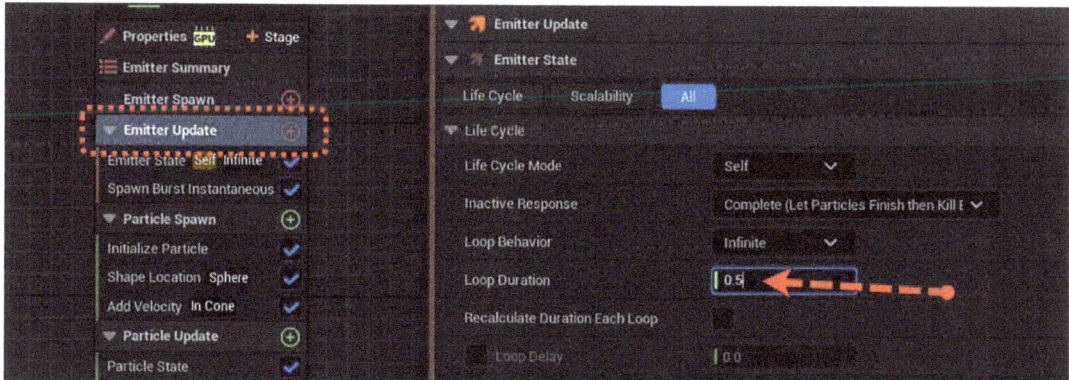

Figure 7-23. *Loop duration of 0.5 seconds*

299

CHAPTER 7   HARNESSING THE POWER OF NIAGARA: PRACTICAL EXAMPLES IN UNREAL ENGINE 5

Figure 7-24 illustrates the preview shows the visual outcome of the particle system with a "Loop Duration" of 0.5 seconds. The result is a denser cluster of particles, due to the more frequent emission of bursts.

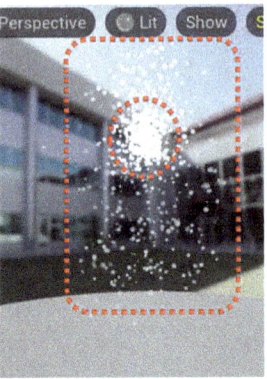

Figure 7-24. *Visual result of change*

The preceding steps collectively illustrate how adjusting the loop duration in a particle system's settings can affect the frequency of particle bursts, thereby changing the visual effect in the particle system's preview within the editor. A shorter loop duration leads to more frequent bursts, resulting in a denser accumulation of particles over time.

Customizing Particle Visuals with Material Swaps in UE's Niagara System

The following figures and steps illustrate the process of changing the material used for a particle system in UE's Niagara visual effects system.

Figure 7-25 illustrates the Niagara system editor shows the "Sprite Renderer" in the Step 1 component of the "Test_Emitter" particle system. Step 2 in this figure also shows the material assigned to the particles is "DefaultSpriteMaterial," which is a basic, often-used material for simple particle effects.

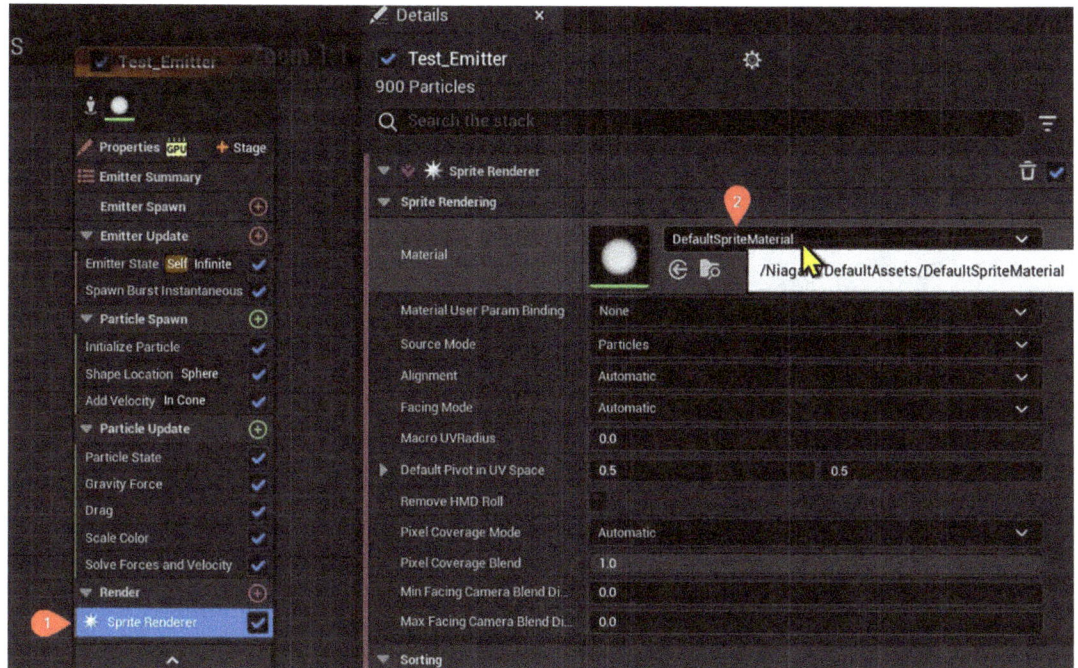

Figure 7-25. *Original material for Sprite Renderer*

Figure 7-26 illustrates the preview window displays the particles with the original material, which renders the particles as soft, white circles, a typical appearance for basic particle effects like smoke or mist.

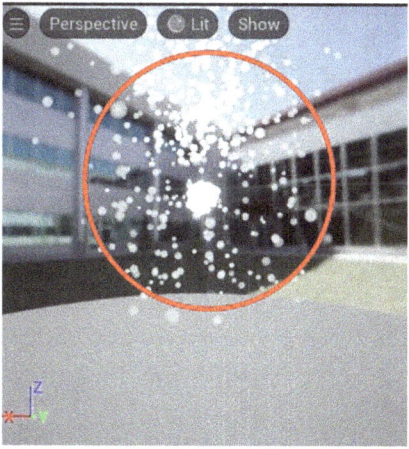

Figure 7-26. *Visual result with default particle materials*

CHAPTER 7   HARNESSING THE POWER OF NIAGARA: PRACTICAL EXAMPLES IN UNREAL ENGINE 5

Figure 7-27 illustrates the material for the "Sprite Renderer" has been changed to "LaserPointerMaterial_Inst," which is a green, glowing material, given its name and appearance in the thumbnail.

Figure 7-27. *Change of the material*

Figure 7-28 illustrates the preview window now shows the particles with the new material, rendering them as green and more vibrant, which could be suitable for effects like a laser show, magical energy, or environmental ambiance.

Figure 7-28. *Visual result from the new change of material*

Changing the material in a particle system can dramatically change the visual style and the feel of the effect, allowing for a wide range of creative possibilities. By simply swapping out the material, the same particle system can be used to represent different elements or phenomena, making it a versatile tool for game development and other real-time applications.

Harnessing the Power of Niagara #1: Smoke from Landing Spaceship

In our project, we initially utilized Niagara to simulate the smoke produced by the combustion in the engine of a landed spaceship. The outcome is illustrated in Figure 7-29, where the smoke effects are visibly emanating from beneath the spaceship.

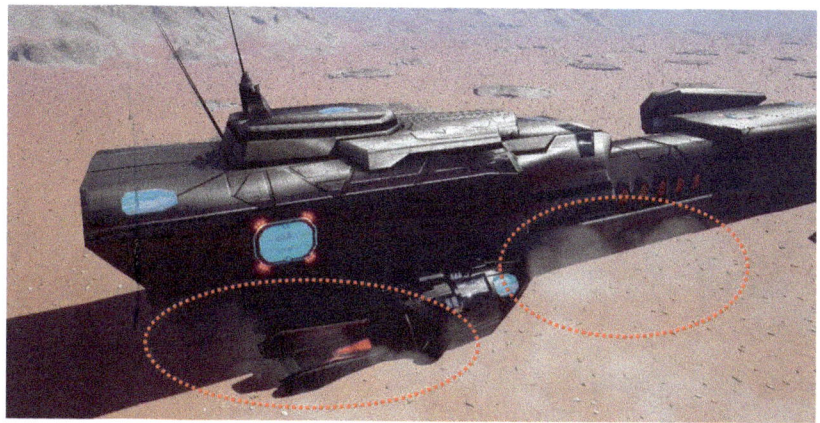

Figure 7-29. End result of bursting smoke with the Niagara system

Import and Set Up the Spaceship

Before creating our first particle system with Niagara, we will prepare a 3D spaceship asset, which is available for free download from turbosquid.com, as shown in Figure 7-30, using the following link:

www.turbosquid.com/3d-models/scifi-space-cruiser-3d-model/868423

CHAPTER 7   HARNESSING THE POWER OF NIAGARA: PRACTICAL EXAMPLES IN UNREAL ENGINE 5

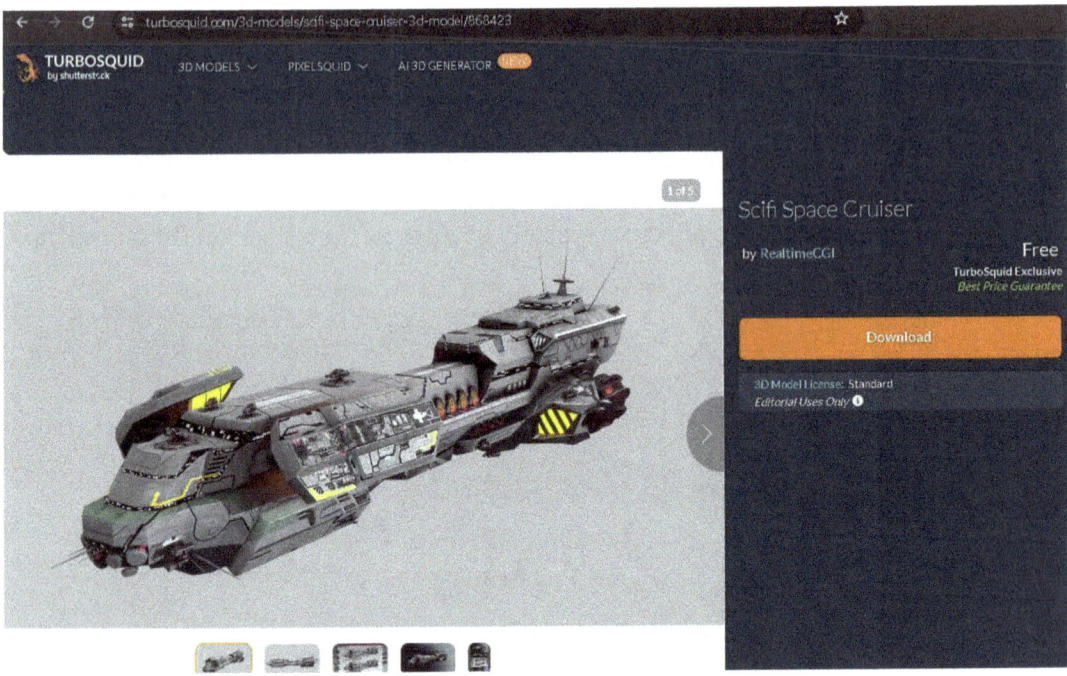

Figure 7-30. Download a space cruiser from TurboSquid

If this asset turns out not to be freely available, we can also opt for an alternative asset if you prefer not to incur additional costs.

There are numerous free spaceship models available that can serve just as well for our purposes, and you're encouraged to source one that fits your project without additional expense.

This way, you can follow along with the tutorial using an asset that is more readily available and budget-friendly.

This spaceship asset has two separate materials, one for the ship and another for the guns, as depicted in Figures 7-31 and 7-32, respectively.

CHAPTER 7　HARNESSING THE POWER OF NIAGARA: PRACTICAL EXAMPLES IN UNREAL ENGINE 5

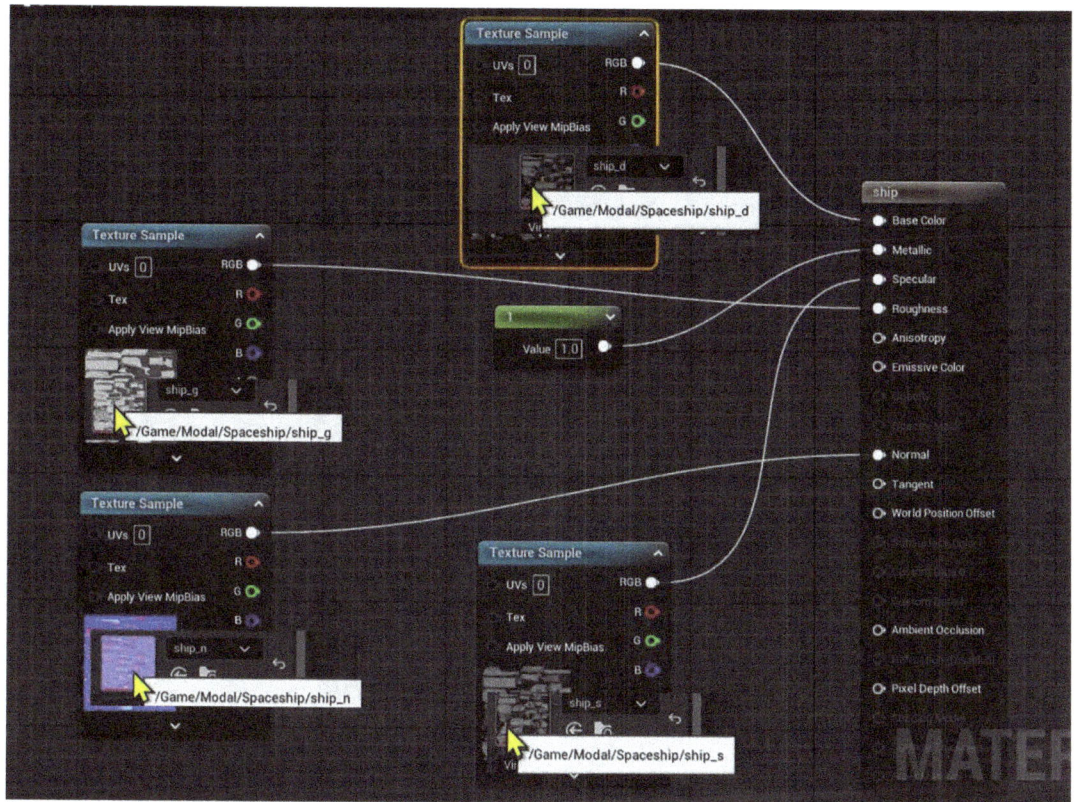

Figure 7-31. *Textures used for the ship material*

CHAPTER 7   HARNESSING THE POWER OF NIAGARA: PRACTICAL EXAMPLES IN UNREAL ENGINE 5

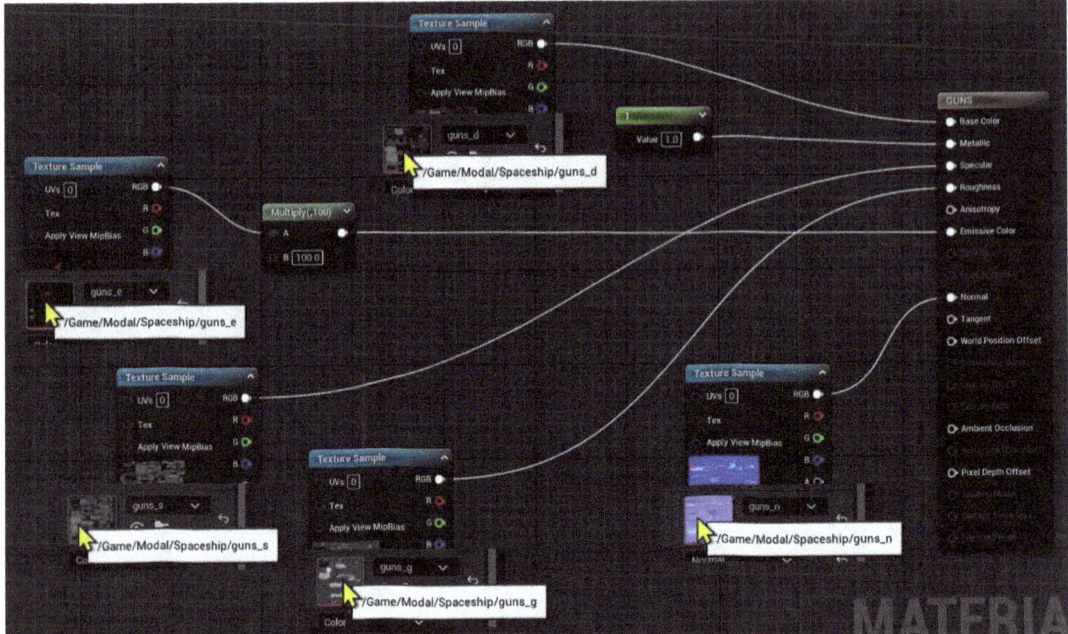

Figure 7-32. *Texture and values used for the gun material*

Figure 7-33 presents a properties panel where the location of the spaceship is set within the 3D space of the level as shown in the following:

- Location: X = 29,170.0, Y = –39,890.0, Z = –70,240.0

You may choose any location that you find suitable for the spaceship.

CHAPTER 7 HARNESSING THE POWER OF NIAGARA: PRACTICAL EXAMPLES IN UNREAL ENGINE 5

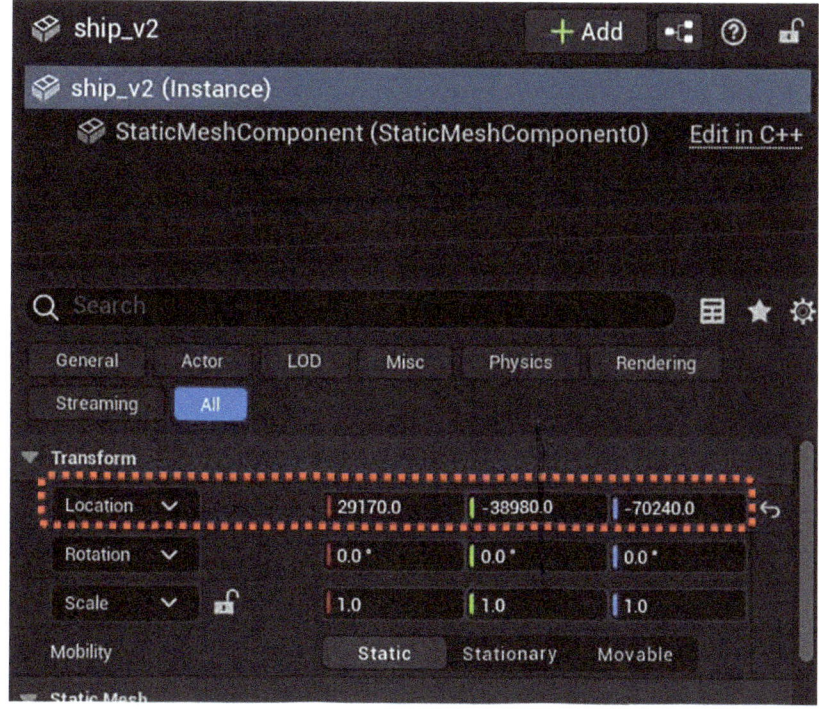

Figure 7-33. *Location of the spaceship*

Lastly, Figure 7-34 displays the result of the applied materials and location settings, offering a view of how the spaceship model would appear within the level, complete with lighting and environmental effects.

Figure 7-34. *The visual result of the spaceship in the level*

Set Up Material for the Smoke Niagara System

We begin by creating or importing a texture that resembles smoke, which has varying degrees of transparency to simulate the wispy nature of smoke. In Figure 7-35, we see that Step 1 involves creating a folder named "*VFX*" within the "*Materials*" folder, and Step 2 involves creating a material named "*Smoke_MAT*".

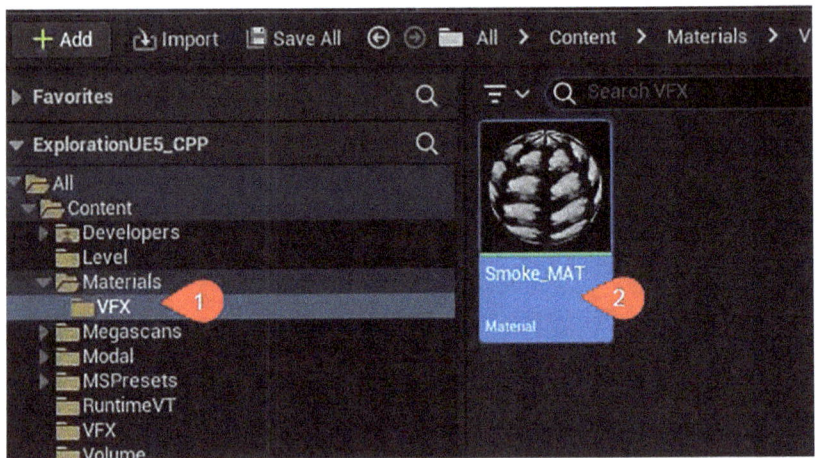

Figure 7-35. *Create a new material named Smoke_MAT inside the folder named VFX under the existing Materials folder*

The setup for smoke materials in such an environment typically involves creating a material that simulates the appearance of smoke.

Here are some general steps on setting up materials for smoke in a VFX software like UE5's Niagara.

Start with a Texture: We start by generating a smoke-like texture through the process of importing a texture from an existing asset from UE contents that can resemble smoke. Figure 7-36 shows a material graph which outlines how various nodes are connected to create the final smoke material. The nodes are typically connected to define how the `base color`, `opacity`, and other material properties are derived from the input textures and parameters.

This graph represents the logic of the smoke's appearance in the software's rendering engine. This texture usually has varying degrees of transparency to simulate the wispy nature of smoke.

Create a Material: Within the material editor, initiate the creation of a new material. It's essential to note that when designing materials for smoke effects, you should set the material's "Blend Mode" to "Translucent." This adjustment is crucial because smoke possesses a semitransparent quality, and the translucent blend mode enables the material to mimic this characteristic accurately. Make sure this setting is applied to achieve the desired visual effect of smoke within your scene.

Apply the Texture: Apply the smoke texture to the material using a texture sampler node as indicated in the following path. This node is connected to the material's base color to provide the smoke's visual content.

The path of the texture sampler for the texture: Content/MSPresets/MSVTTextures/WhitePlaceholder.uasset

---

Adjust Transparency: We need to adjust the transparency of the smoke, which can be done through the material's opacity input. This might involve using a Depth Fade node to ensure the smoke fades out over distance or using the alpha channel of the smoke texture to control transparency.

Particle Color and SubUV: For enhanced detail and diversity in our smoke effect, we incorporate two powerful features: the particle color node and the SubUV node.

- The particle color node gives us the flexibility to alter the color of the smoke dynamically during the particle's life, allowing for a varied and more lifelike appearance.

- On the other hand, the SubUV node is a tool that enables us to animate textures across individual particles, creating the illusion of complex movement within the smoke itself.

By utilizing SubUV animation, we can simulate the intricate, swirling patterns of real smoke. These nodes work together to imbue our smoke with a convincing, dynamic quality that can be fine-tuned to our specific visual needs.

Therefore, particle color nodes allow us to adjust the smoke's color dynamically, and SubUV nodes are used to create more complex animations within the smoke texture as indicated in the following path.

CHAPTER 7   HARNESSING THE POWER OF NIAGARA: PRACTICAL EXAMPLES IN UNREAL ENGINE 5

The path of the texture sampler for the SubUV node: Content/Tutorial/SubEditors/ TutorialAssets/T_SmokeSubUV_8X8.uasset

Fine-tune the Material: Connect various nodes like multiply, add, and clamp to fine-tune how the texture's colors and alpha affect the final appearance. For example, we may want to make the smoke darker or lighter or adjust how it blends with the background.

Material Parameters: We may also create material parameters that can be adjusted within Niagara to control the look of the smoke during runtime. This allows for real-time tweaks to the smoke's appearance without having to go back to the material editor.

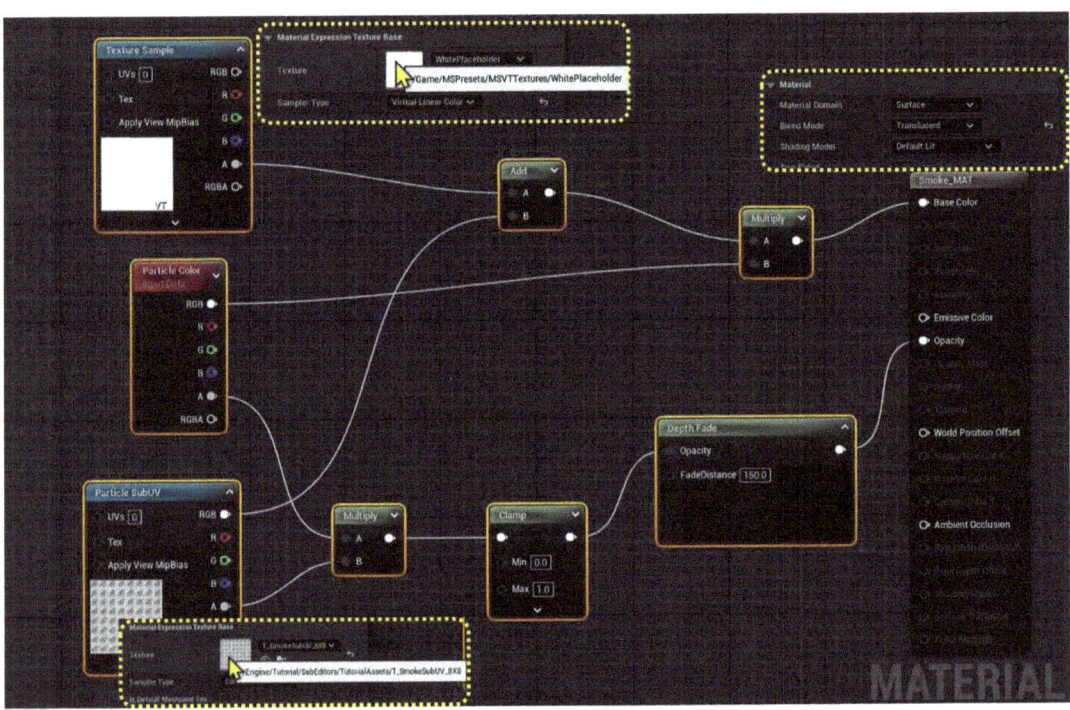

Figure 7-36. *Smoke_MAT*

Set Up the *SmokeFromLandingShip* Niagara System

We create a new Niagara system (based on the Fountain emitter template) named *SmokeFromLandingShip* as shown in Figure 7-37.

CHAPTER 7 HARNESSING THE POWER OF NIAGARA: PRACTICAL EXAMPLES IN UNREAL ENGINE 5

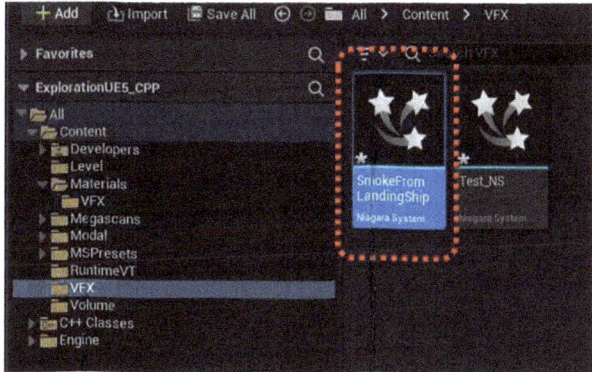

Figure 7-37. *Create a new Niagara system based on the Fountain emitter template*

In Figure 7-38, we see the basic setup of the particle system's basic emitter settings as shown in the following steps:

- Step 1: Rename the emitter to "Smoke."

- Step 2: Set the Sim Target to GPUCompute Sim and the Bounds Mode to Fixed with the following minimum and maximum values. To locate the "Fixed Bounds" values, we may need to expand a section within the interface that could be collapsed by default:

Min: X = –1000, Y = –1000, Z = –1000

Max: X = 1000, Y = 1000, Z = 1000

- Step 3: Set the Loop Duration to 4 seconds.

- Step 4: Set the Spawn Count to 300, as shown in the figure.

These settings are to define the spatial parameters in which the particles will exist and behave, as we described in previous sections.

CHAPTER 7 HARNESSING THE POWER OF NIAGARA: PRACTICAL EXAMPLES IN UNREAL ENGINE 5

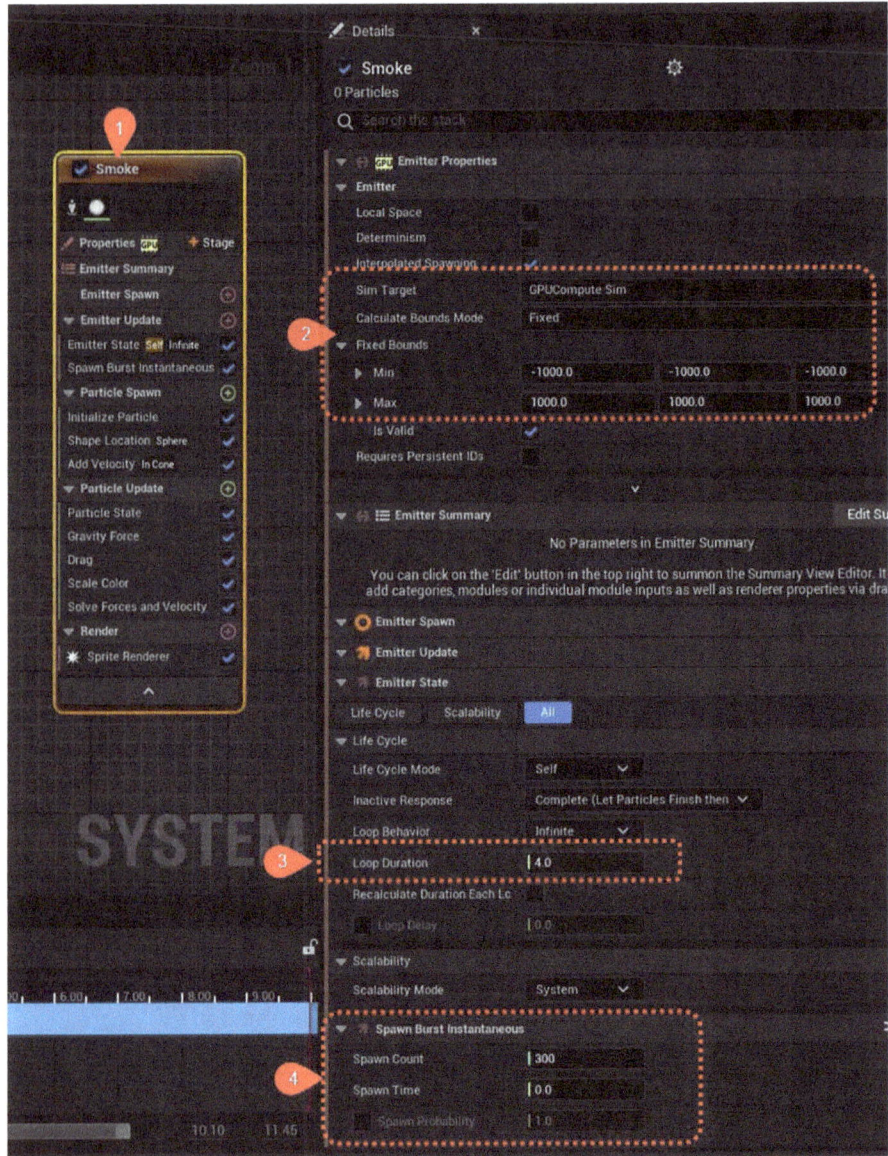

Figure 7-38. Initial setup of the Smoke emitter Niagara system

Initialize Particles

Here, we outline several steps to initialize the properties of smoke particles as shown in Figure 7-39:

- Step 1: To select the "Initialize Particle" module, which will define the initial state of the particles as they are emitted.

- Steps 2 and 3: The Lifetime properties are adjusted, with a minimum and maximum range set to control how long each particle remains active within the simulation. This is essential for ensuring the smoke particles dissipate over a realistic duration. To specify these ranges, look for the drop-down menu adjacent to the "Min" property.

  - This step might not be immediately apparent, as it requires you to interact with the drop-down next to the "Min" property to access and adjust it to a "Random Range." This ensures the particles' lifetimes vary within the defined range, contributing to the realism of the smoke effect.

  - From there, select the "Random Range"; we can enter the desired minimum and maximum values for the "Lifetime Min," for example, 1 and 2.5, respectively. Directly below, for the "Lifetime Max," set the desired values directly, such as 5.0 as the maximum.

- Steps 4–6: The size of the particles is configured using a "Sprite Size" module, with a nonuniform size range established. This likely provides variation to the particle sizes, which contributes to the overall natural look of the smoke effect.

  - We can set our desired range of "Min" and "Max" values, such as 60 and 120, respectively.

- Steps 7–9: The particle rotation is also being adjusted with a direct angle setting, allowing for random rotation of the particles upon spawning. This adds to the chaotic movement characteristic of smoke.

  - We can enter the desired minimum and maximum rotation value, such as −360 and 360 degrees to allow for a full forward and backward rotation range.

CHAPTER 7   HARNESSING THE POWER OF NIAGARA: PRACTICAL EXAMPLES IN UNREAL ENGINE 5

These steps collectively contribute to the visual complexity and realism of the smoke effect by defining how the particles behave from the moment they are generated. The adjustments ensure that the smoke has a natural appearance, with variations in lifespan, size, and rotation that mimic real-world smoke behavior.

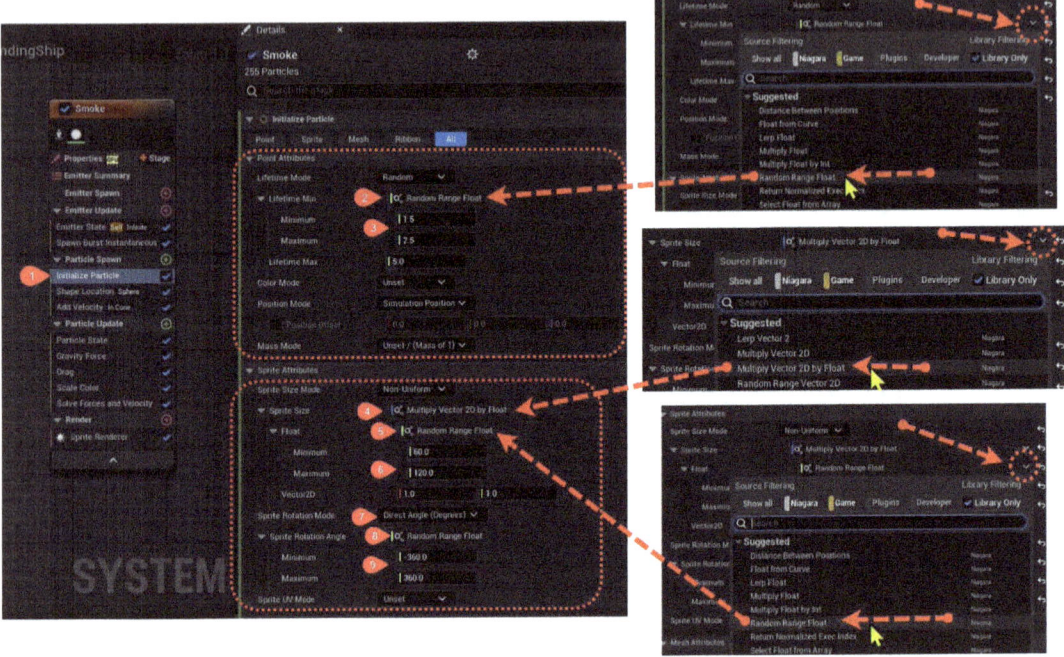

Figure 7-39. *Initialize Particles inside the emitter with several random min-max float values*

Shape Location

Figure 7-40 shows the "Shape Location" module within a particle system editor, which is part of the initial particle setup for a smoke effect.

- As shown in Step 1, we select the "Shape Location" module. The module is used to define the area where the particles will be emitted and to control various aspects of their initial distribution. Shape Location indicates that the initial location of the particles is defined by a spherical volume, which is a common choice for smoke effects as it allows the particles to emerge from a rounded, volumetric space.

314

CHAPTER 7   HARNESSING THE POWER OF NIAGARA: PRACTICAL EXAMPLES IN UNREAL ENGINE 5

- Sphere Radius, as shown in Step 2, is set to 1000 units, establishing the size of the spherical area for particle emission. This large radius suggests that the smoke effect is intended to cover a significant area, perhaps simulating a large-scale event like an explosion or a big fire.

- Non-Uniform Scale, as shown in Step 3, includes a nonuniform scale, which is currently set to uniform scaling (1.0, 1.0, 0.0). This may be used to stretch or squash the spherical volume in particular directions, which can be useful for directing the smoke or fitting the emission volume into a particular space within the scene.

These settings are critical for defining the starting point and overall spread of the smoke effect within the simulation environment. The shape and size of the emission volume have a direct impact on the appearance and behavior of the smoke particles as they rise and disperse.

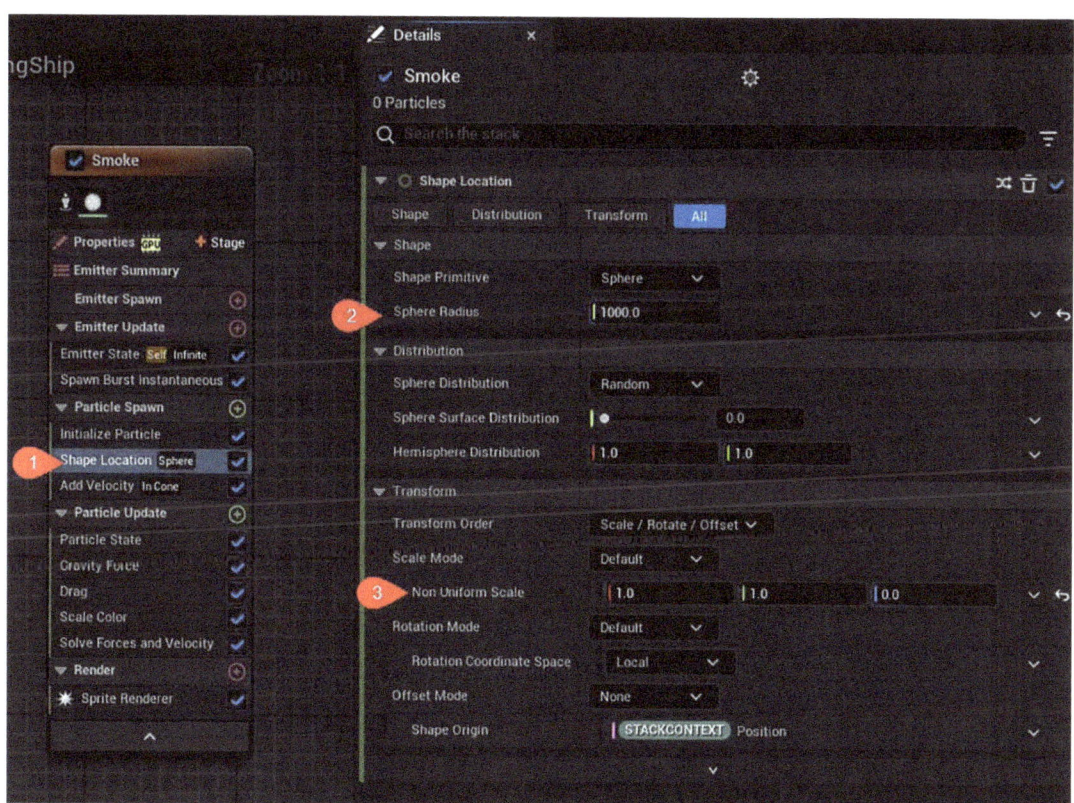

Figure 7-40. Configure the Shape Location

CHAPTER 7　HARNESSING THE POWER OF NIAGARA: PRACTICAL EXAMPLES IN UNREAL ENGINE 5

Add Velocity

Figure 7-41 illustrates the "Add Velocity" module in a particle system editor, which is typically used to define the initial velocity of particles when they are emitted. This particular setup is configured to simulate smoke particles in a particle system called "Smoke." Here's a breakdown of the parameters and their purposes:

- In Step 1, we select the "Add Velocity" module to activate the configuration of the module.

- Velocity Mode: In Step 2, the velocity mode is set to "In Cone," which means that particles will be emitted with a velocity vector that is within a conical shape. This is common for smoke or fire effects where particles need to move upward and spread out as they ascend.

- Velocity Speed, in Step 3, is set with a random range between minimum (800) and maximum (1200) units. This randomness in the speed helps create a more natural and less uniform movement of the smoke particles.

- Distribution Along Cone Axis: Set the value to 0.5 in Step 4, which ensures that the particles will be evenly distributed along the cone axis, neither biased towards the base nor the tip.

- Speed Falloff From Cone Axis, in Step 5, which this option has been disabled to prevent the velocity from decreasing from the center of the cone axis to the edges.

- Cone Axis, in Step 6, defines the direction of the cone. The suggested values of minimum of (-1, -1, 0) and maximum of (1, 1, 1) are set to a random range vector, which allows the cone's direction to vary. In the case of smoke, this could simulate the effect of wind or air currents on the smoke's direction.

- Cone Angle, in Step 8, is set to 10 degrees, which is relatively narrow and would result in a focused stream of particles. A smaller angle would make the smoke rise more straight up, whereas a wider angle would make it disperse more.

CHAPTER 7   HARNESSING THE POWER OF NIAGARA: PRACTICAL EXAMPLES IN UNREAL ENGINE 5

The "Add Velocity" module's configuration in this screenshot is designed to give the smoke a realistic behavior, where particles start with a certain upward momentum and spread out as they rise, simulating the way smoke behaves in a real-world environment. The randomization in speed and direction helps to avoid a too uniform or artificial look, which is crucial for creating convincing smoke effects in visual effects or game development.

Figure 7-41. Configure the velocity of the particles

We can see the result so far in Figure 7-42, where the particles are now spreading outward instead of originally spreading upward from the Fountain template.

Figure 7-42. Particles are now spread outward

Gravity Force

Disabling Gravity Force, as shown in Figure 7-43, in a smoke simulation is an intentional choice to achieve a specific visual effect. In the real world, smoke particles are affected by gravity, but they are also subject to other forces such as buoyancy and air resistance, which can cause them to rise or disperse in the air. Here are some reasons why gravity might be disabled in this simulation:

> Slow Movement: We would like to have a slow-moving smoke; the effect of gravity is minimal compared to the forces of diffusion and slight air currents. Disabling gravity can help replicate this gentle, floating behavior.

> Artistic Direction: The effects may be for a stylized scene where the smoke is required to behave in an unnatural manner, such as lingering in the air or moving horizontally without falling.

In summary, disabling Gravity Force allows the smoke to act in a way that suits the visual requirements of the project, due to the nature of the realistic behavior of hot smoke, a particular artistic vision, and practical considerations in this simulation.

CHAPTER 7   HARNESSING THE POWER OF NIAGARA: PRACTICAL EXAMPLES IN UNREAL ENGINE 5

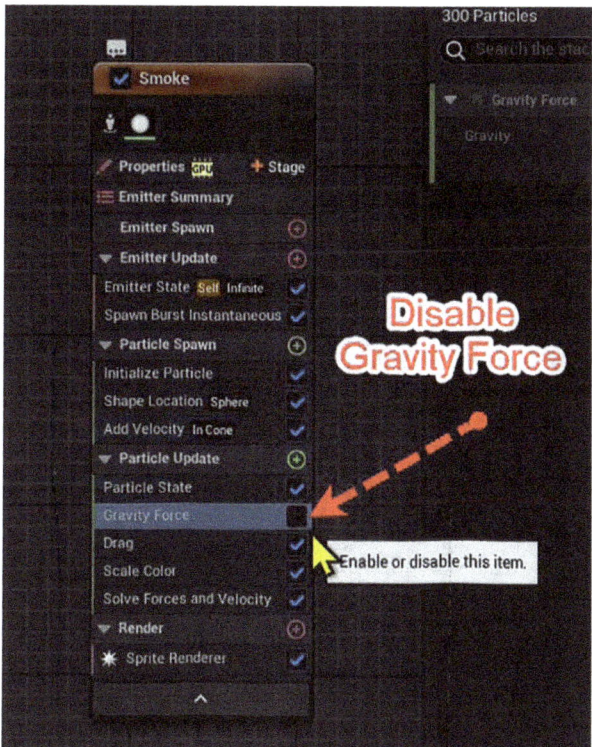

Figure 7-43. *Disable the Gravity Force on the particles*

Drag Value

Figure 7-44 shows the "Drag" module within the particle system editor for a smoke effect, indicating that this feature is active in the simulation. The Drag module is used to simulate air resistance and is essential in creating a realistic motion for particles, like smoke, in the air.

- Step 1: We select the Drag module to activate the properties of the module.

- Step 2: Enable the "Drag" value and set it to 1.0, which determines the amount of air resistance applied to the particles. A drag value of 1.0 implies a moderate level of resistance, slowing down the particles as they move through space. This setting helps to prevent the smoke particles from moving too quickly and instead allows them to disperse in a manner that simulates the effect of smoke interacting with the air.

319

Chapter 7   Harnessing the Power of Niagara: Practical Examples in Unreal Engine 5

Incorporating drag into the simulation helps to ensure that the smoke behaves in a more physically accurate way, reflecting the influence of the environment on the smoke's movement and dispersion. Without drag, smoke particles could move too uniformly and quickly, failing to replicate the languid drift of smoke in the real world.

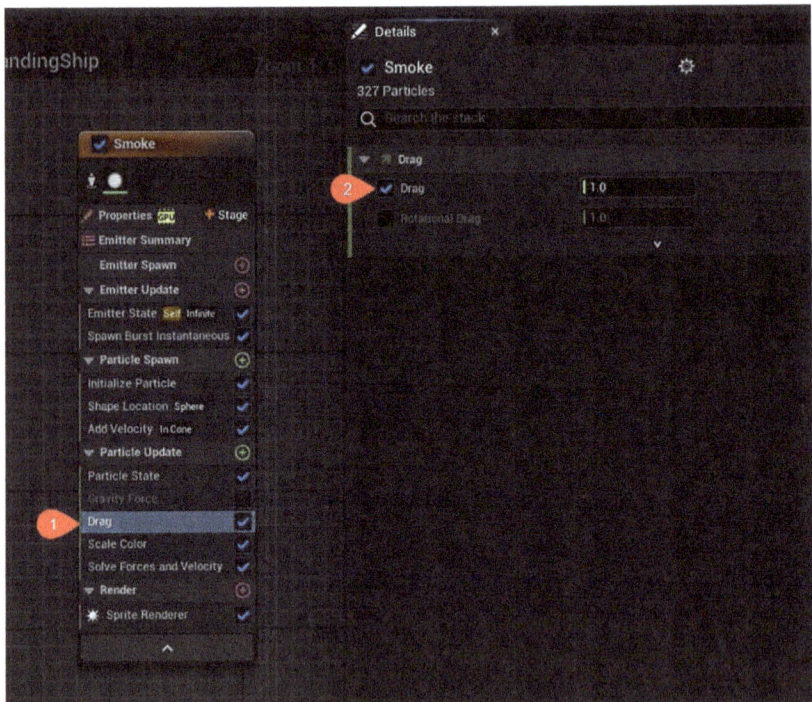

Figure 7-44. *Increase the drag value*

Acceleration Force

The "Acceleration Force" module is the unseen engine driving the behavior of particles within a smoke simulation, providing a sustained force that compounds over time. Unlike a mere initial burst that a velocity parameter might offer, acceleration imbues particles with the capacity for continuous change in speed or direction. This subtle yet profound distinction is key for creating effects that mimic the complexity of real-world physics. As we explore the "Acceleration Force" module, we are not just equipping particles with momentum; we are scripting their evolving journey through the simulated environment, crafting a dance of particles that accelerates with the passage of time.

Figure 7-45 shows the "Acceleration Force" module within a particle system editor, which is configured to apply an additional force to the particles in the smoke simulation. This force is an acceleration, which means it will change the velocity of the particles over time, not just give them an initial push like a velocity parameter would.

- As shown in Step 1, we may need to add the "Acceleration Force" module to our particle system if it is not included by default. This module applies a continuous acceleration to the particles, influencing their velocity over time. To do this, navigate to the appropriate section within the Niagara editor and use the "Add" function to include the "Acceleration Force" module into our effect's stack. Once added, you can then configure the module to define the magnitude and direction of the acceleration that will act upon the particles. This can be used to simulate effects like wind or explosions that would influence the movement of the smoke.

- As shown in Step 2, we set the range of values used in the Random Range Vector feature, which is being applied in a random direction within a specified range. This randomness helps to create a more natural and varied motion in the particles, as smoke in the real world is rarely uniform in its movement.

- As shown in Step 3, we set the minimum and maximum values where the minimum value is (1.2, –1.0, 5.0), and the maximum value is (32.0, 1.0, 1.0). These vectors define the range of the acceleration in each axis (X, Y, Z). The wide range in the values suggests that there's a significant variability in the force applied, which would result in a dynamic and chaotic motion characteristic of smoke.

The Acceleration Force module is a powerful tool for simulating environmental influences on smoke, such as gusts of wind, updrafts, or other forces that could affect the smoke after it has been emitted. By using random ranges, the simulation can ensure that the smoke does not all move uniformly, but rather has variations that make it appear more realistic.

CHAPTER 7 HARNESSING THE POWER OF NIAGARA: PRACTICAL EXAMPLES IN UNREAL ENGINE 5

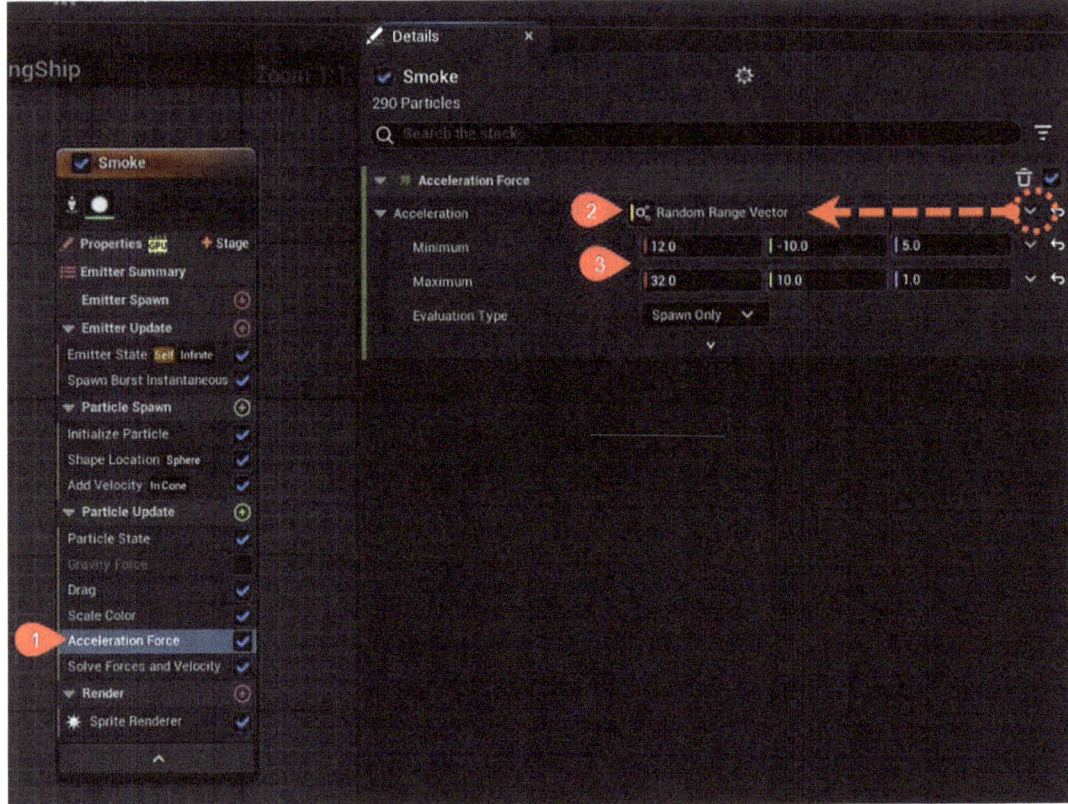

Figure 7-45. *Add some acceleration force*

Color

The Color module is a key player in the artistry of particle effects, providing the tools to dynamically paint each particle with a spectrum of hues across its life cycle. By utilizing a color curve, we gain precise control over the progression of color changes, allowing for a seamless transition from one shade to another as time unfolds. The module embodies the confluence of time and tint, offering a narrative for each particle not just in motion but also in the evolving visual tapestry it weaves within the digital canvas.

Figure 7-46 shows the "Color" module that we can use to configure to adjust the color of the particles over their lifetime using a color curve.

- As shown in Step 1, we need to **add** the Color module and select the Color module (as shown in Step 2) within the Niagara particle system editor for a smoke effect.

CHAPTER 7   HARNESSING THE POWER OF NIAGARA: PRACTICAL EXAMPLES IN UNREAL ENGINE 5

- As indicated in Step 3, we are setting the color value determined by using Color from the Curve feature. The module is set to determine the particle color from a curve, which means the color will change according to the position on the curve that corresponds to the particle's age.

  Color Curve Editor: The color curve is displayed, enabling the user to visually edit the interpolation of color values over the particles' lifetime. This curve signifies the color transition from the beginning (left) to the end (right) of a particle's lifespan. The top axis of the grid within the checkbox bar corresponds to the normalized age of the particles, ranging from 0 (birth) to 1 (death). The bottom axis of the grid within the checkbox bar denotes the color value changes between two points.

- Step 4 involves modifying the timing of the end color value within the particle system's color curve. To make this adjustment, we need to double-click the end point of the curve, which represents the color value as the particles near the end of their lifetime.

  This action allows us to set the value more precisely to an earlier timeframe, thereby changing the color transition that occurs as the particles dissipate. The instruction "double-click and drag the end point to the desired location" provides clarity on how to interact with the curve editor to achieve the intended effect.

  Alternately, instead of dragging the endpoint, which can be less precise, you have the option to double-click the endpoint on the color curve. This will allow us to input the exact value manually, ensuring the color change occurs at the specific intended time. By directly setting the value, we may be able to fine-tune the effect with precision and ease.

- As indicated in Steps 5 and 6, the curve has key points (such as 0.5 and 0.15, as set) that define the color at specific moments in the particle's life. Adjusting these points allows for fine-tuning of the color transition. For example, if the point at 0.5 is set to a lighter color, particles will transition through that color halfway through their life.

323

CHAPTER 7 HARNESSING THE POWER OF NIAGARA: PRACTICAL EXAMPLES IN UNREAL ENGINE 5

The color value for the curve's key points (Steps 5 and 6) can be set by double-clicking the key point. The process may be somewhat tricky, and it might take several attempts to activate the dialog box for entering the value.

By adjusting the color curve, the user can create a more dynamic and realistic smoke effect, where the color may start darker when the particles are emitted and become lighter or fade out completely as they dissipate. This mimics the way smoke changes color in the real world due to factors like concentration, lighting, and mixing with the air. The use of a curve provides a smooth transition between colors rather than a sudden shift, which would be less realistic.

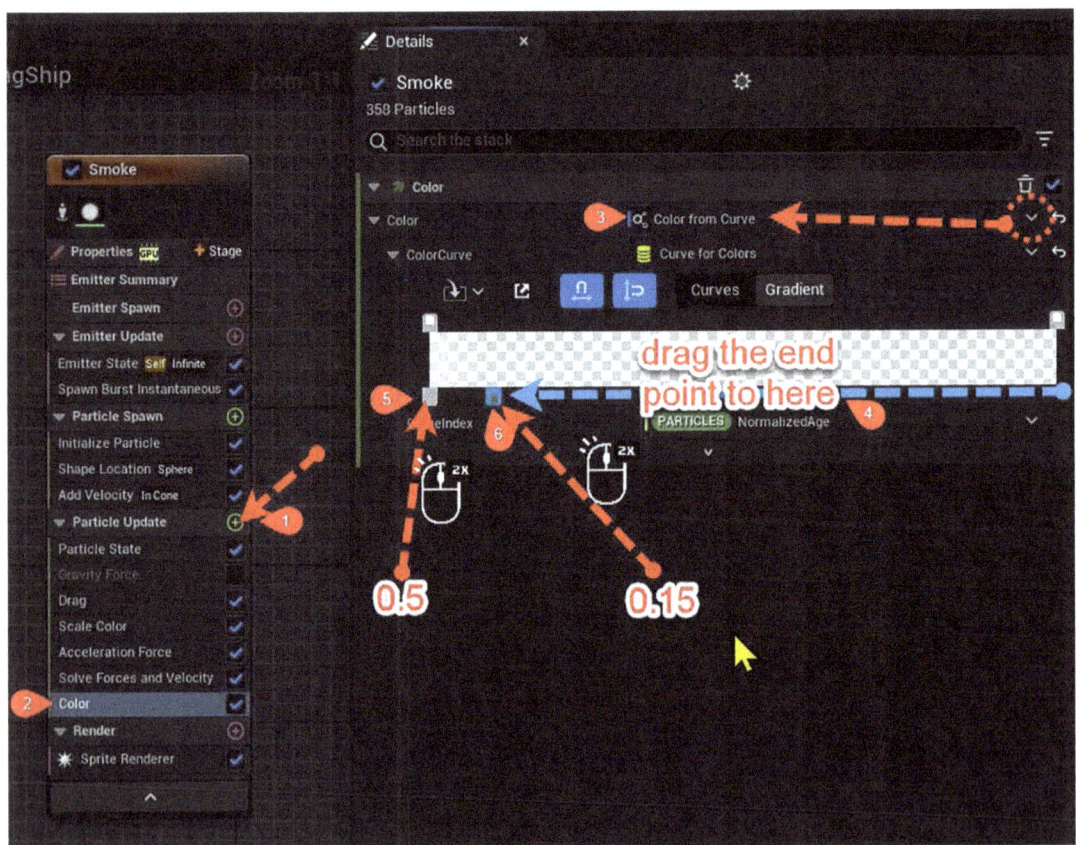

Figure 7-46. Set up color changes for each particle

Scale Sprite Size

The "Scale Sprite Size" module is a vital component within the Niagara particle system editor, particularly showcased through the lens of a smoke effect. This module serves as the control point for the dimensional aspects of the particles, allowing for the meticulous calibration of sprite sizes that constitute the visual structure of the smoke. It's here that the essence of each particle's presence is fine-tuned, with size adjustments crafted over the duration of the particle's life. Whether swelling to billowing proportions or diminishing to a wisp, the "Scale Sprite Size" module equips creators with the ability to shape the narrative of the smoke effect, ensuring that each particle plays its role in the greater choreography of the simulation.

Figure 7-47 indicates the "Scale Sprite Size" module within the Niagara particle system editor for a smoke effect. This module is responsible for adjusting the size of the individual sprites that make up the smoke particles, adjusting how the size of the particles will change over time.

- Scale Sprite Size Selection: As indicated in Step 1, we need to add this extra module "Scale Sprite Size."

- As shown in Step 2, this module is selected.

- Non-Uniform Scaling: As indicated in Step 3, the scaling is set to "Non-Uniform," which allows the user to scale the width and height of the sprites independently. This can create more varied and realistic particle shapes, as smoke particles do not always scale uniformly in all directions.

- As indicated in Step 4, we are setting the Scale Factor to use the "Multiply Vector 2D by Float" feature, which suggests that the sprite size is being manipulated by a factor (a float value) that affects both dimensions (Vector2D) of the sprite, providing a way to adjust the overall size dynamically.

- Under this Scale Factor, as Step 5 indicated, we are setting the Float value to use the Float From Curve feature, which is being used to define how the sprite size changes over the particle's lifetime. This allows the smoke particles to start small, grow larger, and possibly shrink back down, simulating the puffing and dispersal of smoke.

CHAPTER 7   HARNESSING THE POWER OF NIAGARA: PRACTICAL EXAMPLES IN UNREAL ENGINE 5

- Curve Key Points: As shown in Steps 6 and 7, we adjust key points on the curve to define the size of the sprites at different stages of their life cycle.
    - The first key point in Step 6, positioned at a key data value of 0.15 and 0.0, sets the initial size of the particles as they are spawned.
    - The second key point in Step 7, located at a key data value of 0.6 and 5.0, dictates the size of the particles at the midpoint of their existence.
    - By manipulating these key points, we can precisely control the expansion and contraction of the particles over time, allowing for a dynamic size variation that mimics natural growth and shrinkage behaviors.

For Steps 6 and 7, we can set the value for each key point by clicking the point to open the box where we can enter the key data for the axis and the scale value.

Adjusting the "Scale Sprite Size" module is essential for creating realistic smoke effects where the size of the smoke particles changes in a nonlinear fashion, contributing to the natural look and feel of the smoke as it billows and disperses in the environment.

CHAPTER 7   HARNESSING THE POWER OF NIAGARA: PRACTICAL EXAMPLES IN UNREAL ENGINE 5

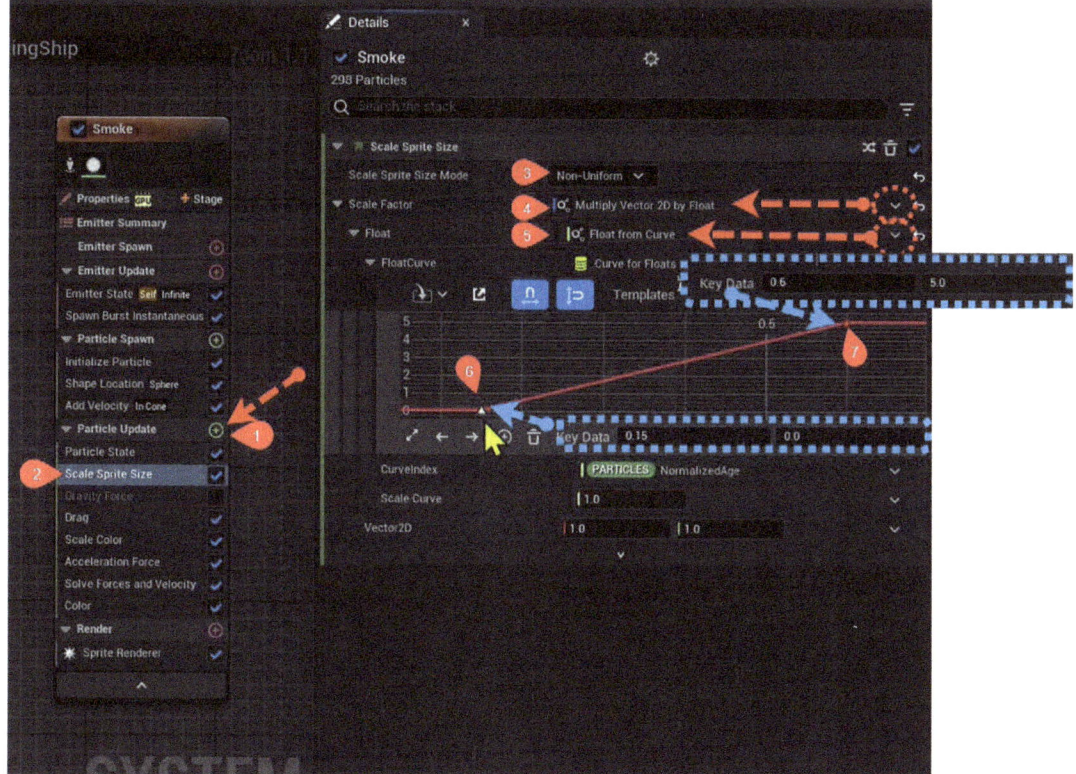

Figure 7-47. *Configure the Sprite Size throughout the lifetime update*

Sprite Renderer

In the orchestration of particle systems, the Sprite Renderer within the Niagara editor is where two-dimensional artistry meets three-dimensional space. This vital module is the final brushstroke in the creation process, translating individual particles into visible sprites – each one a flat image projected within the simulated depth of the scene. Here, designers have the capability to define how these sprites are visually represented, from their texture and material to their orientation and rendering order.

The Sprite Renderer is the nexus where particles are given visual form and texture, enabling them to contribute to the overall aesthetic and feel of the effect. It is the stage where particles are not just emitted and manipulated but are finally rendered for the viewer's eye, completing the illusion of life and motion in the virtual tableau.

Figure 7-48 shows the "Sprite Renderer" module settings within the Niagara particle system editor, which is used for rendering particles as sprites to simulate smoke effects.

- In Step 1, the "Sprite Renderer" module is selected.

- The specific settings for the "Sprite Renderer" module indicate that the "Smoke_MAT" material is being used to render the particles as shown in Step 2. This material was set up at the beginning of this section, which contains a texture that represents smoke.

- In Step 3, we see a parameter labeled "Alignment," which is set to "Unaligned." By using "Unaligned," the simulation can achieve a higher degree of visual complexity and randomness in orientation and appearance, which is crucial for replicating the fluid, ever-changing nature of smoke.

- Step 4 shows the "Facing Mode" set to "Face Camera." This is an important setting for particle systems, especially for effects like smoke. It ensures that no matter where the camera moves, the sprites will always face toward it, maintaining the illusion that they are volumetric. This is crucial for smoke and other similar effects, as it helps to maintain the visibility and consistency of the effect from all viewing angles.

- Sub Image Size: As shown in Step 5, this defines how the texture atlas is divided. For example, an 8x8 grid indicates that the texture atlas is divided into 64 (8 by 8) equal parts. Each particle can then use one of these segments as its texture, allowing for multiple variations of the smoke to be displayed.

  - The key feature highlighted in the settings is the use of SubUVs, particularly the "Sub Image Size" parameter. In particle systems, SubUVs are a technique used to create more complex and detailed visual effects with a single texture atlas. A texture atlas is a large image containing different frames of an animation or different variations of a texture.

  - Efficiency: Using SubUVs is a resource-efficient way to get a lot of visual complexity out of a single texture, as it allows for the reuse of the texture in varied ways without the need for multiple textures or more complex particle systems.

CHAPTER 7   HARNESSING THE POWER OF NIAGARA: PRACTICAL EXAMPLES IN UNREAL ENGINE 5

The outcome of these settings can be seen in the preview image, where the smoke appears to consist of varied textures, suggesting that the particles are using different SubUV frames to achieve a more natural and less repetitive smoke effect. This technique is essential for creating high-quality visual effects that are also performance-friendly, as it reduces the need for additional assets and computations.

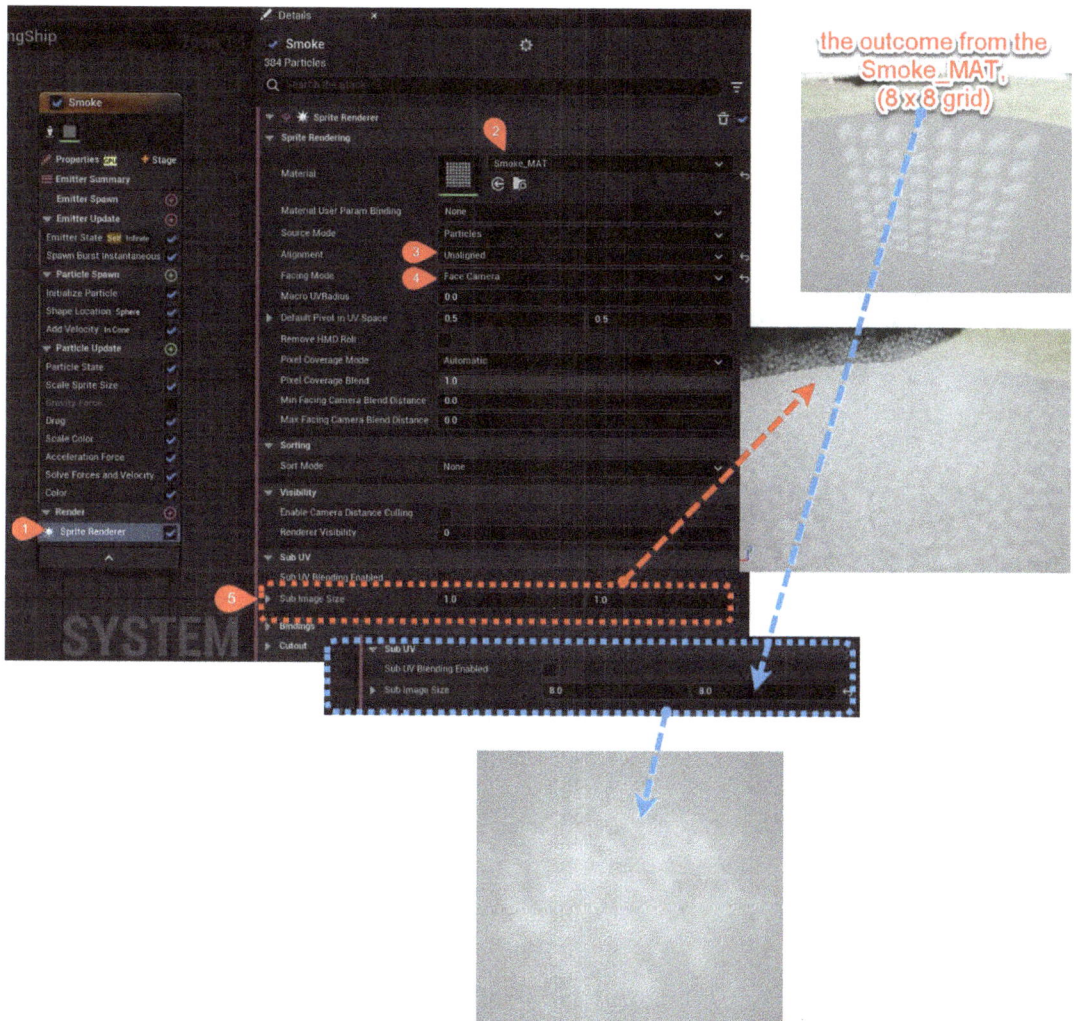

Figure 7-48. Set up the Sprite Renderer to use Smoke_MAT with SubUV

CHAPTER 7 HARNESSING THE POWER OF NIAGARA: PRACTICAL EXAMPLES IN UNREAL ENGINE 5

Placement of the SmokeFromLandingShip Niagara System

After completing the Niagara setup, we can now place the emitter in our level, specifically under our spaceship. The author had chosen the following location value to place this emitter in the level as shown in Figure 7-49. You are welcome to amend the location as you see fit yourself:

Location: X = 30,700.0, Y = –39,000.0, Z = –70,650.0

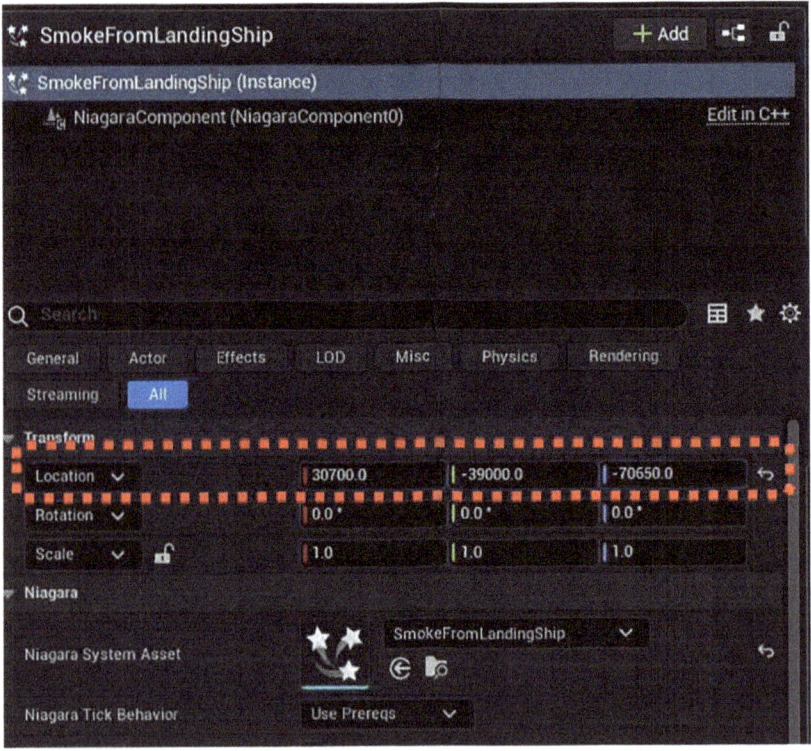

Figure 7-49. *Place the SmokeFromLandingShip Niagara system under the rear of the spaceship*

In Figure 7-50, we see the end result of this setup in the 3D view of the engine. The smoke is bursting out from beneath the landing ship, which is consistent with the expectation of a smoke effect generated by a ship touching down or hovering close to the ground. The smoke seems to spread out from the emitter, providing a realistic visual of dust and debris being disturbed by the ship's descent.

This kind of effect not only adds realism to the scene but also helps to convey the scale and power of the landing ship by visually representing its interaction with the environment.

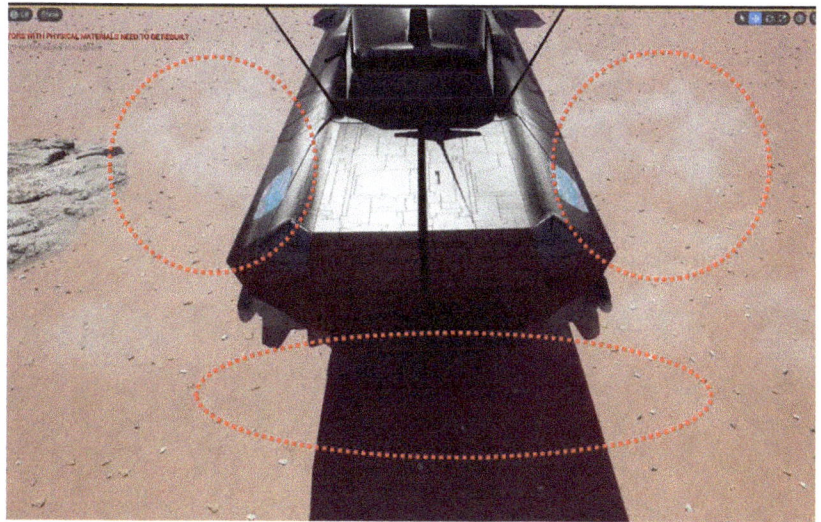

Figure 7-50. *The visual result of the bursted smoke coming from the Niagara system*

Harnessing the Power of Niagara #2: Spiral Effects on Magical Ball

In our project, we are going to begin creating the concept for a Magical Ball with spiral effects, envisioned as a defensive weapon. The design of this Magical Ball utilizes special effects through a visual effects system such as Niagara to produce an aura of energy or a protective shield around the Magical Ball.

The outcome is illustrated in Figure 7-51, where the spiral effect emanates from a Magical Ball. The spiral effect, delineated by the dotted red line, suggests a dynamic and visually compelling use of particle simulation to add a mystical or magical quality to the object in focus. The green arrow seems to indicate a specific direction or flow of the effect, which could be a part of the visualization or editing process to highlight the movement and behavior of the effect within the scene.

CHAPTER 7   HARNESSING THE POWER OF NIAGARA: PRACTICAL EXAMPLES IN UNREAL ENGINE 5

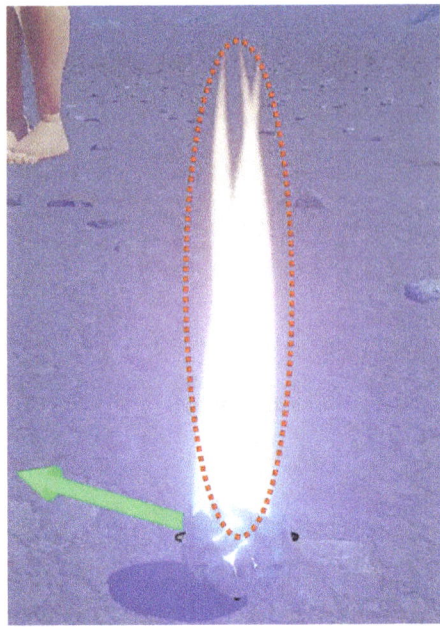

Figure 7-51. *Spiral effects from the Magical Ball*

Import and Set Up the Magical Ball

We will prepare our Magical Ball using this free downloadable asset from turbosquid.com, as shown in Figure 7-52:

www.turbosquid.com/3d-models/object-abstract-3d-model-1707700

The author plans to make this "Magical Ball" appear as shiny metal; therefore, we are using the available material (from Unreal Editor content itself) named "BrushedMetal," shown in Figure 7-53:

- Step 1: Place this asset in a newly created folder named MagicalBall.
- Step 2: Assign *BrushMetal* to this Magical Ball asset from the following path: Engine/Plugins/Enterprise/DatasmithContent/Content/Materials/FBXImporter/VRED/BrushedMetal.uasset

CHAPTER 7 HARNESSING THE POWER OF NIAGARA: PRACTICAL EXAMPLES IN UNREAL ENGINE 5

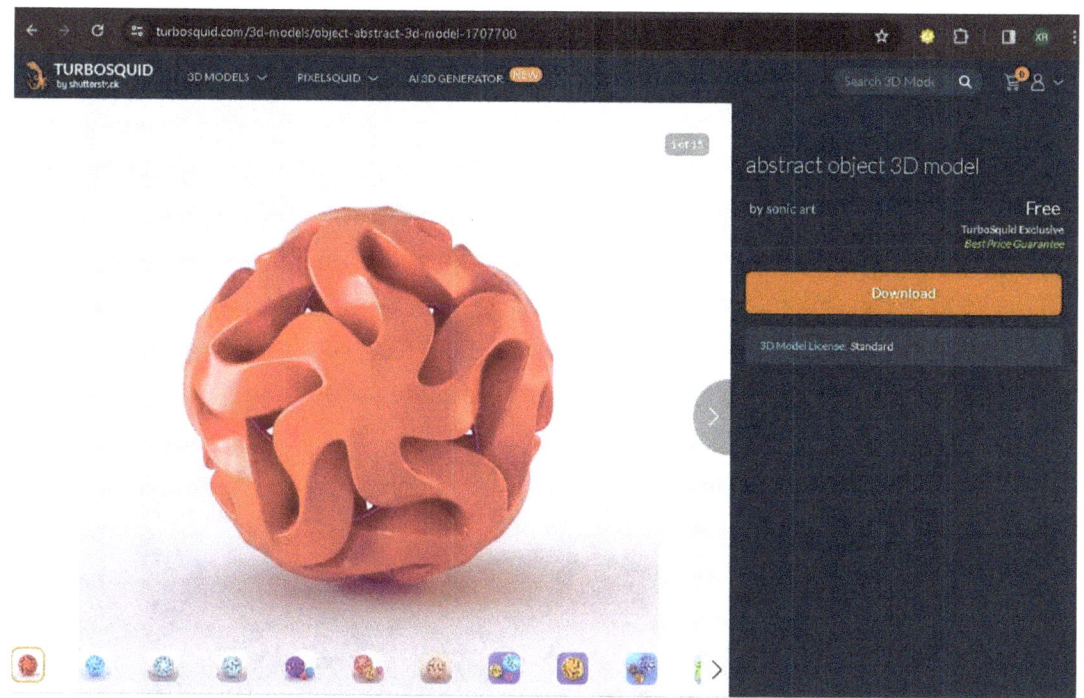

Figure 7-52. *Magical ball asset from turbosquid.com*

Figure 7-53. *Assign the BrushedMetal material to this magical ball*

CHAPTER 7   HARNESSING THE POWER OF NIAGARA: PRACTICAL EXAMPLES IN UNREAL ENGINE 5

Initialize Particle

As with the previous example, we will configure some initial settings for our Niagara system asset. For this setup, we begin by using the "Fountain" template as a starting point. This template provides a solid foundation with preconfigured values that we can then customize to suit our project's requirements. In this example, we set the name of the asset as Spiral_Effect, which is displayed in Figure 7-54. It highlights a VFX asset named "Spiral_Effect" which is categorized as a Niagara system. This indicates that a special visual effect called Spiral_Effect has been created and is accessible within the project for further editing or implementation in the game.

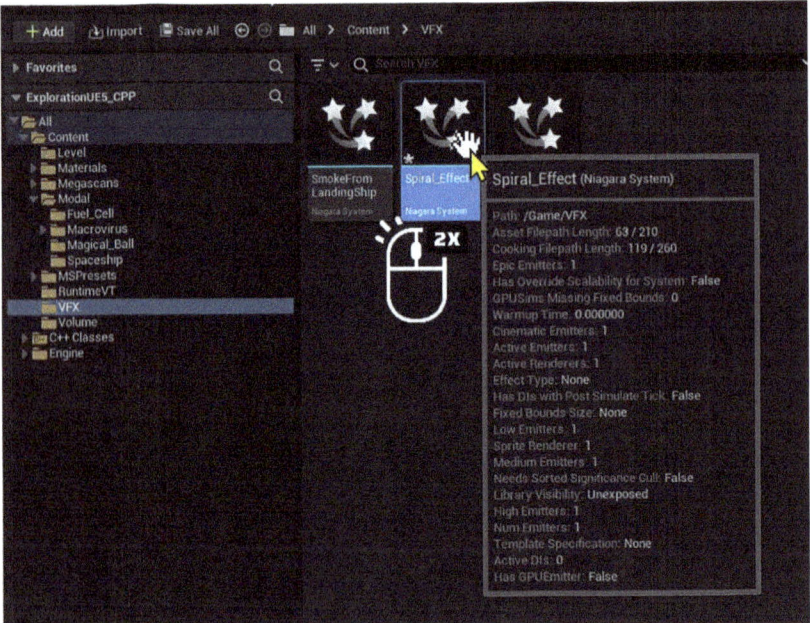

Figure 7-54. Open up this Spiral_Effect Niagara system

In Figure 7-55, the outlined sections indicate various parameters of the effect, such as emitter properties, emitter state, and emitter spawn.

- Step 1 indicates the change of name to Spiral.
- In Step 2, we set the effect in local space.

334

CHAPTER 7   HARNESSING THE POWER OF NIAGARA: PRACTICAL EXAMPLES IN UNREAL ENGINE 5

- With GPUCompute Sim, as shown in Step 3, we can leave the FixedBounds min-max bounds value as min: –100 and max: 100 in X, Y, Z bounding values.

- In Step 4, we focus on the "Loop Duration" setting within the "Emitter State" module. Here, you can set the duration for each loop of the particle effect. The loop duration is currently set to 5.0 seconds, which defines the length of time one cycle of the particle effect will play before repeating.

- Moving to Step 5, the attention shifts to the "Spawn Rate" under the "Emitter Spawn" module. The spawn rate is crucial because it dictates how many particles are generated per second. In this step, the "Spawn Rate" is set to a high value of 8000.0, indicating a very dense stream of particles being emitted, which can be suitable for creating intense effects like a spiral of smoke or a swarm of sparks. Adjusting this value allows for finer control over the visual density and frequency of the particles within your scene.

CHAPTER 7   HARNESSING THE POWER OF NIAGARA: PRACTICAL EXAMPLES IN UNREAL ENGINE 5

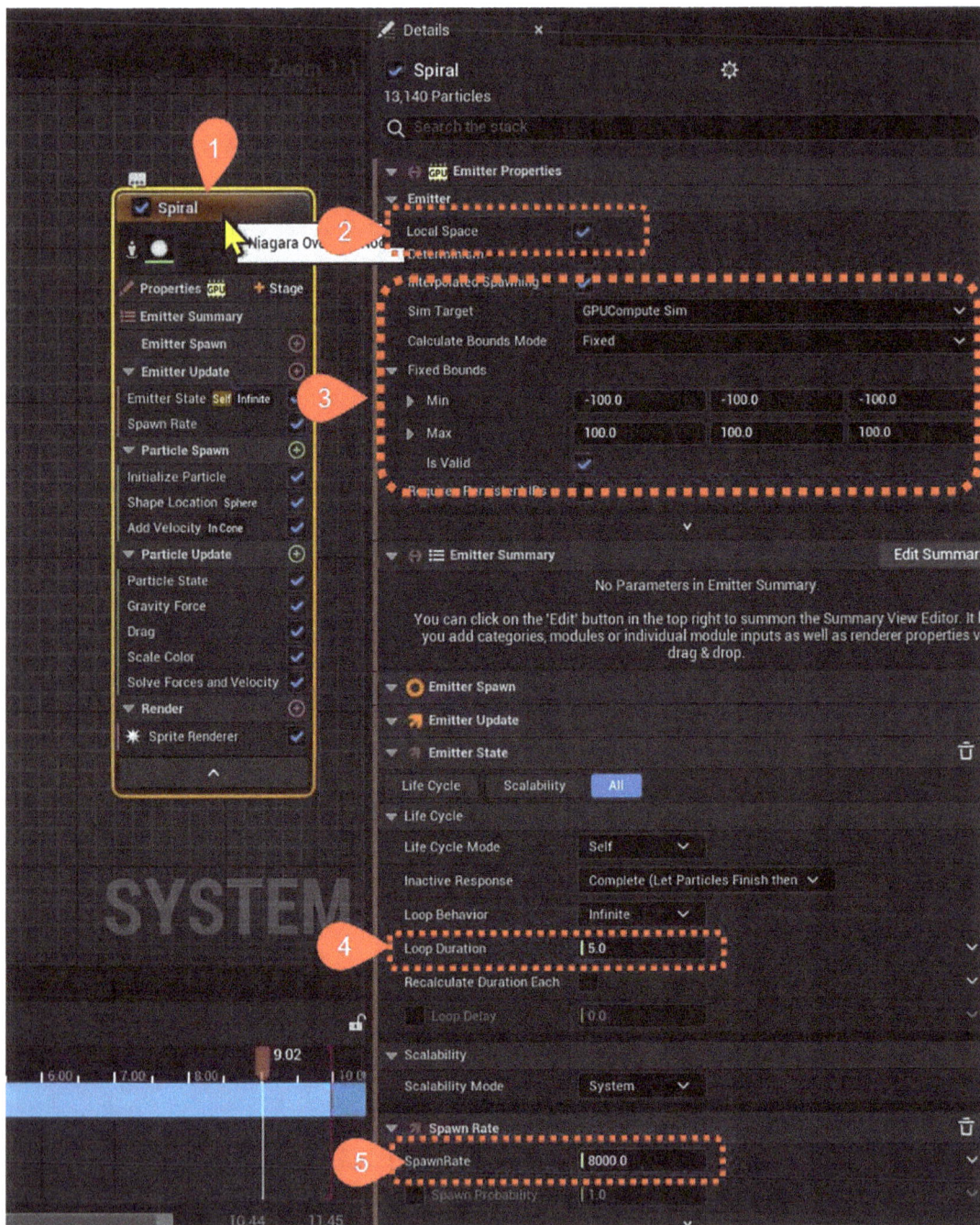

Figure 7-55. *Initial settings for this Niagara system, named Spiral*

Figure 7-56 continues to show the Niagara editor, focusing on the "Initialize Particle" section:

- Step 1: This step is crucial for defining the initial properties of particles when they are first created. In the 'Initialize Particle' module section, we set up the initial properties for each particle as it is created.

- Step 2: Lifetime configuration

  - "Lifetime Mode": Set to "Random" to give each particle a random lifetime within a specified range.

  - "Lifetime Min": Set to 1.0 seconds as the minimum amount of time a particle will exist.

  - "Lifetime Max": Set to 1.5 seconds as the maximum amount of time a particle will exist before it disappears.

- Step 3: Sprite size configuration

  - "Sprite Size Mode": Set to "Random Uniform" to ensure each particle's size is randomly determined but maintains the same aspect ratio.

  - "Uniform Sprite Size Min": Set to 1.0 to specify the minimum size of the particles.

  - "Uniform Sprite Size Max": Set to 5.0 to specify the maximum size of the particles.

- Step 4: Mesh scale configuration

  - "Mesh Scale Mode": Set to "Random Uniform" which means the scale of each instance will be uniformly scaled by a random value within a defined range.

  - "Mesh Uniform Scale Min": Set to 1.0 as the minimum scale multiplier for the particles.

  - "Mesh Uniform Scale Max": Set to 2.0 as the maximum scale multiplier for the particles.

Chapter 7   Harnessing the Power of Niagara: Practical Examples in Unreal Engine 5

These settings collectively contribute to the randomness and variety of the particles, creating a more dynamic and visually interesting effect. The minimum and maximum values define the ranges within which each particle's individual properties, such as lifetime and size, can vary. Adjusting these settings would affect the visual characteristics of the Spiral_ Effect, such as how large the particles appear and how long they remain visible on screen.

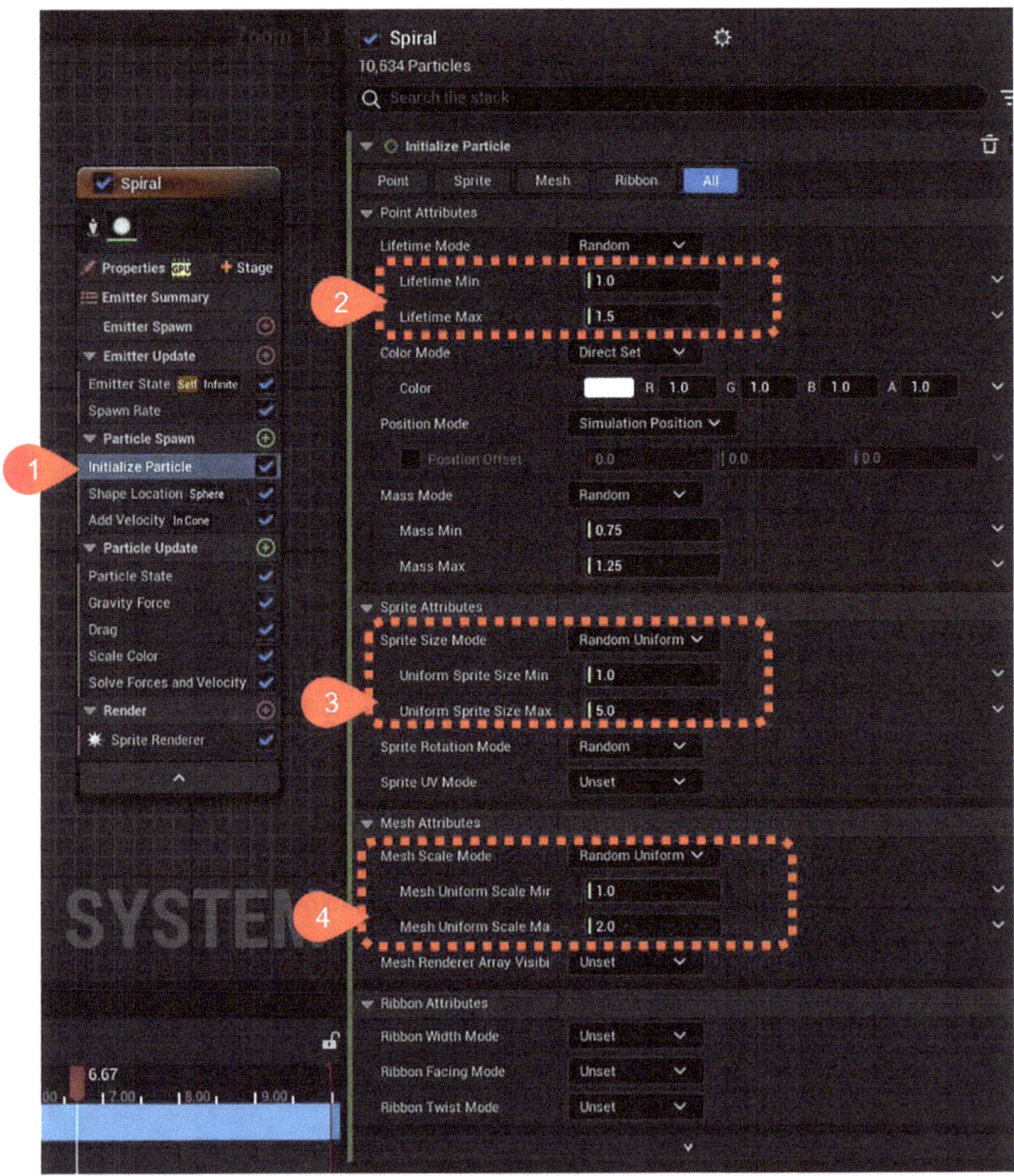

Figure 7-56. Set the initialize value for each of the particles

Rotate Around Point

In the intricate ballet of particles that enlivens a digital scene, the ability to pivot and swirl around a central axis adds a dimension of mesmerizing realism. The "Rotate Around Point" feature within Niagara's comprehensive toolkit enables precisely this effect. It's a feature that empowers creators to anchor their particles to an invisible fulcrum, allowing them to trace circular paths that can mimic anything from the majestic twirl of a galaxy to the intimate dance of embers caught in a draft. As particles adopt a rotational dance around a point, they bring forth a dynamic element to simulations, instilling a sense of order and rhythm within the visual chaos. This module is not just about rotation; it's about adding layers of movement that elevate a simple effect into a spectacle.

Figure 7-57 shows the configuration for setting up the particles to rotate around one point.

- Step 1 illustrates the addition of the "Rotate Around Point" module. This module is not part of the default particle setup, so it needs to be manually added.

  - Including this module is essential for any effect where particles need to exhibit rotational behavior around a point, adding a sense of orbiting motion that can be visually compelling and central to effects like tornadoes or galaxy simulations.

- Step 2 points to the "Rotate Around Point" module where we specify the characteristics of the rotation.

  - Adjusting settings here allows you to dictate how particles will move around a central axis, which is fundamental for creating effects such as spiraling smoke or twirling debris. This module gives life to the particles, transforming them from static to dynamic elements within the scene.

- Step 3 is tied to the "Rotation Phase" parameter, which is configured to sync with the engine's time, using a multiplier of 0.3.

  - This linkage means that as time progresses in the game, the phase of the particle rotation changes accordingly, introducing a time-based dynamic element to the effect. This ensures that the spiral movement evolves during gameplay, contributing to the realism and keeping the visuals interesting and less repetitive.

CHAPTER 7 HARNESSING THE POWER OF NIAGARA: PRACTICAL EXAMPLES IN UNREAL ENGINE 5

- Step 4 involves setting the "Radius" to 100, which is the distance from the central point to the path of the particles.
  - This value is critical as it defines the scope of the spiral effect, influencing the spatial extent and overall size of the pattern. A larger radius would result in a broader, more expansive spiral, whereas a smaller value would keep the spiral tight and confined.

Together, these settings are used to create a spiraling motion for particles, which can simulate effects like magical auras, energy fields, or other visually dynamic phenomena in a game environment. The relationship between the rotation phase and the engine time suggests an ongoing, possibly perpetual, movement that could add a layer of complexity and realism to the visual effect.

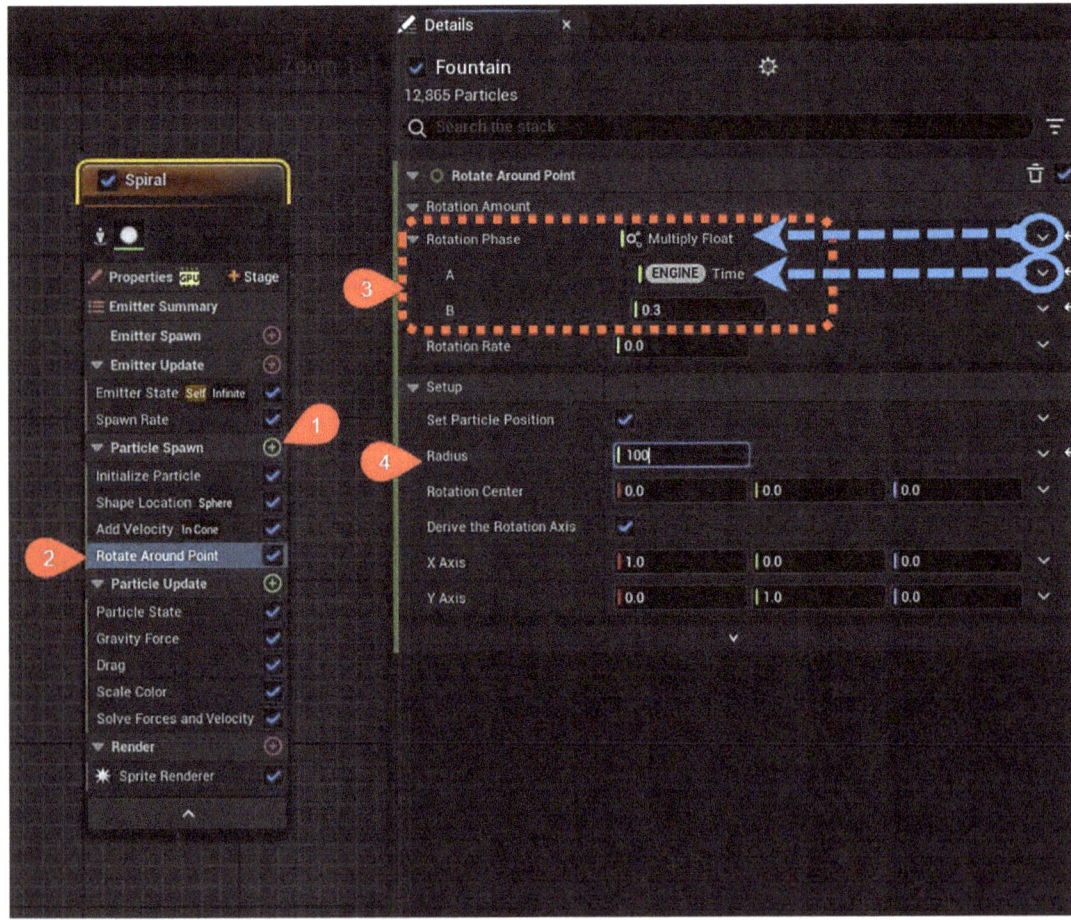

Figure 7-57. Rotate Around Point

CHAPTER 7   HARNESSING THE POWER OF NIAGARA: PRACTICAL EXAMPLES IN UNREAL ENGINE 5

Disable the Shape Location

In Figure 7-58, we see the interface of the Niagara visual effects system within UE5, and here we're focusing on managing the "Shape Location" module. Instead of removing it, we are disabling it, which means that the module remains in the system and can be re-enabled if needed. Disabling, rather than deleting, allows for nondestructive editing – we can test the effects of not using this module without permanently losing its settings.

Disabling the Shape Location is particularly useful in scenarios where the "Shape Location" might conflict with other modules like "Rotate Around Point" or when the intent is for particles to emerge from a singular point rather than a shape-defined area. We can ensure that the initial emission point of the particles does not adhere to a shape, granting us the flexibility to define the emission behavior through other means.

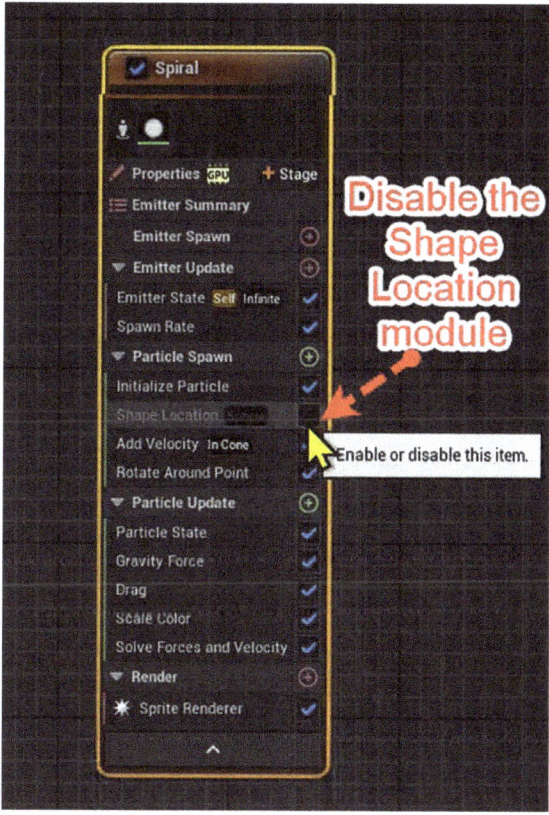

Figure 7-58. *Disable the Shape Location module*

CHAPTER 7 HARNESSING THE POWER OF NIAGARA: PRACTICAL EXAMPLES IN UNREAL ENGINE 5

Add Velocity

Applying velocity to particles is a fundamental aspect of creating dynamic and realistic effects. In the context of a spiraling effect, adding linear velocity could be used to make the particles move outward or upward from the center of the spiral, giving the impression of energy being released or particles being propelled by a central force. Adjusting the velocity parameters can significantly impact the visual style and behavior of the particle effect, contributing to the overall feel and function of the effect within the game.

Figure 7-59 illustrates the "Add Velocity" module within the particle system stack.

- As shown in Step 1, we select this module, which is responsible for adding movement to the particles after they are spawned. In this case, it appears the velocity is set to be linear, which would give the particles a consistent motion in a specified direction.

- Step 2 shows the detailed settings for the "Add Velocity" module. Here, the velocity vector is set to (0, 0, 50), which means that the particles will be moving linearly along the Z-axis if we assume a standard right-handed 3D coordinate system. The "Velocity Speed Scale" is set to 50, which likely multiplies the base velocity to give the final speed at which the particles will move.

CHAPTER 7   HARNESSING THE POWER OF NIAGARA: PRACTICAL EXAMPLES IN UNREAL ENGINE 5

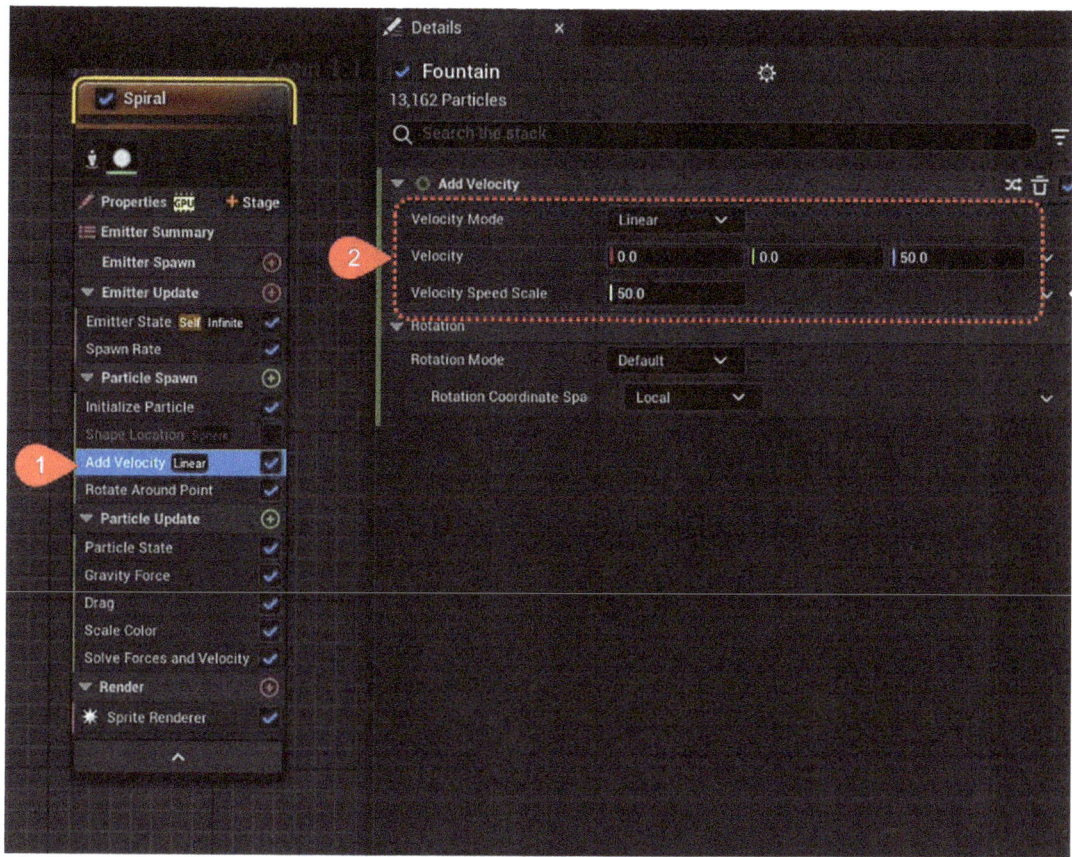

Figure 7-59. *Add velocity*

CHAPTER 7   HARNESSING THE POWER OF NIAGARA: PRACTICAL EXAMPLES IN UNREAL ENGINE 5

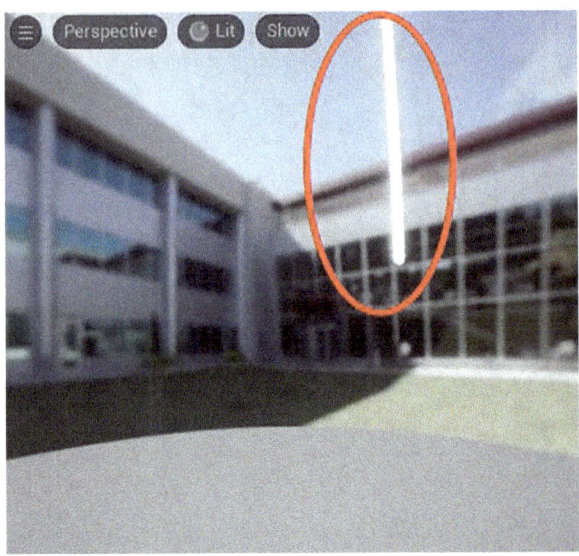

Figure 7-60. *The visual of the result so far*

We can see the result so far, illustrated in Figure 7-60 where we can see the effects as line shape spinning around the center point, as Rotate Around Point in previous section.

Scale Sprite Size

In Figure 7-61, the "Scale Sprite Size" module within a "Spiral" particle system is selected. This module is used to define the properties and behaviors of particle effects in a 3D environment. The purpose of the "Scale Sprite Size" module is to add visual variety and realism to the particle effect. By scaling particles over their lifetime, the effect can more closely mimic real-world behaviors, adding to the immersion and aesthetic quality of the scene. For a spiral effect, such scaling might simulate a magical energy that intensifies and fades, contributing to the visual storytelling within the game.

- Step 1 shows that we need to add a new module named "Scale Sprite Size" module.
- Select the "Scale Sprite Size" module, shown by Step 2, to change the size of the individual sprites (or particles) over time or based on other factors.

- The line curve, indicated by data in Steps 3 and 4, visually represents the size scaling over time.

  - In Step 3, we're setting the key data for the "Scale Sprite Size" module. The curve begins with key data of 0.0 and 1.0. This defines a scaling behavior where the particle grows from a scale of 0 to its full size throughout its lifetime.

  - For Step 4, the second key data point on the size scale curve is set to decrease back to 1.0 and 0.0, which means that as the particles approach the end of their lifespan, they scale down to disappear. This value indicates that the particles will shrink away completely before they expire.

  - These points allow the user to define how the size of the sprite changes over the lifetime of the particles. By adjusting these points, the user can create complex size variation patterns. While the process of changing the key data point value is a little tricky, we can single-click the little point itself to activate the Key Data dialog box.

This curve suggests that the sprite size will start small, increase to a larger size, and then reduce back to small before the particle expires. This kind of scaling is commonly used to simulate natural phenomena like explosions, where particles start small, expand quickly, and then dissipate.

- The "Uniform Curve Scale" parameter, shown in Step 5 with the Uniform Curve Scale value of 0.5, modifies the overall scale of the curve effect, allowing the user to fine-tune the intensity of the size change without altering the shape of the curve itself.

CHAPTER 7 HARNESSING THE POWER OF NIAGARA: PRACTICAL EXAMPLES IN UNREAL ENGINE 5

Figure 7-61. Scale Sprite Size

Gravity Force

In the intricate dance of particles within a visual effect, gravity plays a pivotal role, often grounding the animation in the familiar behavior of the physical world. The purpose of configuring the gravity vector in this manner might be to create an effect where particles are pushed upward, opposing the natural downward force of gravity. This setup of the gravity values could simulate effects such as anti-gravity, upward-moving energy streams, or objects that naturally rise, such as bubbles in water or embers in a fire. Figure 7-62 highlights the "Gravity Force" module with the settings if the particles should be affected by the gravity.

CHAPTER 7   HARNESSING THE POWER OF NIAGARA: PRACTICAL EXAMPLES IN UNREAL ENGINE 5

- As shown in Step 1, we select this module to activate the properties of the "Gravity Force" module.

- As shown in Step 2, the vector field for gravity is set to (0.0, 0.0, 50.0), indicating that the force of gravity is being applied in the positive Z-axis direction. This would mean the force is acting upward, which is unconventional as gravity usually pulls downward.

The strength and direction of this force are crucial for achieving the desired movement of the particles within the "Spiral" effect, contributing to the overall realism and dynamic feel of the scene.

Figure 7-62. Gravity force

CHAPTER 7   HARNESSING THE POWER OF NIAGARA: PRACTICAL EXAMPLES IN UNREAL ENGINE 5

Drag

In any dynamic environment, the interplay of forces brings animation to life, and in the realm of particle simulation, the "Drag" module plays a crucial role. Here, "Drag" isn't just about slowing things down; it's about infusing movement with realism. Figure 7-63 indicates the use of the module named Drag to configure to apply a drag force to the particles, which simulates resistance that slows down their motion over time, akin to air or fluid resistance in the physical world.

- As indicated by Step 1, we activate the properties of the Drag module.

- The details panel on the right, marked by Step 2, shows the properties of the "Drag" module. The drag value is set to 0.5 to create a realistic particle movement. By adjusting it, the particles can have a more natural deceleration, making them appear as if they are moving through a medium like air or water.

The reason to configure drag in a particle system is to simulate the natural slowing down of objects as they move through a substance, which adds to the realism of the effect. For a spiral effect, particularly one that may represent energy or a magical force, drag could be used to make the particles appear to dissipate energy as they move away from the source, creating a more believable and visually appealing effect.

CHAPTER 7　HARNESSING THE POWER OF NIAGARA: PRACTICAL EXAMPLES IN UNREAL ENGINE 5

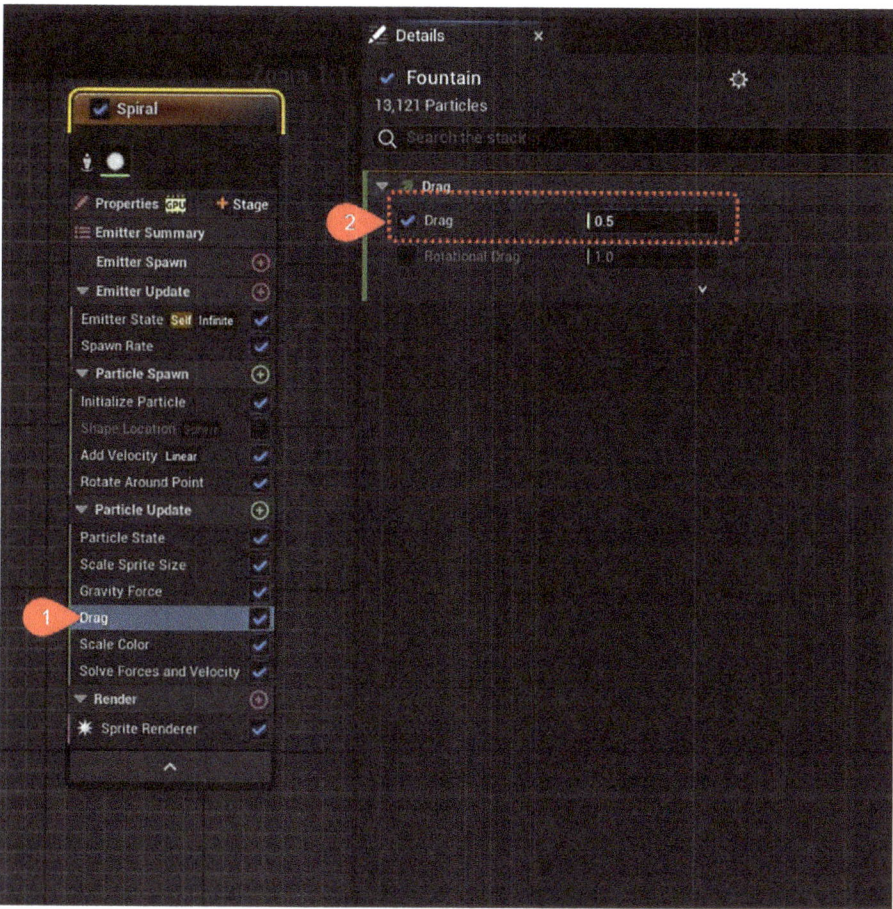

Figure 7-63. *Drag*

Scale Color

In the vibrant tableau of visual effects, color commands attention, shaping the narrative of each particle's journey from inception to dissipation. Figure 7-64 shows a detailed view of the "Scale Color" module that allows the color properties of the particles to be adjusted over time or based on certain conditions. This module is crucial for creating visually dynamic effects where particle colors can change during their lifespan.

- As shown in Step 1, we activate the configuration of the Scale Color module.

CHAPTER 7   HARNESSING THE POWER OF NIAGARA: PRACTICAL EXAMPLES IN UNREAL ENGINE 5

- Step 2 demonstrates how the color is being modified. It appears that the color is being determined by a gradient, as indicated by the "Color from Curve" and "Curves" options. This provides a way to smoothly transition between colors over the particle's lifetime or based on other factors.

- The curves, pointed out by Step 3, represent the change in color and opacity over time. One curve appears to control the alpha (transparency) of the particles, showing that particles will fade out over time. The key data points allow for precise control over how these properties change.

Clicking on top of the gradient key point can activate the color picker dialogs shown, which are used to select the colors that will be applied to the particles at different points in their life cycle.

Right-clicking the bottom opacity key point can activate the value dialog box to set the opacity value.

The purpose of these color adjustments in the "Spiral" particle system is to add visual complexity and emotional impact to the effect. Color transitions can suggest changes in energy, temperature, or other narrative elements in the game, enhancing the player's immersion and the overall aesthetic of the effect.

CHAPTER 7   HARNESSING THE POWER OF NIAGARA: PRACTICAL EXAMPLES IN UNREAL ENGINE 5

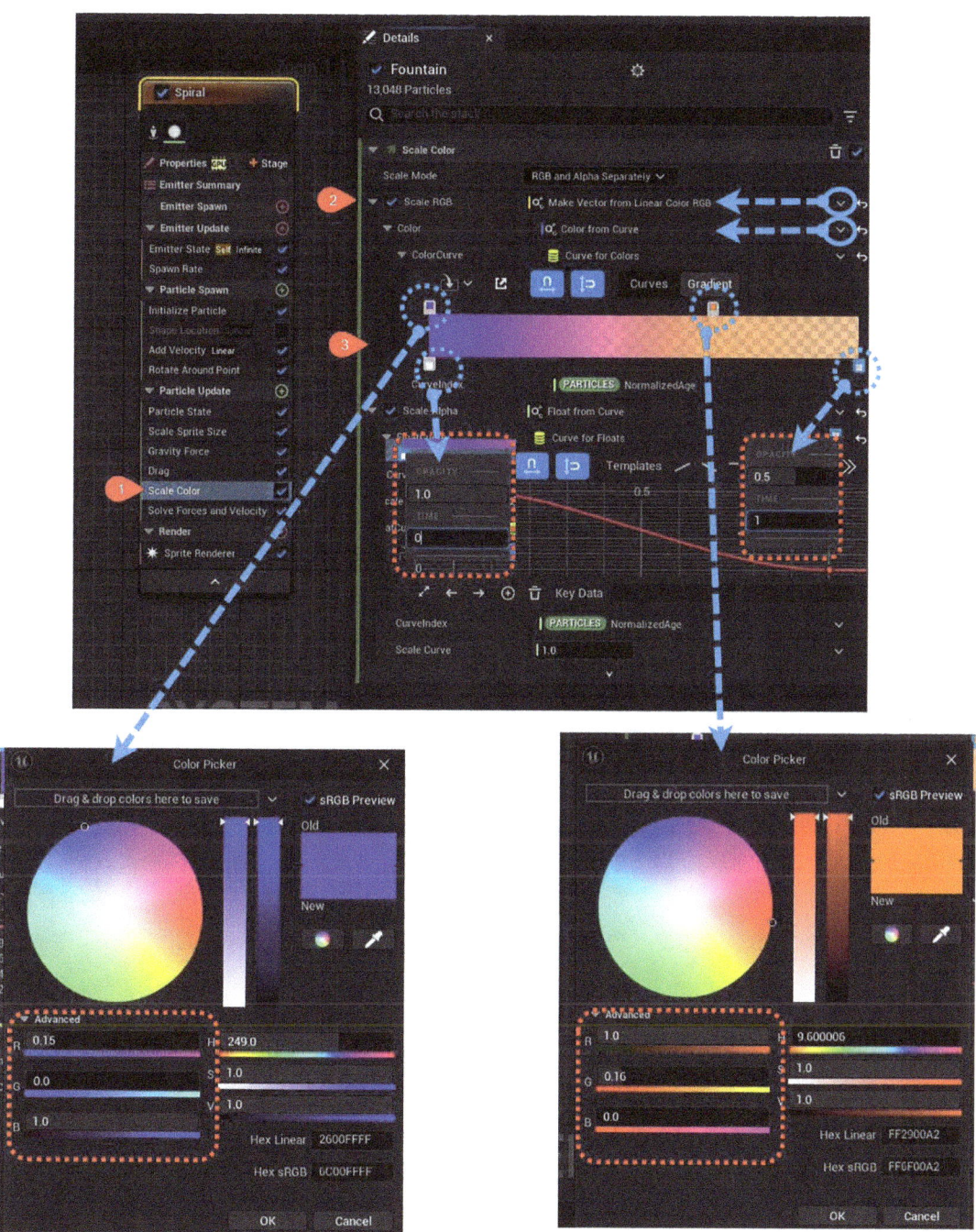

Figure 7-64. *Scale Color*

Sprite Renderer

Through the "Sprite Renderer," (see Figure 7-65), we dictate how each particle is visually manifested and perceive it in relation to the camera and the world we've crafted.

Unlike the previous example, we are not using any specific customized material assets, as we stick with the *DefaultSpriteMaterial*.

- We select the Sprite Renderer module as shown in Step 1 to activate the properties of the module.

- The central panel, shown in Step 2, indicates the "Source Mode" of the sprites. It is set to "Particles," meaning that the rendering of the sprites is based on the individual particles generated within the system.

- In Step 3, the "Facing Mode" is set to "Face Camera." This setting ensures that no matter where the camera moves within the 3D space, the sprites will always face toward it, maintaining their visibility to the viewer.

The "Sprite Renderer" is a crucial component for visual effects in game development, as it dictates how the individual particles will be visualized. The settings shown here would be typical for a particle system that is intended to be seen from all angles, ensuring that the sprites are always oriented for optimal viewing, contributing to the immersive experience in the game environment.

CHAPTER 7   HARNESSING THE POWER OF NIAGARA: PRACTICAL EXAMPLES IN UNREAL ENGINE 5

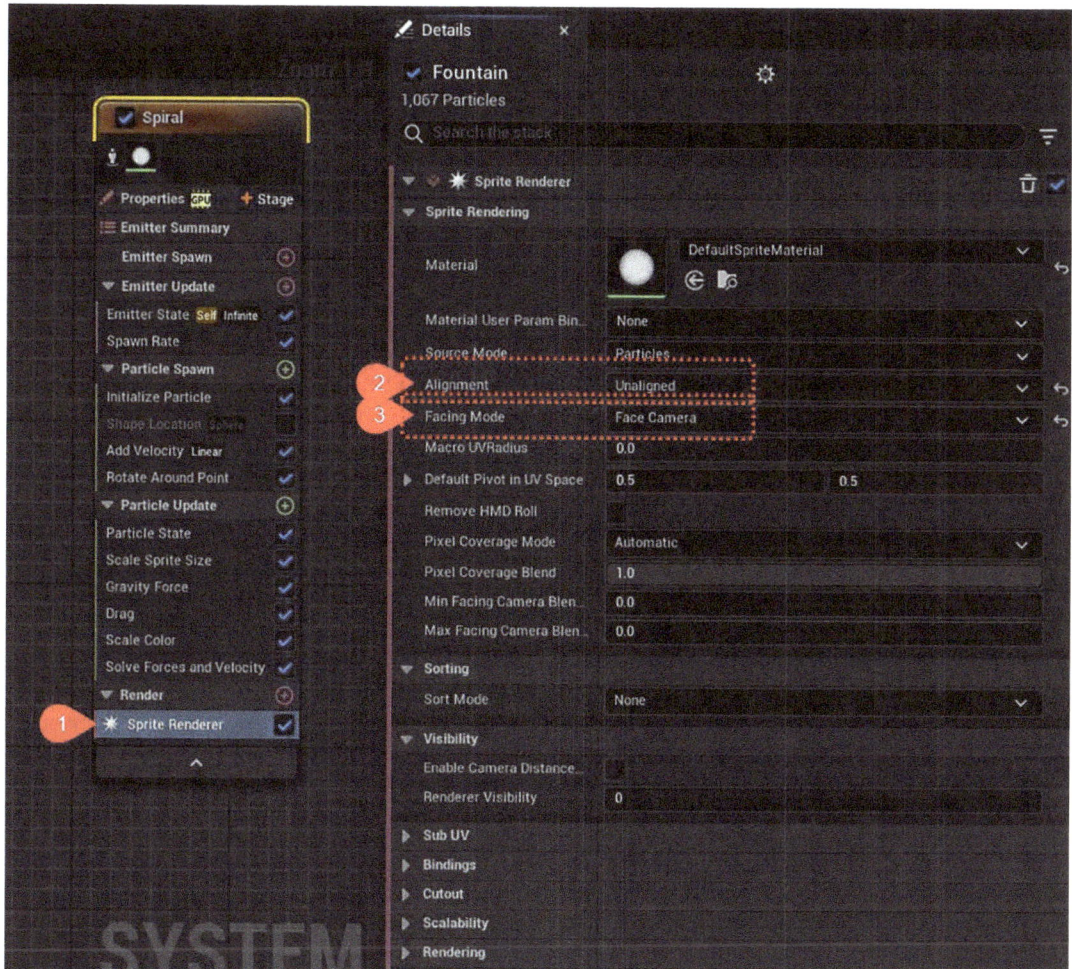

Figure 7-65. Sprite Renderer

Spiral in Action

Bring In the Magical Ball

It's time now for us to set up our Magical Ball with our built spiral effects, (see Figure 7-66), where

- We place the Magical Ball asset as indicated in Step 1.

- The Location is set to X = –68,860, Y = –40,470, Z = –69,400 as shown in Step 2.

- The scale value is set to 0.03125 as shown in Step 3.

353

CHAPTER 7 HARNESSING THE POWER OF NIAGARA: PRACTICAL EXAMPLES IN UNREAL ENGINE 5

Please note that the author has chosen a location close to the TPP character. However, you are welcome to place the Magical Ball at any location you deem appropriate. As we will have the character pick up this Magical Ball at a later stage, the author decided to scale down the asset to fit the character's hand, using the given scale value.

Additionally, if there's any chance that your Magical Ball does not have the material being set up from the previous step, we can apply the available material from the library named "BrushedMetal," which corresponds to the object's shiny appearance in the 3D view.

- As indicated in Step 4, the "Mobility" status is set to "Movable," indicating that it can be animated or manipulated during gameplay, which would be important for an object with an active role in the game, such as a magical artifact that players can interact with at a later stage.

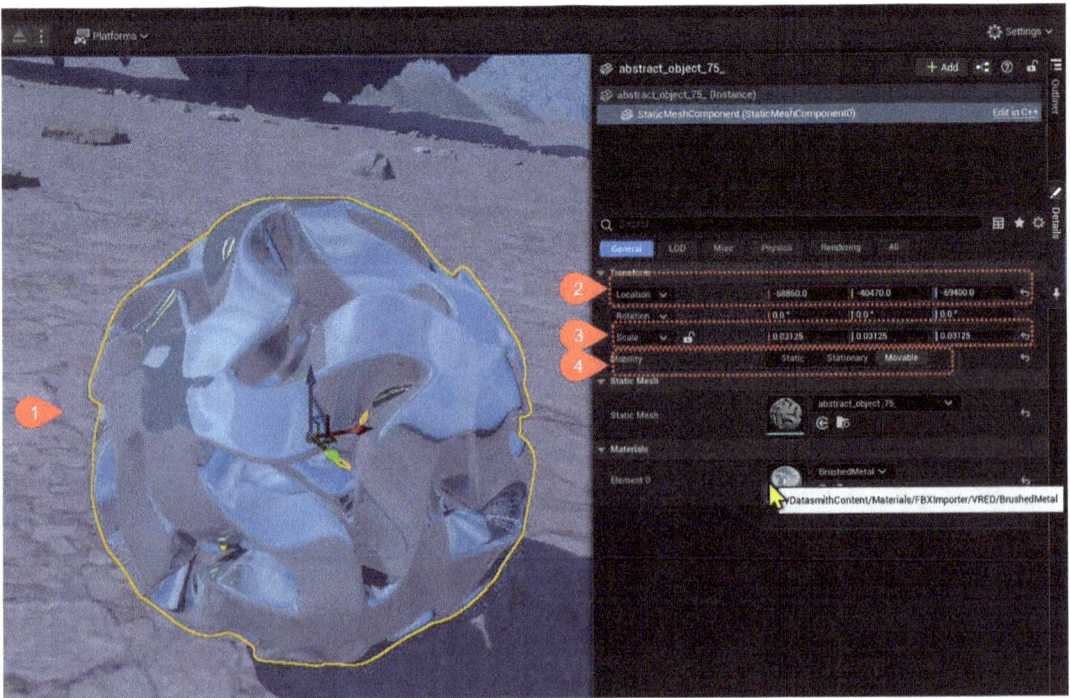

Figure 7-66. Placement of the Magical Ball

CHAPTER 7   HARNESSING THE POWER OF NIAGARA: PRACTICAL EXAMPLES IN UNREAL ENGINE 5

Bring In the Niagara System Particle Effect

The Niagara system particle effect can be positioned similarly to placing an Actor: by single-clicking to select the asset (as shown in Figure 7-67) and right-clicking to select "Place Actor" in the level itself (as shown in Figure 7-68).

However, at this stage, the particle effect operates independently from the Magical Ball, meaning that if we move the Magical Ball, we must also move the particle system separately to align it with the new location of the Magical Ball. This process can be quite inconvenient, particularly if we plan to associate a group of three spiral effects with this Magical Ball. To streamline this process, we can incorporate the spiral effect as one of the components of the Magical Ball itself.

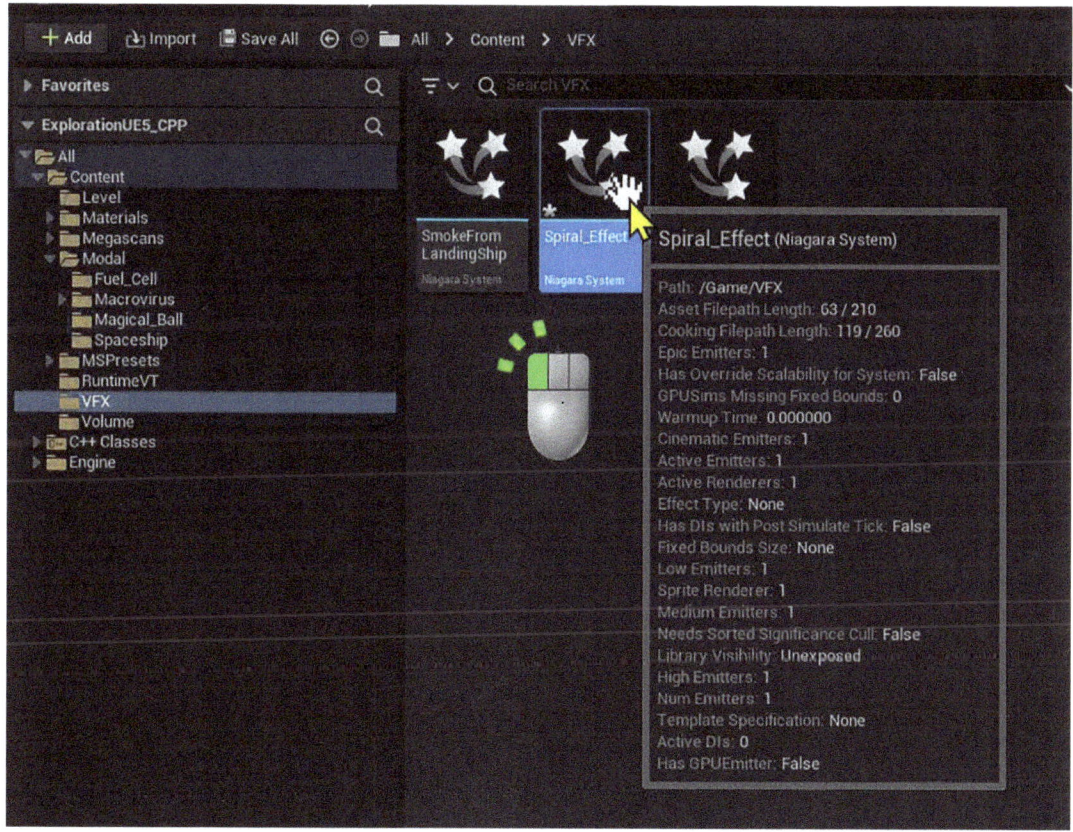

Figure 7-67. Select the Spiral_Effect asset

CHAPTER 7   HARNESSING THE POWER OF NIAGARA: PRACTICAL EXAMPLES IN UNREAL ENGINE 5

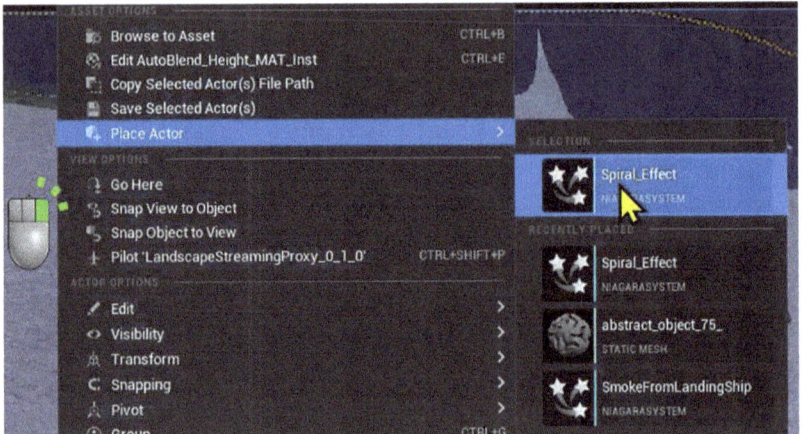

Figure 7-68. *Place the Spiral_Effect asset*

Attaching Niagara Particles

The sequence for attaching Niagara particles to the Magical Ball resembles the copy-paste concept. As demonstrated in Figure 7-69

- We can right-click the spiral effects in the level (Step 1).
- Access the Edit menu (Step 2).
- Select the Cut option (Step 3) to store the information on the clipboard.

CHAPTER 7   HARNESSING THE POWER OF NIAGARA: PRACTICAL EXAMPLES IN UNREAL ENGINE 5

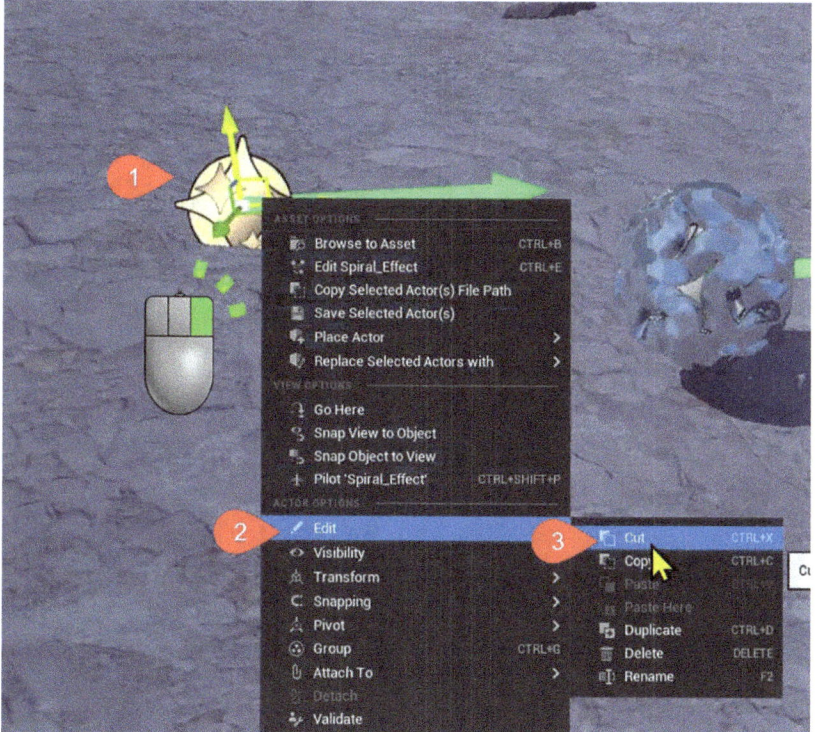

Figure 7-69. *Edit to cut this asset into the clipboard*

Figure 7-70 shows that we will paste the spiral effect to the Magical Ball by following these steps:

- As shown in Step 1, we are to select the Magical Ball.
- Right-click the *StaticMeshComponent* (Step 2).
- Choose the Paste option (Step 3) to integrate the spiral effect into the components of the Magical Ball.
  - Since the pasted spiral effect is quite large, we resize the spiral component, as shown in Figure 7-71, using 0.5 as the uniform value for the scale in Scale-X, Scale-Y, and Scale-Z.

CHAPTER 7　HARNESSING THE POWER OF NIAGARA: PRACTICAL EXAMPLES IN UNREAL ENGINE 5

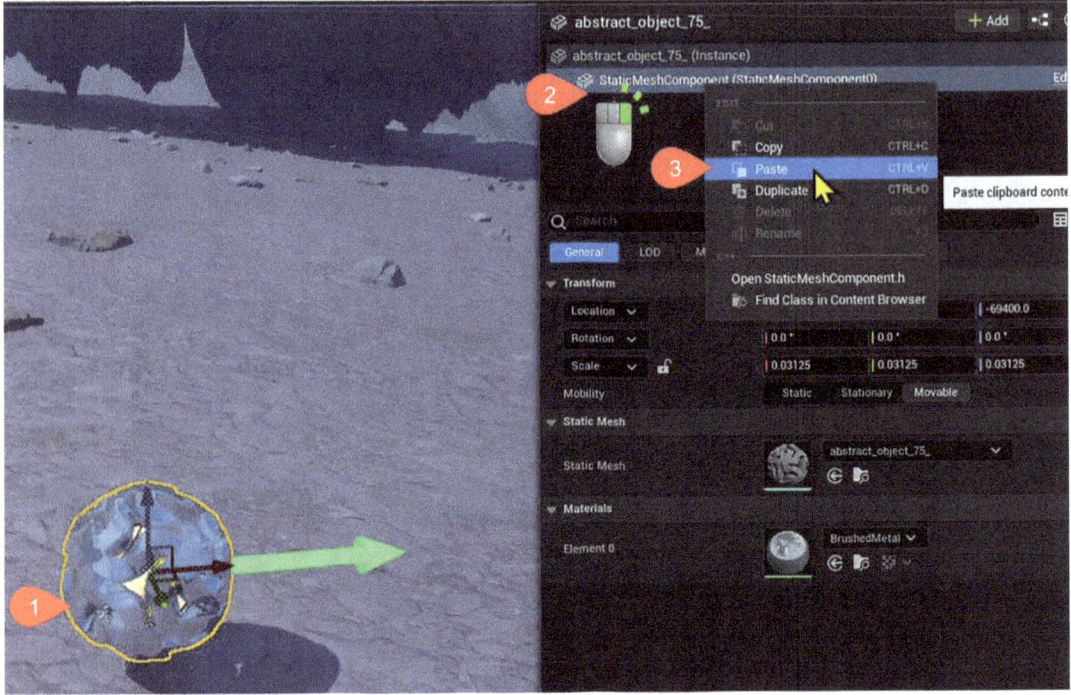

Figure 7-70. *Paste the asset as a component under the Magical Ball static mesh*

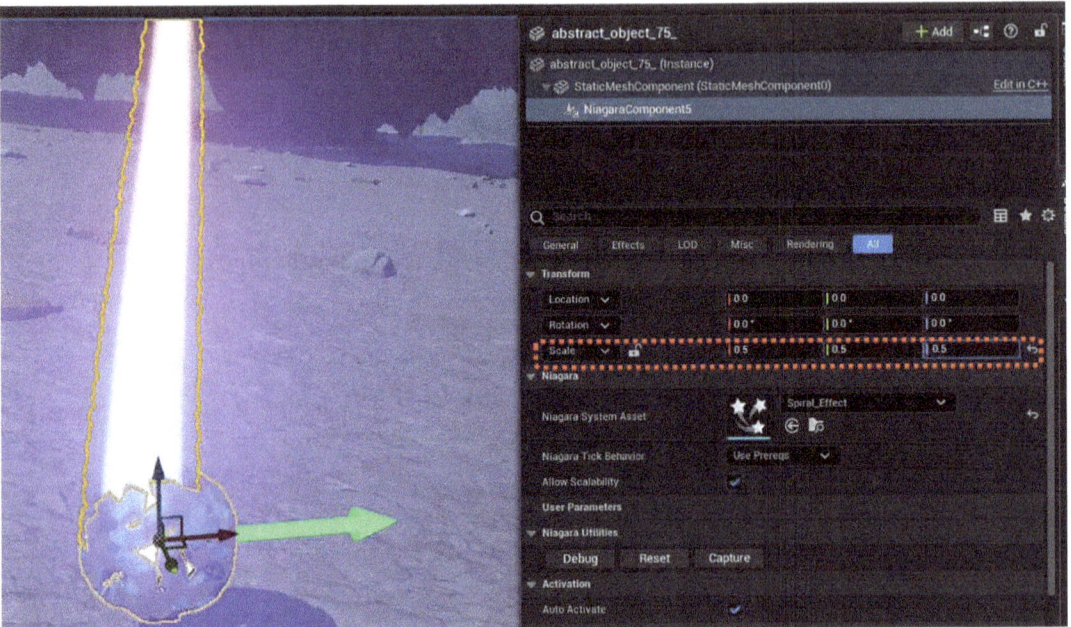

Figure 7-71. *Scale down the Spiral_Effect*

CHAPTER 7 HARNESSING THE POWER OF NIAGARA: PRACTICAL EXAMPLES IN UNREAL ENGINE 5

To add two more spirals, the simplest method is to use the Duplicate option, as shown in Figure 7-72, which creates additional scaled-down spiral effects.

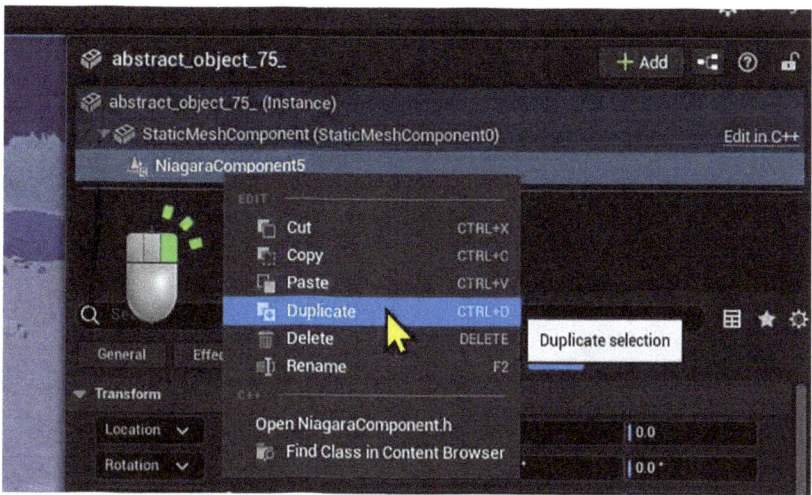

Figure 7-72. *Duplicate the Spiral_Effect component*

To position each spiral to start at different points, we adjust the Rotation-Z value, as shown in Figure 7-73 for the second spiral component and Figure 7-74 for the third.

CHAPTER 7 HARNESSING THE POWER OF NIAGARA: PRACTICAL EXAMPLES IN UNREAL ENGINE 5

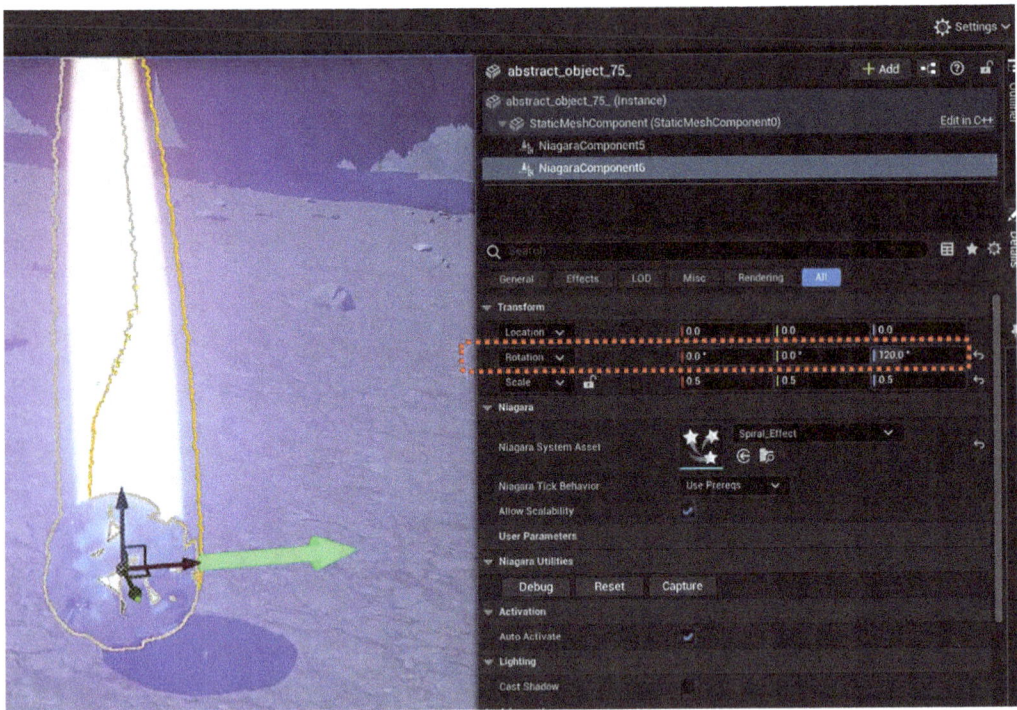

Figure 7-73. *Set the rotation for the second Spiral_Effect component*

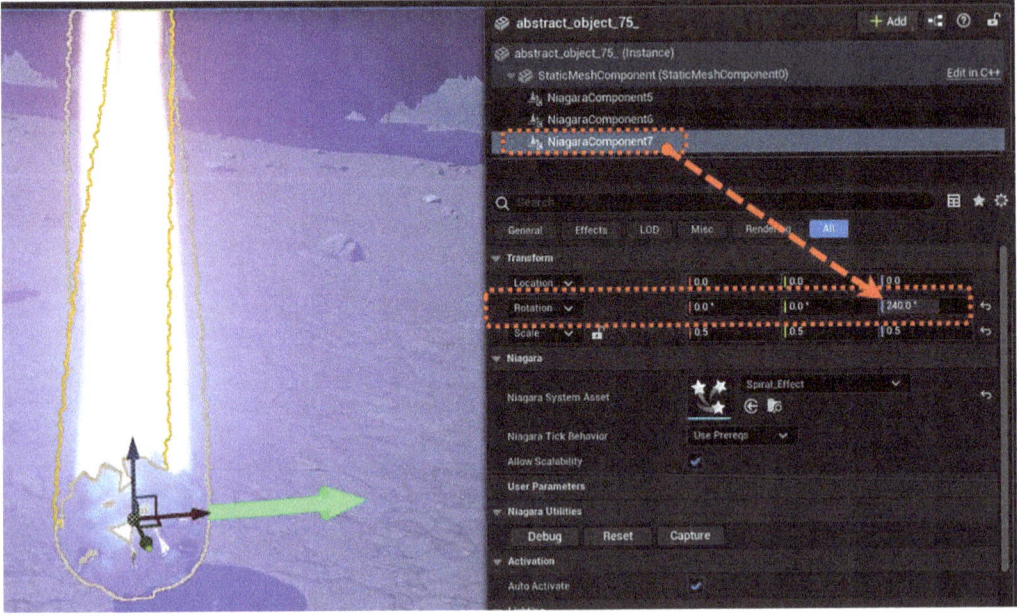

Figure 7-74. *Set the rotation for the third Spiral_Effect component (the numbering of the NiagaraComponent is irrelevant)*

CHAPTER 7   HARNESSING THE POWER OF NIAGARA: PRACTICAL EXAMPLES IN UNREAL ENGINE 5

Finally, the completed assembly of the Magical Ball with the three spiral effects rotating around its center point is visible in Figure 7-75.

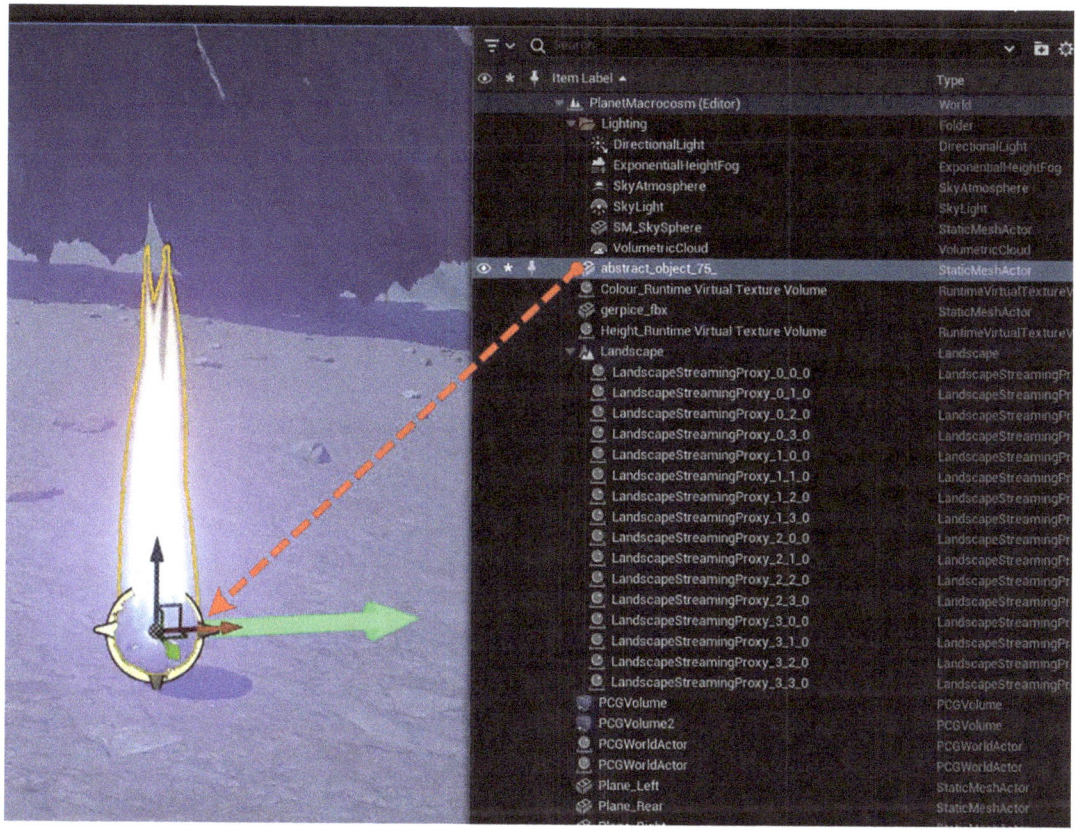

Figure 7-75. *Final result of this Magical Ball with three Spiral_Effect components*

Harnessing the Power of Niagara #3: Dust Storm

The forthcoming Niagara system we plan to develop will simulate environmental effects such as dust storms. This system will comprise two emitters that create a dust storm incorporating elements like dust rock and smoke storm. Figure 7-76 illustrates the various components that are combined to produce a realistic dust storm effect.

CHAPTER 7 HARNESSING THE POWER OF NIAGARA: PRACTICAL EXAMPLES IN UNREAL ENGINE 5

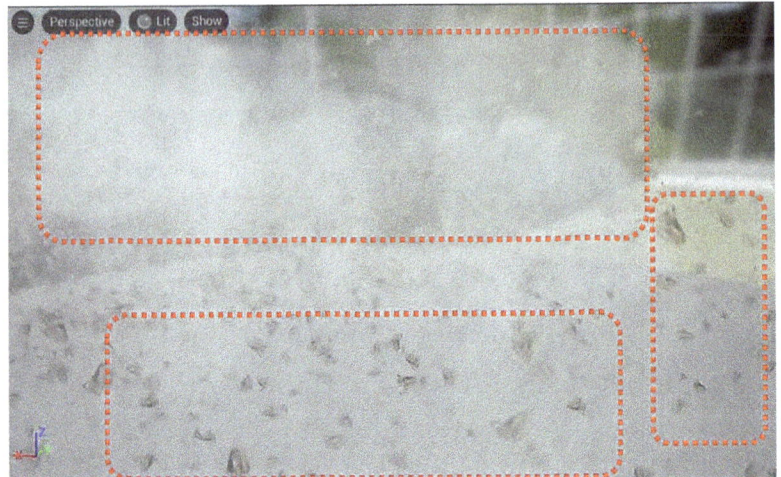

Figure 7-76. The combo emitter: dust storm and smoke storm

Set Up Materials for the Dust-Rock Emitter

Similar to Example #1, we need to import a series of rock textures to set up our dust-rock emitters. These rock textures can be downloaded from a texture atlas, which is freely available at the following link, as shown in Figure 7-77, and imported into our project under the Texture folder as shown in Figure 7-78.

www.pngaaa.com/detail/3260971

Figure 7-77. Download the 2D rock texture from a set of texture atlas

CHAPTER 7   HARNESSING THE POWER OF NIAGARA: PRACTICAL EXAMPLES IN UNREAL ENGINE 5

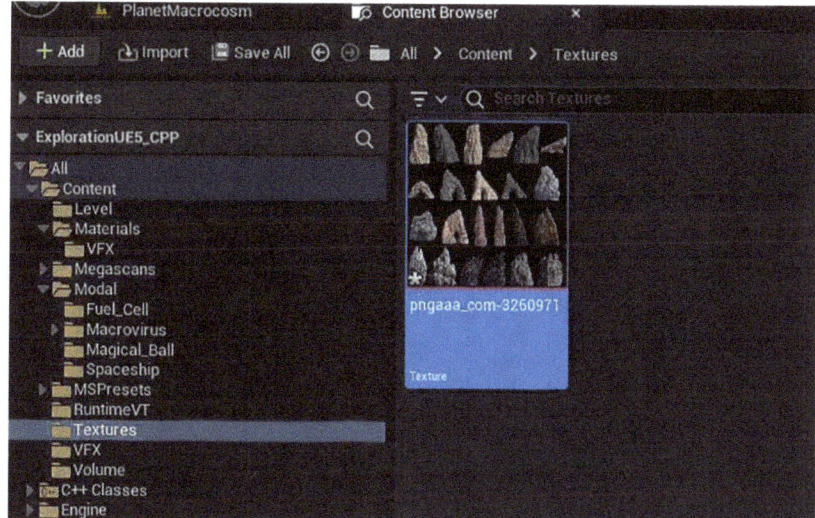

Figure 7-78. *Import the texture png file into a newly created folder named Textures*

To set up the material for this dust-rock emitter, we set up a material named Dust_Storm_Rock inside the VFX folder under the Materials folder, as shown in steps in Figure 7-79.

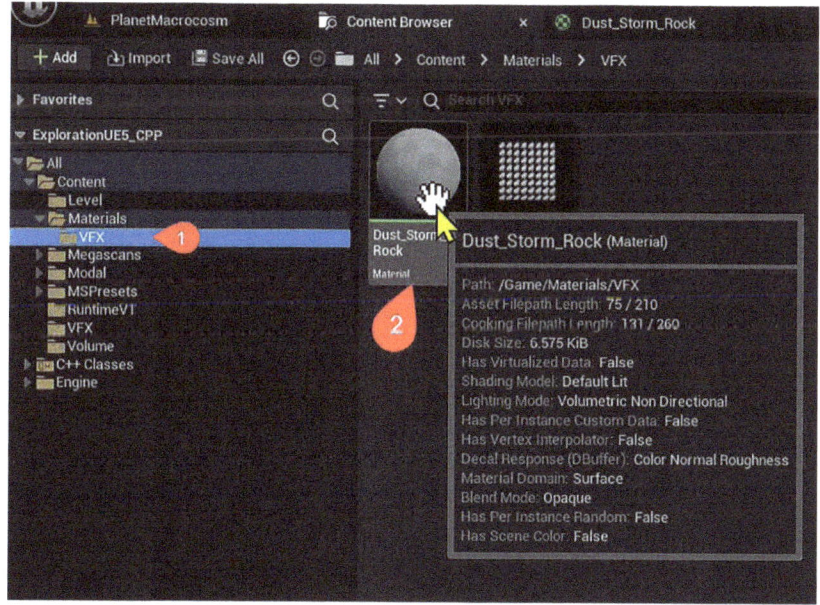

Figure 7-79. *Create a material named Dust_Storm_Rock in the VFX folder under the Materials folder*

CHAPTER 7   HARNESSING THE POWER OF NIAGARA: PRACTICAL EXAMPLES IN UNREAL ENGINE 5

As shown in Example #1, the configuration for this material is illustrated in Figure 7-80. We are now setting the texture to the downloaded rock textures. The purpose of these operations is to create a visually complex and dynamic material for particles that will be used to simulate a dust storm, with varying colors and textures that give the illusion of swirling dust and rock debris.

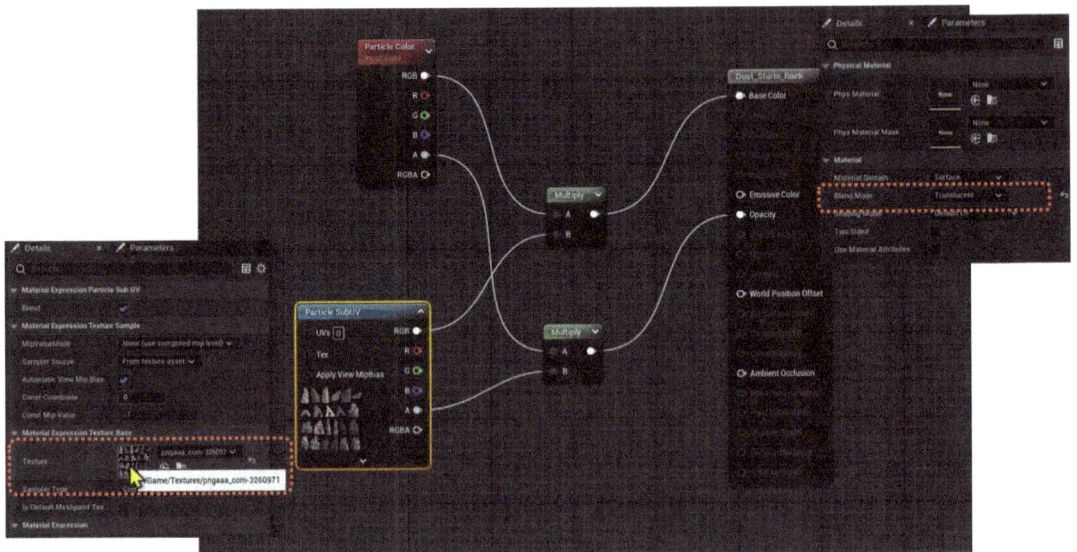

Figure 7-80. *The setup for the particle materials*

Set Up the Dust-Storm Niagara System

Again, like previous examples, we are setting up a new Niagara system based on the Fountain template (named Dust_Storm as shown in Figure 7-81).

Chapter 7 Harnessing the Power of Niagara: Practical Examples in Unreal Engine 5

Figure 7-81. *Set up the Dust_Storm Niagara system based on the Fountain template*

In Figure 7-82, we can see the basic modification of this emitter, such as

- In Step 1, we rename the emitter as "Dust_Rock" emitter within the Niagara system.

- In Step 2, we change the simulation target to "GPU Compute Sim" and define fixed bounds for the particle system.

This time, we are setting a significantly larger value for the bounds to encompass the entire level, as it is necessary for the particles to cover the entire area, with the minimum bounding values of –10,000 and maximum bounding values of 10,000.

CHAPTER 7   HARNESSING THE POWER OF NIAGARA: PRACTICAL EXAMPLES IN UNREAL ENGINE 5

- In Step 3, the emitter's life cycle is managed under the "Life Cycle" section, with options for setting both the life cycle mode and scalability mode to "System," instead of our previous "Self" settings.

Setting the Life Cycle to "System" indicates that the emitter is managed collectively at the system level rather than on an individual particle basis or at the component level. The rationale behind this approach is the presence of a large number of particle systems that require oversight. Therefore, it is more efficient and consistent to have the system collectively manage the particles.

- Furthermore, Step 4 shows the "Spawn Rate" parameter is set to 25,000, which is considerably higher than in our previous examples. This increase is necessary because this emitter is designed to release a larger number of particles to cover the entire level.

These settings are crucial for creating realistic particle simulations, such as rocks in a dust storm, allowing for precise control over their behavior and appearance.

CHAPTER 7   HARNESSING THE POWER OF NIAGARA: PRACTICAL EXAMPLES IN UNREAL ENGINE 5

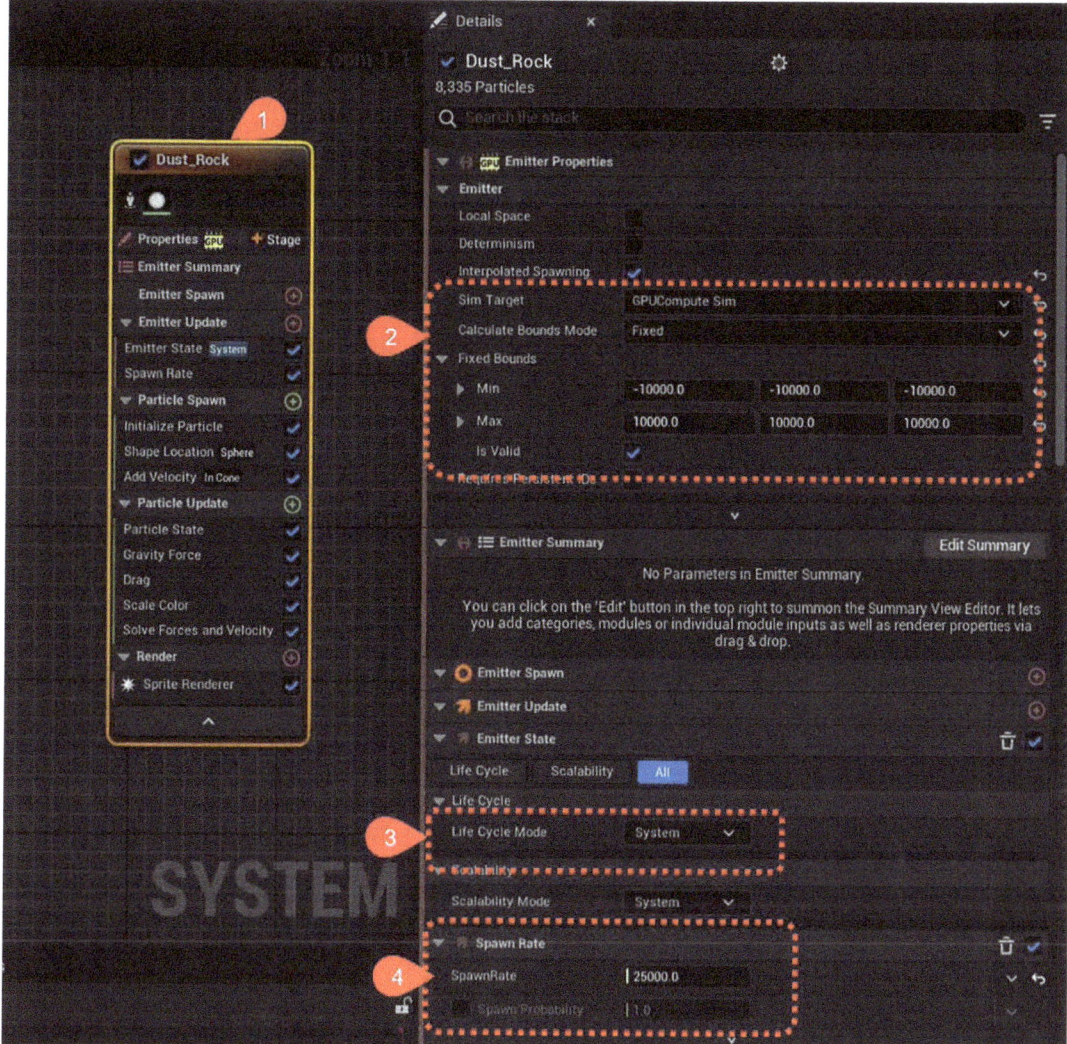

Figure 7-82. Initial settings for the Dust_Rock emitter

Initialize Particles

Here are some key points noted in Figure 7-83 to initialize the particles:

- In Step 1, select the Initialize Particle module.
- Lifetime: In Step 2, the particle's lifetime's minimum value is set to a random value between a minimum of 5.0 seconds and a maximum of 10.0 seconds, with the randomness contributing to a more natural and varied effect as each particle will last for a different amount of time.

367

- In Step 3, the maximum range is set slightly higher at 250.0 to maximize the chances that we can see the particles before they disappear from the system.

- Mass: In Step 4, the mass of the particles is also randomized between 0.75 and 1.25, which could affect how the particles move and interact with forces within the environment, like wind or gravity.

- Sprite Size: In Step 5, the size of the particle sprites is nonuniform and determined by multiplying a base size vector by a random float value, providing variation in the size of the particles to simulate a more realistic dust-rock effect. The minimum size is set at 50.0, and the maximum can go up to 70.0.

- Sprite Rotation Angle: In Step 6, this setting allows for a random rotation of the particle sprites between –360 and 360 degrees, giving the particles a randomized orientation upon spawning, which contributes to the chaotic nature of a dust storm.

The detailed settings here are essential for creating a dynamic and visually convincing dust and rock particle effect that behaves realistically in relation to the virtual environment it occupies.

CHAPTER 7   HARNESSING THE POWER OF NIAGARA: PRACTICAL EXAMPLES IN UNREAL ENGINE 5

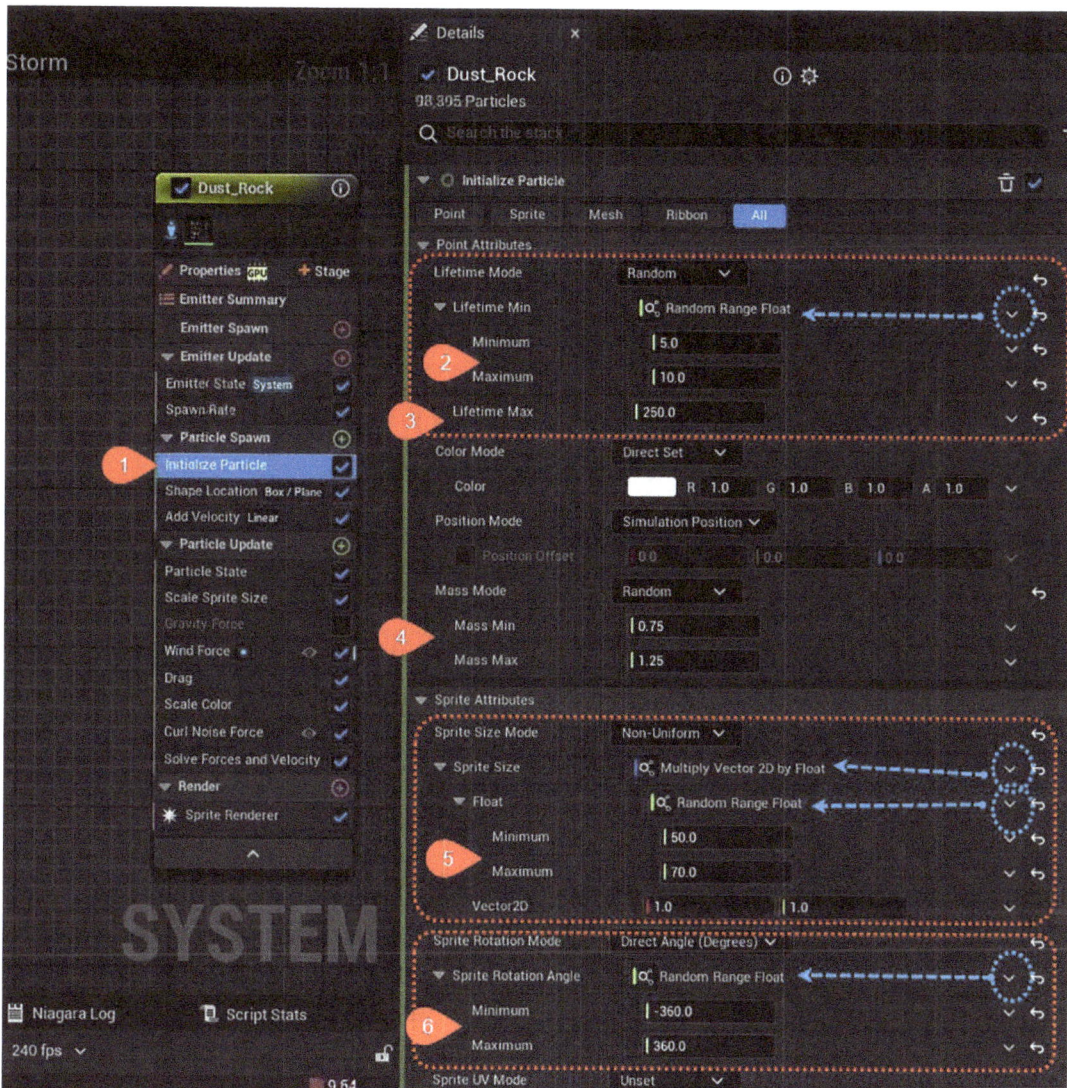

Figure 7-83. Initialize a particle

Shape Location

Here, we leave the "Shape Location" enabled to allow particles to be emitted with specific defined shapes. The Shape Location module stands as a cornerstone for defining the birthplace of particles. It's here that we can dictate the initial geometry from which their particles emerge, whether it be from the points of a star, the surface of a sphere, or the vertices of a custom mesh.

CHAPTER 7   HARNESSING THE POWER OF NIAGARA: PRACTICAL EXAMPLES IN UNREAL ENGINE 5

This module is instrumental in crafting the starting point of a particle's journey, setting the stage for how they will interact with the world around them. The choice of shape can drastically alter the behavior and visual outcome of the particle effect, making it a fundamental step in the creation of anything from a delicate puff of smoke to a chaotic explosion. It's the module that gives the first hint of structure to the otherwise formless potential of a particle system.

In Figure 7-84, we can see the "Shape Location" settings within the particle system editor. This specific panel configures where and how particles are emitted within the 3D space of the scene.

- In Step 1, we activate the Shape Location Box/Plane configuration.

- In Step 2, we set the Box Size and Midpoint. The parameters within the red dashed area show the specific settings for the box shape. The "Box Size" has been set to 20,000 in width, 20,000 in depth, and 2750 in height, which defines a very large area from which particles can emit, suggesting a vast effect like a dust storm that covers a large portion of the level.

- The "Box Midpoint" is set at 0.5 for X, Y, and Z, placing the center of the box at a normalized position within the given size parameters. This would typically ensure that the emission of particles is centered and evenly distributed within the box volume.

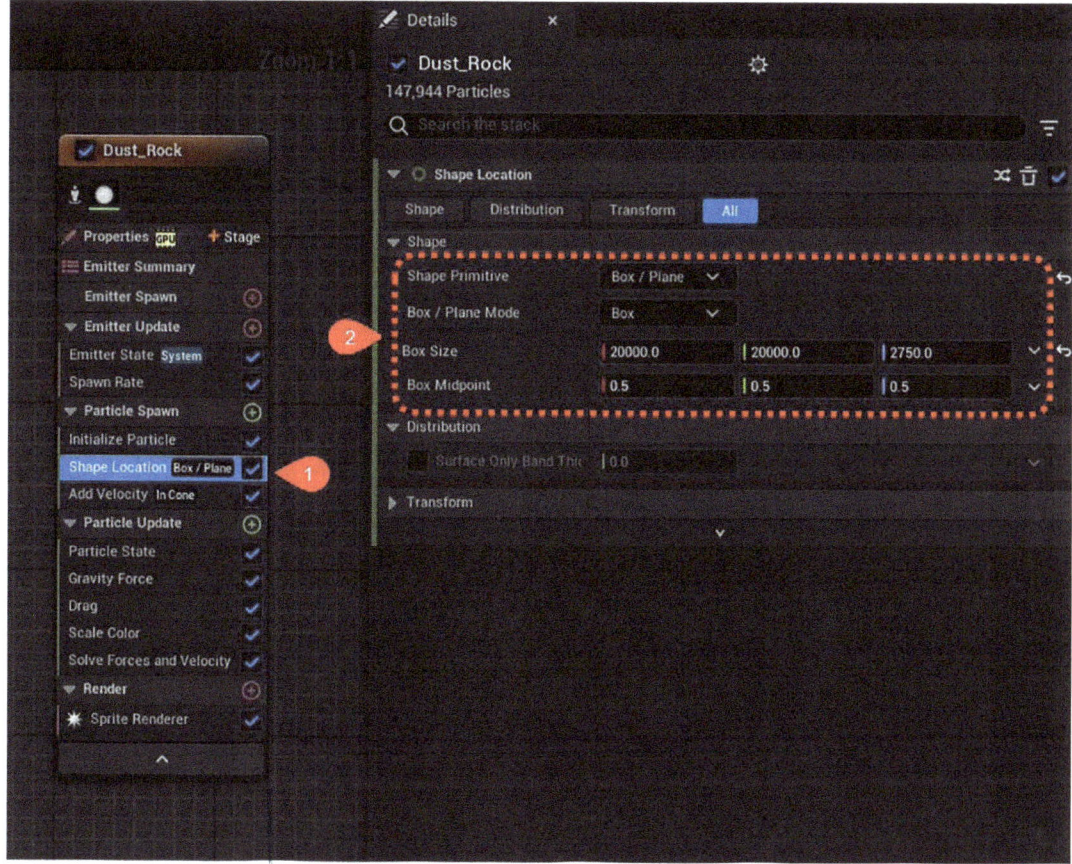

Figure 7-84. Shape Location

Add Velocity

The "Add Velocity" module is akin to the gust of wind that sets a weather vane spinning; it imparts motion to particles, offering them a vector of speed and direction. This module is essential for dictating the initial impetus of particles within a scene, allowing them to soar, drift, or cascade based on the velocity vectors provided. Whether simulating the gravitational pull on a waterfall or the chaotic dispersal of debris in an explosion, "Add Velocity" is the tool that breathes kinetic life into particles.

It's where creators can inject the energy of motion, transforming static scenes into dynamic canvases of movement and interaction. Figure 7-85 shows the "Add Velocity" module in a particle system editor, which is part of the settings for a "Dust_Rock" particle emitter. This module is responsible for defining the initial velocity of the particles when they are emitted.

CHAPTER 7 HARNESSING THE POWER OF NIAGARA: PRACTICAL EXAMPLES IN UNREAL ENGINE 5

- In Step 1, we select the Add Velocity module.

- In Step 2, we set Add Velocity as Linear; particles will move in a straight line from their point of creation.

  - Velocity Minimum and Maximum: These settings determine the range of velocities that particles will be emitted with. The "Minimum" vector is set to 200 units in the X direction, –5 in the Y direction, and –10 in the Z direction. The "Maximum" values are set to 300 units in the X direction, 5 in the Y direction, and 10 in the Z direction. This range allows for variation in the speed and direction of the particles, making the effect more dynamic and realistic.

- Velocity Speed Scale: Set to 50.0, which likely acts as a multiplier to the base velocity, scaling up the speed at which particles move.

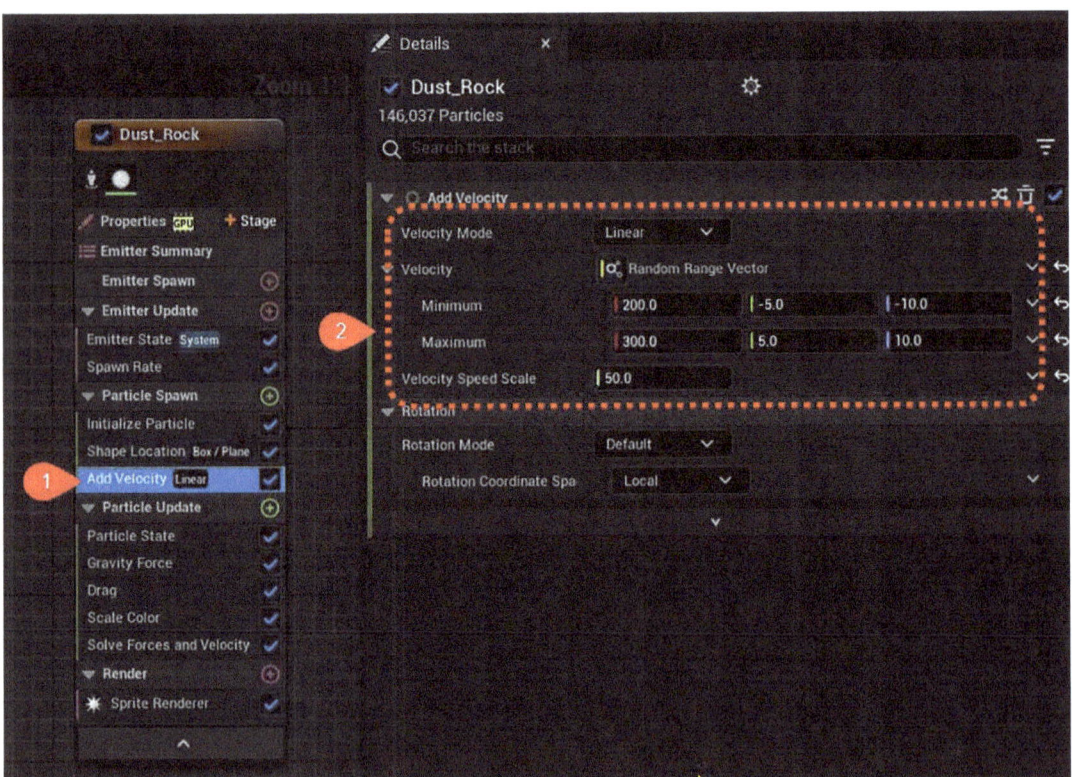

Figure 7-85. Add Velocity

Scale Sprite Size

The "Scale Sprite Size" module within the Niagara particle system editor is designed to control the size of individual sprites that comprise the particles. Figure 7-86 displays the settings for the "Scale Sprite Size" within the particle system editor, highlighting how these parameters can be adjusted to scale the particles over their lifespan. This setting is used to adjust the size of the particle sprites over time or based on certain conditions.

- Step 1 shows that we need to add this extra module.

- Highlighted Section: In Step 2, the "Scale Sprite Size" module is selected, as indicated by the side panel.

- Non-Uniform Scaling: In Step 3, the parameters within the red dashed area show that the scaling is nonuniform, allowing for different scaling factors in the X and Y directions. This can create more varied and dynamic shapes for the particles, as they won't necessarily remain square or circular as they scale.

- Float Curve: In Step 4, this is a graphical representation of how the size of the particles will change over time or based on a certain parameter, with the curve influencing the scale factor during the particle's life cycle.

  - The blue arrow points to a template selection for the curve, in this case, a "Ramp Up Down" template which suggests that the particle size will increase and then decrease in a symmetrical fashion, likely to simulate a puffing effect where particles expand and then contract as part of their behavior.

Overall, the "Scale Sprite Size" settings will define the dynamic changes in the particle size throughout the particles' existence, contributing to the realism and complexity of the visual effects, such as rocks in a dust storm that might appear larger as they get closer or as they are first emitted and then shrink as they disperse.

CHAPTER 7   HARNESSING THE POWER OF NIAGARA: PRACTICAL EXAMPLES IN UNREAL ENGINE 5

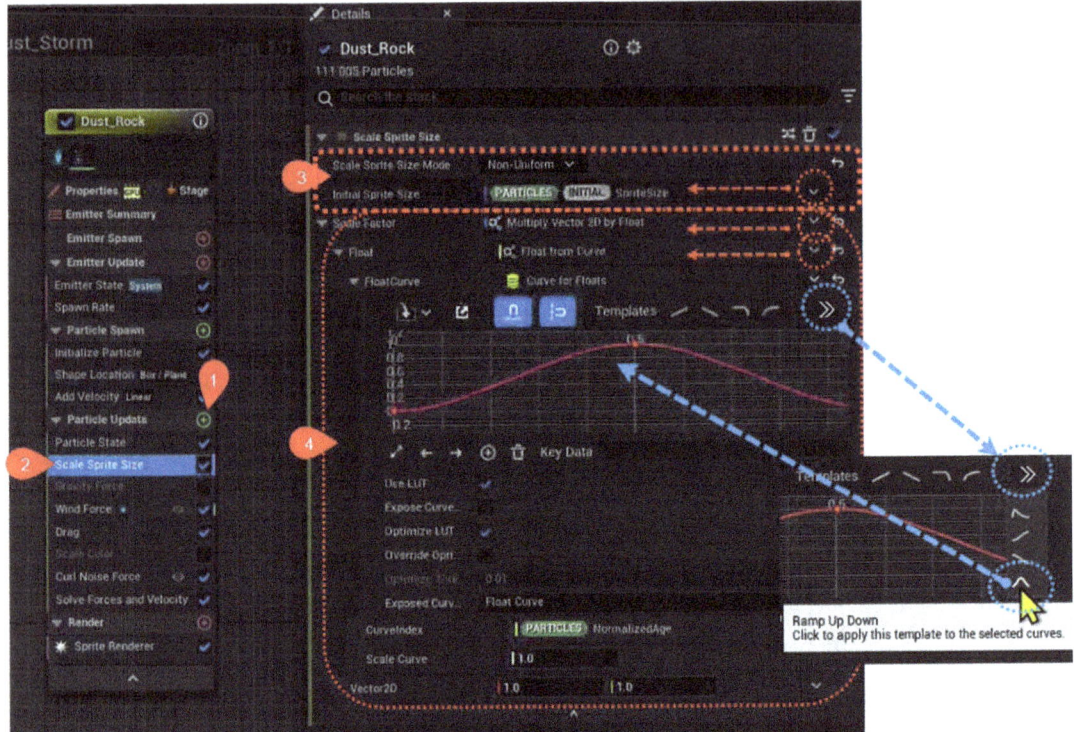

Figure 7-86. *Scale Sprite Size*

Disable Gravity Force

Figure 7-87 shows that "Gravity Force" being disabled would mean that particles will not be influenced by any gravity-like force that pulls them in a specific direction, which is typically downward within a 3D environment. Since this is about a small dust rock, the particles are meant to represent debris or dust in space or another zero-gravity environment; gravity should be disabled to ensure that the particles float freely without falling.

CHAPTER 7   HARNESSING THE POWER OF NIAGARA: PRACTICAL EXAMPLES IN UNREAL ENGINE 5

Figure 7-87. *Disable Gravity Force*

Drag

The "Drag" module is essential in animating forces that bring particle simulations to life. Figure 7-88 exhibits the settings of the "Drag" module within a particle system editor, which influence the behavior of particles post emission.

- Drag Module: As shown in Step 1, the module is selected in the particle system editor, suggesting that we can adjust the resistance applied to the particles as they move through space.

375

- Drag Parameters: As shown in Steps 2 and 3, the values within the red outlined area specify the amount of drag or air resistance applied to the particles. Both "Drag" and "Rotational Drag" are set to 1.0:

  - This parameter applies linear drag to the particles, which will slow them down over time, simulating the effect of resistance due to the particles moving through a medium like air or water. A value of 1.0 implies a certain level of resistance, but without further context of the system's units or scale, it's not clear how strong this resistance is.

  - Rotational Drag: Similarly, this applies to the rotation of the particles, slowing down their spin over time. This can be used to simulate how rotating objects gradually come to a stop due to the resistance they experience.

By adjusting the drag, the user can simulate realistic movements of particles in an environment, as objects in real life typically slow down due to air resistance or friction. If the particles represent dust or debris, adding drag would make their movement more natural as they would not continue at a constant speed indefinitely but would instead decelerate over time. The exact values would need to be fine-tuned based on the desired visual effect and the environmental conditions being simulated.

CHAPTER 7   HARNESSING THE POWER OF NIAGARA: PRACTICAL EXAMPLES IN UNREAL ENGINE 5

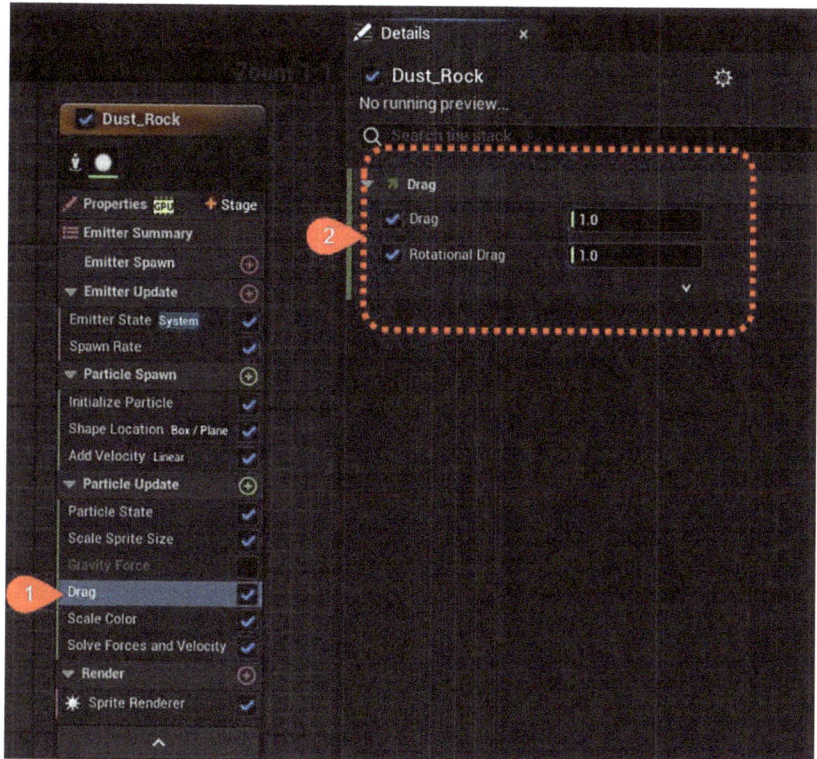

Figure 7-88. Drag

Wind Force

The "Wind Force" module within the Niagara particle editor introduces a breath of natural dynamics to the virtual environment, simulating the unseen but ever-present force of wind. This module allows creators to infuse their particle effects with the fluid, directional movement characteristic of wind, from gentle zephyrs that sway grass and leaves to fierce gales that whip through scenes with tangible force.

By adjusting the strength and direction of the wind within this module, particles can be made to drift, swirl, or surge in ways that mimic real-world weather patterns, adding layers of realism and immersion to digital landscapes. It's a tool that not only animates particles but also breathes life into the world they inhabit, making the invisible winds of virtual realms visible through their influence on the environment.

Figure 7-89 presents the "Wind Force" module settings within a particle system editor. Wind force is critical in creating realistic movements by applying directional force to simulate the effect of wind in the environment.

- As shown in Step 1, we need to add this extra module.

- Step 2 shows that this module is selected.

- Wind Speed: Step 3 shows that within the module, the "Wind Speed" parameter has a random range vector set for the maximum value with 1.0 units in the Z direction, suggesting that there is some variability to how the wind will affect the particles vertically. The minimum is set to 0.0 in all directions, which means there might be moments when the wind has no effect on the particles.

- Turbulence: In Step 4, the "Turbulence" settings below allow for the simulation of chaotic, irregular changes in the wind's direction and speed, which can make the effect more realistic, as natural wind is rarely uniform.

  - The "Turbulence Mode" is set to "Curl Noise," which is a common technique for simulating fluid-like motion.

  - The "Scale Min/Max" and "Speed Min/Max" provide a range for the scale and speed of the turbulence, offering control over how pronounced the turbulence effect will be on the particles.

The "Wind Force" module is vital for emulating how particles would behave in real-world conditions, where wind and turbulence can greatly affect the trajectory and speed of objects like dust and debris. By fine-tuning these settings, the user can create a more immersive and believable particle simulation for their visual effects project.

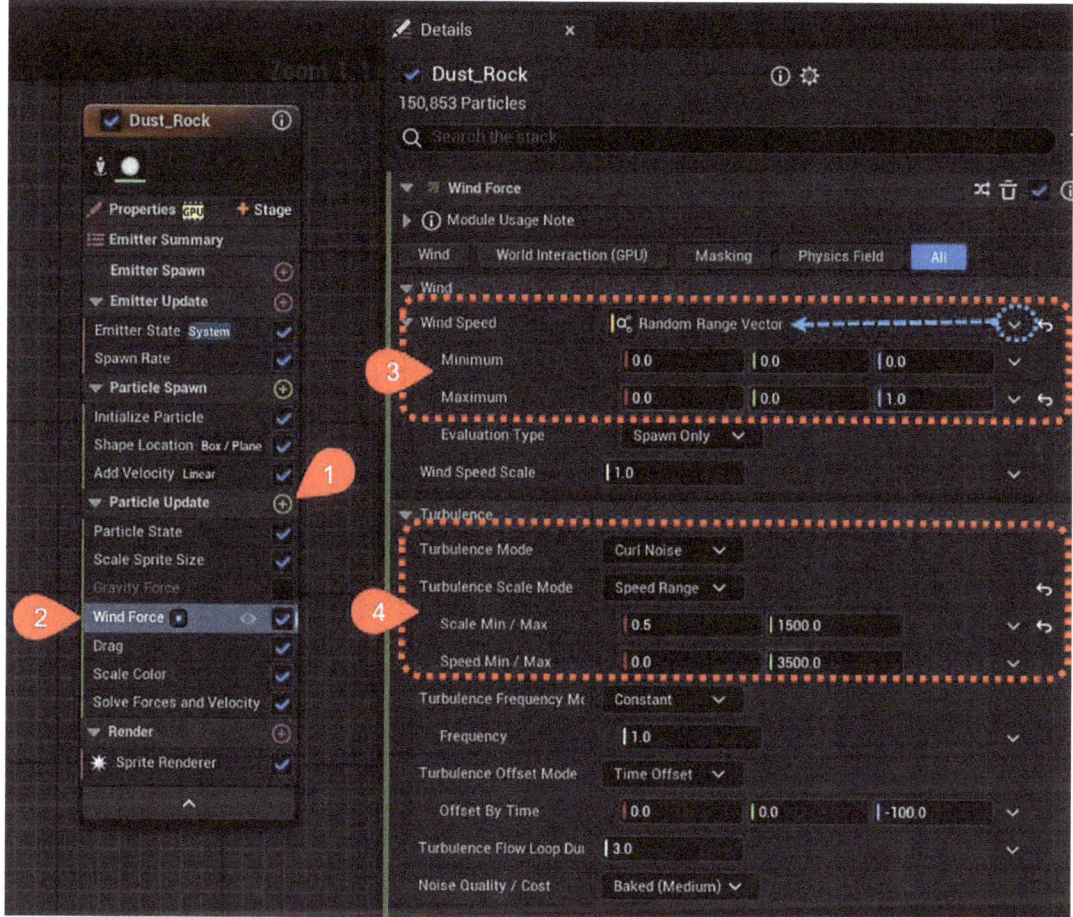

Figure 7-89. Wind Force

Scale Color

The Scale Color module plays a pivotal role in defining the visual journey of each particle, from its creation to its fade into oblivion, bringing a spectrum of dynamism to digital creations. Figure 7-90 showcases the "Scale Color" module, which is utilized to alter the color properties of particles over time or in response to specific conditions.

- Scale Color Module: In Step 1, we select the module to adjust how the color of the particles will change during their life cycle.

CHAPTER 7   HARNESSING THE POWER OF NIAGARA: PRACTICAL EXAMPLES IN UNREAL ENGINE 5

- Color Scaling Curves: Step 2 features curves that represent how the particle color attributes scale over time. There are two curves visible:
  - The upper curve in red likely represents the scaling of one color channel (perhaps red, green, or blue), showing how its intensity changes.
  - The curve suggests that the color intensity starts at a lower value, ramps up to full intensity, then decreases again, which could simulate particles heating up and then cooling down.
  - The lower curve in blue represents the alpha transparency of the particles. This curve follows a similar pattern to the red one, indicating that the particles may become more opaque at the midpoint of their life cycle and then fade out toward the end.

    The blue arrows point to the template options for the curves, and the "Ramp Up Down" template is highlighted, suggesting that the user can apply this preset shape to the curves, which is useful for creating symmetrical fade-in and fade-out effects for the particle colors and transparency.

Overall, the "Scale Color" module is crucial for adding visual depth and realism to the particle effects by allowing the particles' colors to change dynamically throughout their existence. This is particularly useful for simulating natural phenomena like fire, smoke, or dust, where the appearance of the particles can significantly change from the moment they are emitted to when they dissipate.

CHAPTER 7   HARNESSING THE POWER OF NIAGARA: PRACTICAL EXAMPLES IN UNREAL ENGINE 5

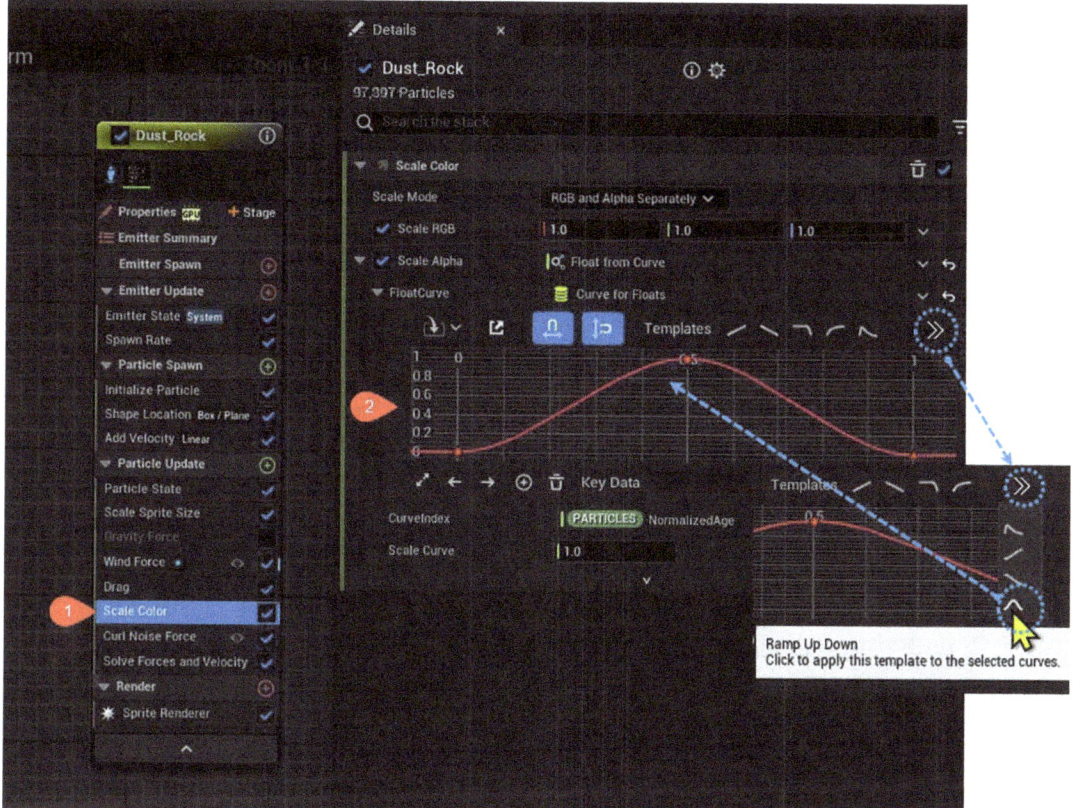

Figure 7-90. Scale Color

Curl Noise Force

The "Curl Noise Force" module within the Niagara editor stands as a testament to the complex beauty of chaos in motion, blending the art of simulation with the unpredictability of nature. This module is adept at imbuing particles with a lifelike, turbulent force that echoes the capriciousness of elements such as wind gusts and swirling water currents. By applying this intricate force, particles acquire a movement that is as random and organic as the forces governing our natural world.

It's this module that enables creators to transcend the boundaries of digital environments, crafting effects that breathe with the essence of chaos, yet move with the purpose and fluidity of nature itself. Through the "Curl Noise Force" settings, particles are not merely animated; they are endowed with the unpredictable grace of natural motion, enhancing the realism and dynamism of digital creations.

Figure 7-91 features the "Curl Noise Force" module settings within a particle system editor, which is commonly used to apply a complex, turbulent force to the particles, giving them a more natural and chaotic movement that simulates the random nature of forces like wind or water currents.

- Curl Noise Force: In Step 1, we need to add this extra module.

- In Step 2, we need to select this module to configure forces that will affect the particles to create a more dynamic behavior.

- Noise Strength: Step 3 shows the parameters to define the strength of the Curl Noise Force. The "Minimum" value is set to 1000.0, and the "Maximum" value is set to 5000.0, based on a random range float.

  - This means that each particle will experience a different level of force within this range, contributing to the variety in their motion.

- Additional Settings: As shown in Step 4, the "Pan Noise Field" settings with values at 0.1 might control the overall movement of the noise field, influencing how the noise pattern changes or moves over time.

The "Curl Noise Force" module is pivotal in adding lifelike motion to particles, making them appear as if they are being influenced by natural, swirling forces. By adjusting these settings, artists and visual effects designers can simulate a wide range of environmental effects, from the gentle drifting of airborne dust to the violent tumult of a stormy sea.

CHAPTER 7 HARNESSING THE POWER OF NIAGARA: PRACTICAL EXAMPLES IN UNREAL ENGINE 5

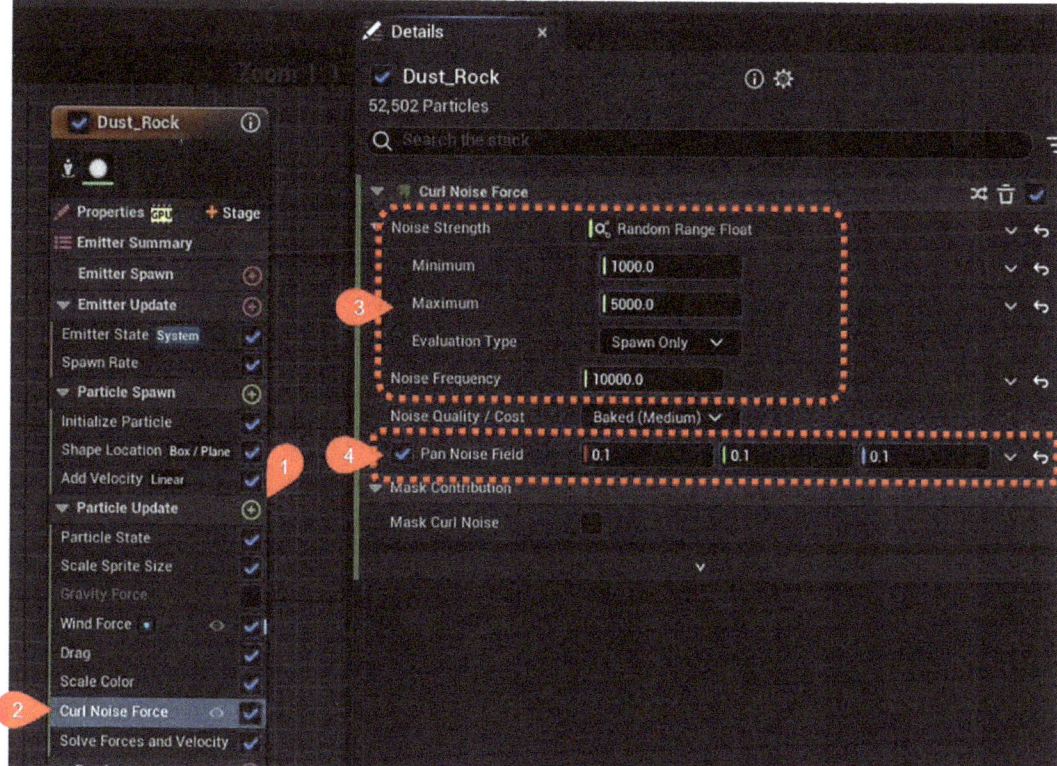

Figure 7-91. Curl Noise Force

Sprite Renderer

Within the particle system editor, Figure 7-92 showcases the settings of the "Sprite Renderer" module, a pivotal tool for defining how particles are visually represented as sprites in the scene.

- As shown in Step 1, we select this Sprite Renderer module to allow us to adjust how the individual particles are rendered as sprites in the system. Sprites are 2D images that represent the particles in 3D space.

- Material: As shown in Step 2, we applied the "Material" field with the "Dust_Storm_Rock" material to the sprites. This material will determine how the particles look, including their texture, color, and any visual effects like transparency or glow.

383

- Alignment and Facing Mode: As shown in Step 3, which we had seen from previous examples, we can set "Alignment" and "Facing Mode."
    - The alignment is set to "Unaligned," which means the sprites do not automatically align themselves based on any particular axis or direction.
    - The "Facing Mode" is set to "Face Camera," ensuring that the sprites always face toward the camera, which is a common technique to maintain the illusion that a 2D sprite is part of the 3D scene.
- SubUV: As shown in Step 4, the "SubUV" section is highlighted at the bottom of the red dashed area. "SubUV" refers to the technique of using a single texture (atlas) that contains multiple frames of animation or variations for the sprite.
    - The "Sub Image Size" settings of 8.0 by 8.0 indicate how many frames or variations are in the texture atlas. This is often used to animate the particles or provide a variety of appearances without needing multiple textures.

The outcome of these settings can be seen in the preview image as shown in Figure 7-93.

CHAPTER 7   HARNESSING THE POWER OF NIAGARA: PRACTICAL EXAMPLES IN UNREAL ENGINE 5

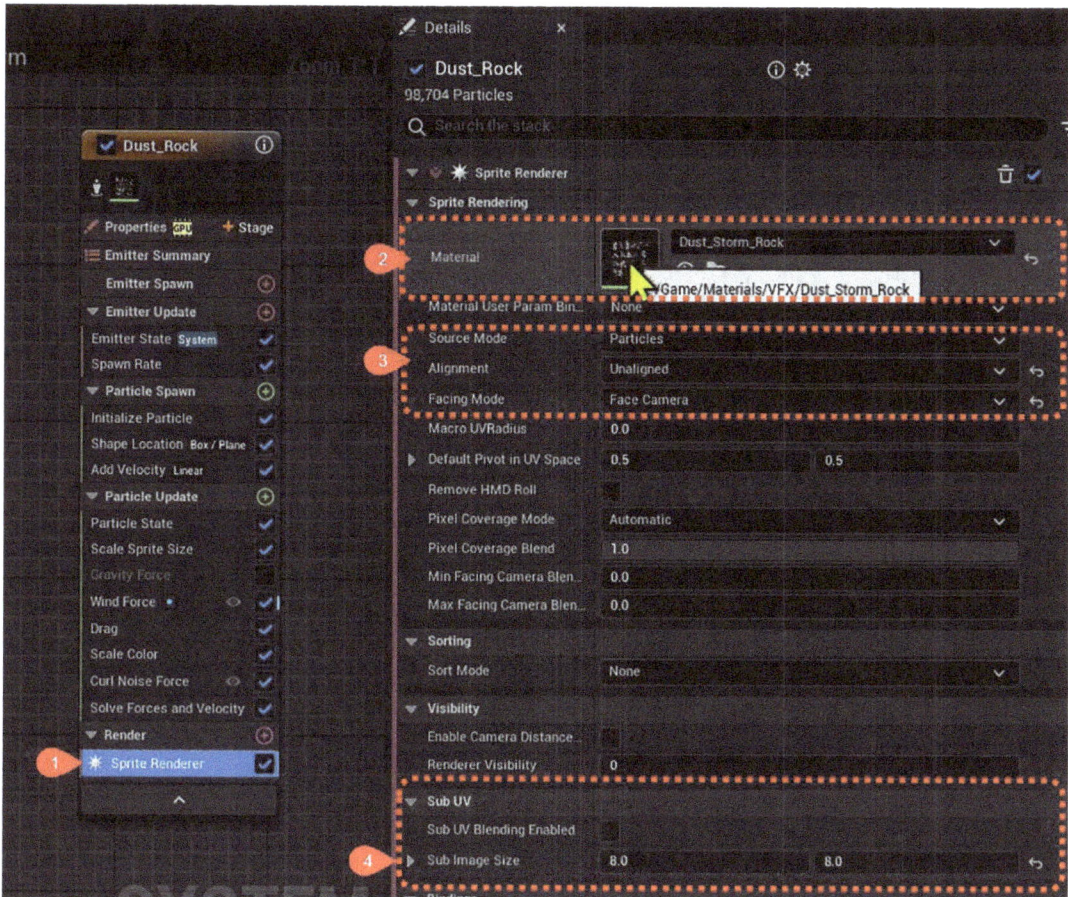

Figure 7-92. Sprite Renderer

CHAPTER 7 HARNESSING THE POWER OF NIAGARA: PRACTICAL EXAMPLES IN UNREAL ENGINE 5

Figure 7-93. *Outcome of the dust-rock emitter*

Add Another Emitter

As mentioned earlier at the beginning of this example, our Dust_Storm Niagara system is composed of more than one emitter, Dust_Rock and Smoke_Storm. We can follow Figure 7-94 to add additional emitter module with the context menu for a particle system and select the option to "Add Emitter." This is a step where we can add a new emitter to the particle system to produce additional effects.

Emitter Templates: Figure 7-95 shows a list of emitter templates that we can choose from. These are preconfigured emitters with specific behaviors, such as "Fountain," which simulates a looping fountain spray. We will select Fountain as the template to create our new particle effect.

Final Effect Preview: Figure 7-96 shows a preview of the particle effect within the environment. The "Dust_Storm" effect is visible, and the newly added "Fountain" emitter is highlighted, suggesting that it has been configured and is now part of the overall particle effect.

These steps illustrate how we can enrich a particle system by adding different types of emitters to create complex and layered effects. By combining different emitters with varied properties, we can achieve a more dynamic and visually appealing scene.

CHAPTER 7   HARNESSING THE POWER OF NIAGARA: PRACTICAL EXAMPLES IN UNREAL ENGINE 5

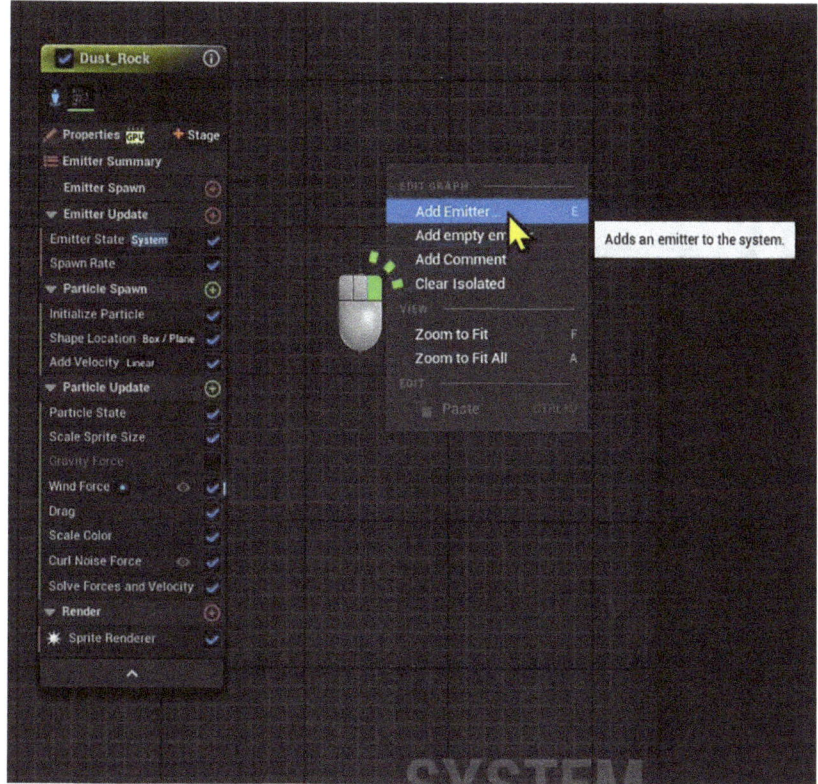

Figure 7-94. Right-click to add another emitter

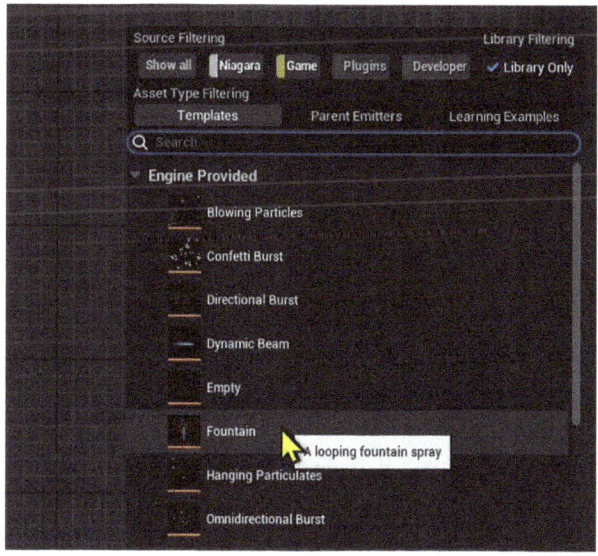

Figure 7-95. Select the Fountain as template

CHAPTER 7   HARNESSING THE POWER OF NIAGARA: PRACTICAL EXAMPLES IN UNREAL ENGINE 5

Figure 7-96. *Fountain emitter shown alongside the Dust_Rock*

Initial Settings for Smoke Storm

Similar to the previous Smoke from the Landing Spaceship example, Figure 7-97 showcases a section of the particle system editor for this emitter.

- In Step 1, we rename the emitter to Smoke_Storm.
- Sim Target: In Step 2, we set this to "GPUCompute Sim," indicating that the simulation is intended to be computed on the GPU for better performance.
  - Calculate Bounds Mode: Fixed, which means the bounds of the emitter are manually set to contain the particles.
  - Fixed Bounds: These values define the spatial limits for the particle simulation. The "Min" and "Max" values suggest a considerable size, allowing for a large area of effect.

CHAPTER 7   HARNESSING THE POWER OF NIAGARA: PRACTICAL EXAMPLES IN UNREAL ENGINE 5

- The author suggests to use these values:

  Min values of X = –10,000, Y = –10,000, Z = –1375

  Max values of X = 10,000, Y = 10,000, Z = 1375

- Life Cycle Mode: In Step 3, we set this to "System," which implies that the life cycle of the emitter is managed on a system level. This means that the emitter will operate according to the system's settings, ensuring consistent behavior across all particles within this emitter.

- Spawn Rate: In Step 4, the "Spawn Rate" is set to 50.0, determining how many particles are generated per unit of time. This rate influences the density and appearance of the smoke effect.

This configuration demonstrates the setup for a smoke effect that is part of a larger particle system named "Dust_Storm."

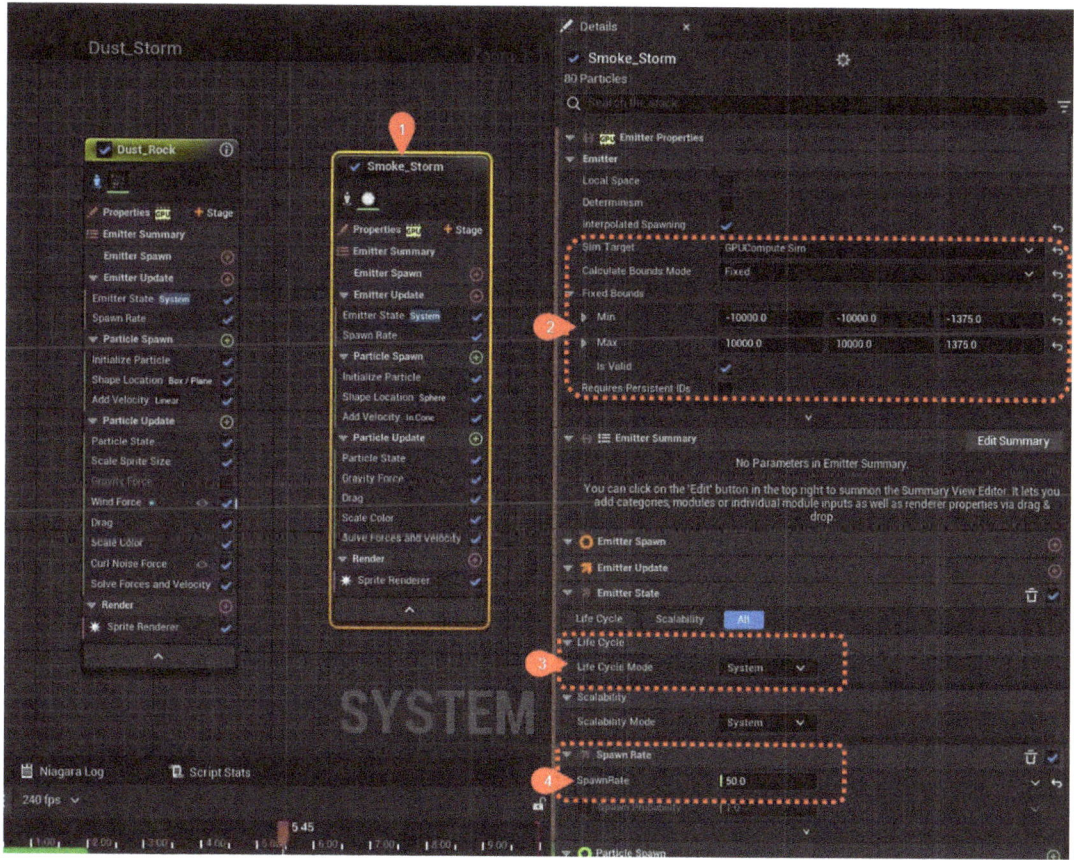

Figure 7-97. *Initial settings for Smoke_Storm*

Initialize Particle for Smoke Storm

Figure 7-98 illustrated the "Initialize Particle" settings within the "Smoke_Storm" emitter module of a particle system editor, where the fundamental attributes like lifespan, color, size, and mass are defined.

- In Step 1, we select the Initialize Particle module to see the configuration.

- Lifetime: Steps 2 and 3 indicate the random values between a minimum of 10.0 and a maximum of 20.0, with a maximum range of 200. This randomness in lifespan ensures that not all particles will disappear at the same time, providing a more natural dissipation of the smoke.

- Color Mode: In Step 4, below the lifetime settings is the color mode, which is set to "Random Range." This setting allows each particle to have a color value within a specified range, leading to a variation in color among the smoke particles. The color minimum and maximum values are defined, which will likely give the smoke a range of shades for added realism.

- Sprite Size: In Step 5, the sprite size settings within the red dashed area are set to "Non-Uniform," allowing different scaling in the X and Y directions. The size of the sprites is determined by a random range between 4000.0 and 6000.0, which could represent a large variation in particle size, contributing to the volumetric appearance of the smoke.

These settings in the "Initialize Particle" module are critical for establishing the initial look and behavior of the smoke particles, which will evolve based on these initial properties throughout the particles' life cycles. The randomization of attributes like lifespan, color, and size is essential for creating a convincing smoke effect that mimics the chaotic and diverse nature of real-world smoke behavior.

CHAPTER 7   HARNESSING THE POWER OF NIAGARA: PRACTICAL EXAMPLES IN UNREAL ENGINE 5

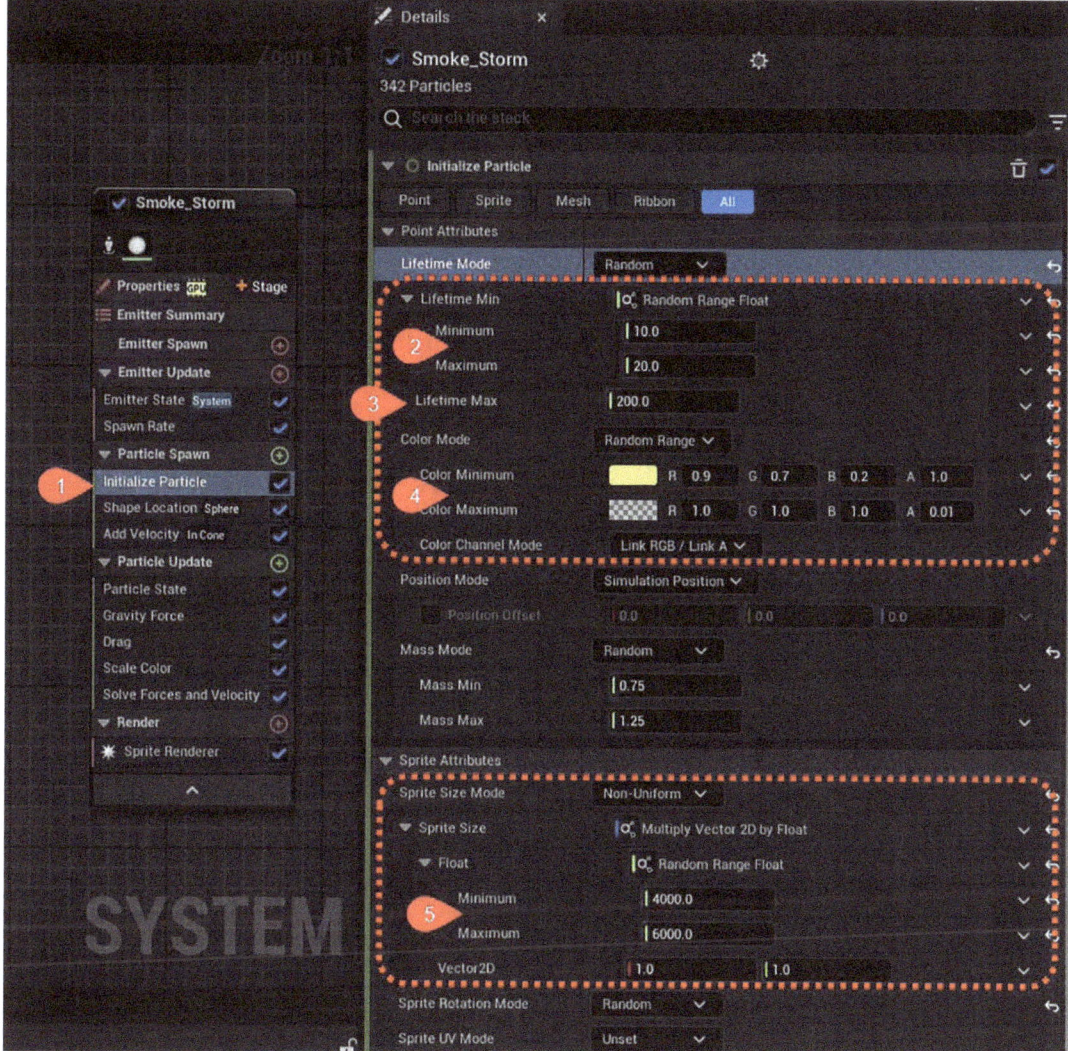

Figure 7-98. Initialize Particle for Smoke_Storm

Shape Location for Smoke_Storm

The Shape Location module is pivotal in defining the initial position from which particles emerge, and Figure 7-99 displays the "Shape Location" settings for the "Smoke_Storm" emitter.

- In Step 1, we select the "Shape Location" module where we can define the initial area or volume from which particles are emitted. This setting is crucial because it determines the starting point and distribution of the particles in 3D space.

- Box/Plane Emitter Shape: In Step 2, we set particle emission as a "Box/Plane Mode."

  - The dimensions of the box are set to 20,000.0 by 20,000.0 by 2750.0, which define a very large volume for the smoke particles to be emitted from, suggesting a significant area that the smoke will cover in the scene.

  - The "Box Midpoint" values are all set to 0.5, which likely centers the box within the local coordinate space of the emitter.

These settings within the "Shape Location" module are critical for creating the initial boundary and distribution area for the smoke effect. In a visual simulation, such as creating a smoke effect over a large area, these dimensions would contribute to the scale and the spread of the effect, impacting how it interacts with other elements in the scene and how it appears to the viewer.

CHAPTER 7   HARNESSING THE POWER OF NIAGARA: PRACTICAL EXAMPLES IN UNREAL ENGINE 5

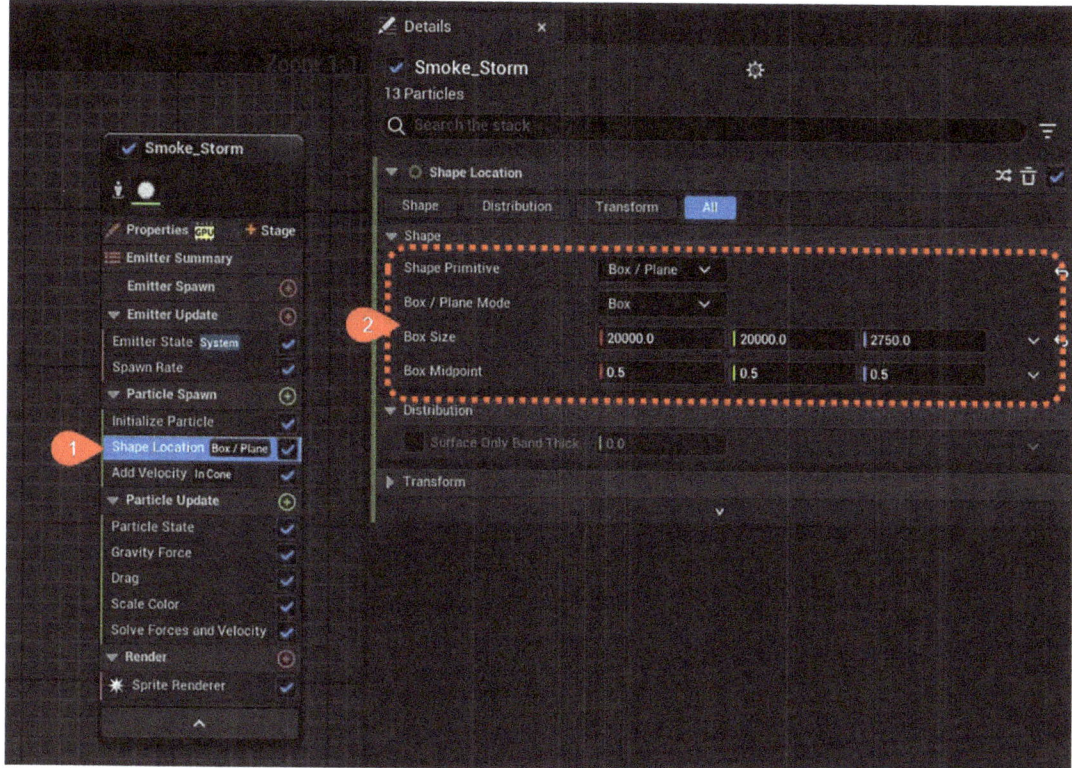

Figure 7-99. *Shape Location for Smoke_Storm*

Add Velocity for the Smoke Storm

The "Add Velocity" module in Niagara functions like a gust of wind that propels a weather vane; it bestows motion upon particles by providing them with a vector of speed and direction. Figure 7-100 illustrates the "Add Velocity" module settings for the "Smoke_Storm" emitter, highlighting how adjustments to these settings can significantly influence the behavior and visual appearance of the particle effect.

- In Step 1, we select this Add Velocity module to adjust the setting of the velocity.

- Velocity Mode: In Step 2, the mode is set to "Linear," meaning that the velocity added to the particles will cause them to move in a straight line from their point of emission.

- Velocity Values: As shown in Step 3, the velocity is set using a "Random Range Vector," which will assign each particle a velocity within the specified minimum and maximum values. The "Minimum" fields are set to 50.0, 5.0, and 1.0, presumably for the X, Y, and Z axes, respectively. The "Maximum" fields are set to 100.0, –5.0, and 10.0 for the X, Y, and Z axes, respectively. This range provides variability in the speed and direction of the particles, making the smoke movement appear more natural and less uniform.

- Velocity Speed Scale: As shown in Step 4, there is a "Velocity Speed Scale" parameter set to 100.0, which likely acts as a multiplier to the velocity values, enhancing the effect of the motion imparted on the particles.

The "Add Velocity" settings are crucial for simulating the dynamic motion of smoke as it would naturally rise and disperse in the environment. By tweaking these parameters, users can create a variety of smoke behaviors, from gentle drifts to rapid billows, depending on the needs of the scene.

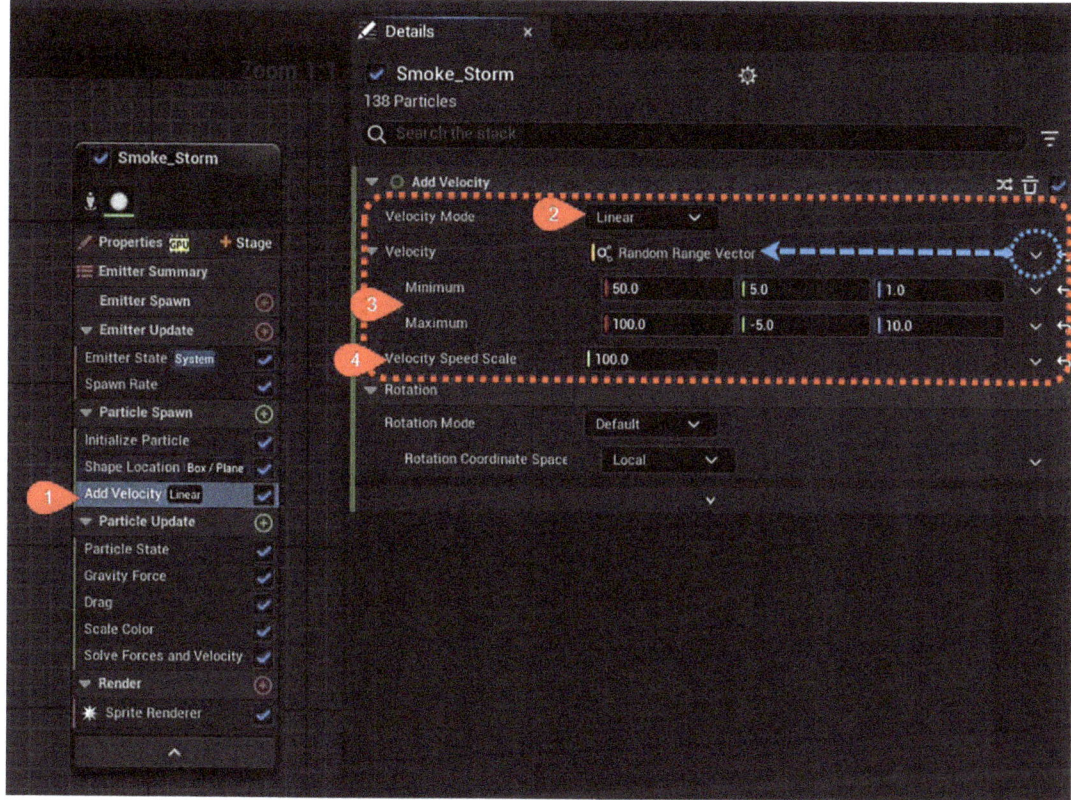

Figure 7-100. Add Velocity for the Smoke_Storm

Disable Gravity and Drag Modules for Smoke_Storm

In the realm of particle simulation, the interplay between forces can define the essence of motion and realism. Among these, the "Gravity" and "Drag" modules are fundamental forces, grounding particles in a simulation that mimics the physical world. However, there are moments in creative design where defying these natural laws can open the door to extraordinary visual narratives.

Disabling the "Gravity" and "Drag" modules allows particles to break free from the chains of downward pull and air resistance, respectively, enabling them to float, swirl, or dart with unbound freedom. This act of suspension or acceleration, unencumbered by gravity or drag, invites a canvas where the only limit is imagination, allowing for the creation of ethereal scenes, weightless environments, or particles that move with supernatural grace. Through this deliberate manipulation of natural forces, designers can sculpt motion in ways that elevate the visual experience to realms beyond the ordinary.

Figure 7-101 indicates that we disable both the "Gravity Force" and "Drag" modules for the following reasons:

> Gravity Force: Disabling this module would prevent the particles from being affected by a force that simulates gravity. In the context of smoke, we typically want the smoke to rise or float through the air, not fall to the ground as it would under the influence of gravity. By disabling gravity, the smoke particles can move in a more realistic manner consistent with smoke behavior in the real world.

> Drag: This module simulates air resistance. Disabling drag would allow the particles to move unhindered by any resistance, which can be useful if the desired effect is for the smoke to spread quickly and smoothly. In some scenarios, especially where the smoke needs to represent a fast-moving or forcefully expelled substance (like steam or pressurized gas), not having drag can make the motion appear more forceful and less impeded.

In summary, disabling these two modules suggests a desire for the smoke particles to have an unrestricted, possibly upward or omnidirectional, movement pattern, without being pulled down by gravity or slowed by air resistance. This would be typical for simulating smoke in a zero-gravity environment or depicting a more ethereal and less dense form of smoke.

Figure 7-101. Disable Gravity Force and Drag for Smoke_Storm

Sprite Rotation Rate for Smoke_Storm

In the dynamic world of visual effects, the ability to control every aspect of a particle's behavior is key to crafting compelling imagery. The "Sprite Rotation Rate" module within Niagara offers just that control by dictating the angular velocity of sprites. This feature allows creators to imbue each particle with a distinct rotation, adding a layer of realism or fantastical motion that can dramatically enhance the visual impact of an effect.

Whether it's simulating the lazy drift of autumn leaves, the rapid spin of fireworks, or the steady rotation of floating debris in water, adjusting the sprite rotation rate can bring a sense of depth and dynamism to the scene. It's a tool that transforms static images into moving parts of a larger, animated tapestry, contributing significantly to the overall feel and flow of the particle system.

CHAPTER 7　HARNESSING THE POWER OF NIAGARA: PRACTICAL EXAMPLES IN UNREAL ENGINE 5

Figure 7-102 showcases the "Sprite Rotation Rate" module settings within the "Smoke_Storm" emitter's particle system editor. This module controls the rate at which individual particle sprites rotate around their pivot point, adding dynamic visual effects to the particles.

- As shown in Step 1, we need to add the extra module.

- Module Selection: As shown in Step 2, we need to select the "Sprite Rotation Rate" module to configure the rotation rate of each sprite.

- Rotation Rate: As shown in Step 3, the rotation rate is configured to have a randomized value, as indicated by the "Random Range Float" option.

  - The minimum and maximum values are set to 15.0 and 25.0, respectively. This range means that when the particles spawn, they will each be assigned a rotation speed between these values, contributing to the variation in the appearance and motion of the smoke particles.

The use of a random range for the sprite rotation rate can give the smoke a more natural and less uniform look, as each particle will rotate at a slightly different speed, mimicking the chaotic nature of smoke where different parts may twist and turn at different rates due to varying air currents and other environmental factors.

CHAPTER 7   HARNESSING THE POWER OF NIAGARA: PRACTICAL EXAMPLES IN UNREAL ENGINE 5

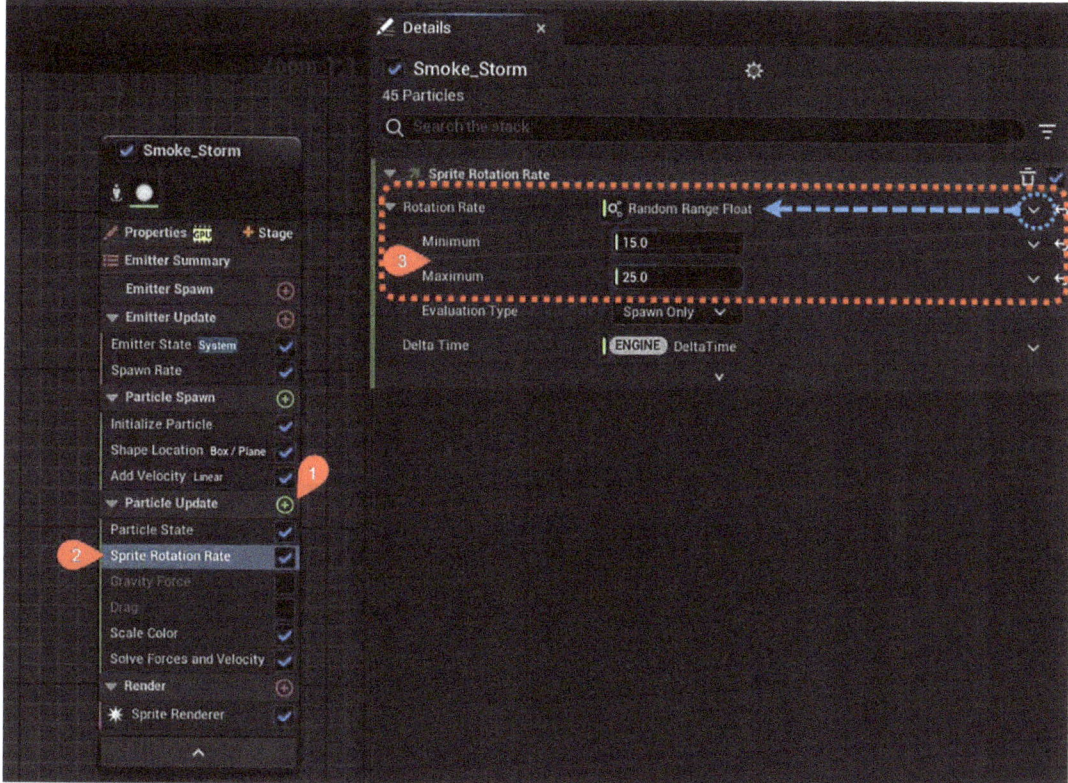

Figure 7-102. *Sprite Rotation Rate*

Scale Sprite Size for Smoke_Storm

The "Scale Sprite Size" module is essential for dynamically adjusting particle sizes, crucial for adding realism and depth to effects. It allows particles to grow or shrink, enhancing visual storytelling within a scene. Figure 7-103 shows the "Scale Sprite Size" module settings within the "Smoke_Storm" emitter configuration of a particle system editor, so that we can configure the size of the particle sprites over time or in response to certain conditions. This is a key factor in creating visually dynamic and varied particle effects.

- As shown in Step 1, we need to add the extra module.

- Scale Sprite Size Module: As shown in Step 2, we need to select the module to make the adjustment.

- Scale Factor: As shown in Step 3, the "Scale Sprite Size Mode" is set to "Non-Uniform," which allows the particle sprites to scale differently in each dimension, providing a more organic and less rigid appearance.

  - The "Scale Factor" is determined by a "Multiply Vector 2D by Float" operation, which suggests that the size of each particle sprite is being multiplied by a scalar value to vary its size.

  - Float Curve: The "Float Curve" panel within the red dashed area displays a graph that represents how the sprite size scales over the normalized age of the particles. The curve shows a progression where the size increases and then decreases, which could simulate a puffing effect where smoke particles expand as they rise and then contract.

  - The "Curve for Floats" button and the "Templates" drop-down provide tools for shaping this curve, allowing users to apply predefined or custom curves to achieve the desired effect. In this case, we can start with the Linear Ramp Up template and modify the second key data point as 1.0 and 2.0.

The ability to manipulate sprite size over the lifetime of the particles using curves is vital for simulating realistic behaviors in particle-based effects. For smoke, this can represent the natural expansion and dissipation of smoke particles as they interact with the environment, contributing to the realism of the smoke simulation.

CHAPTER 7   HARNESSING THE POWER OF NIAGARA: PRACTICAL EXAMPLES IN UNREAL ENGINE 5

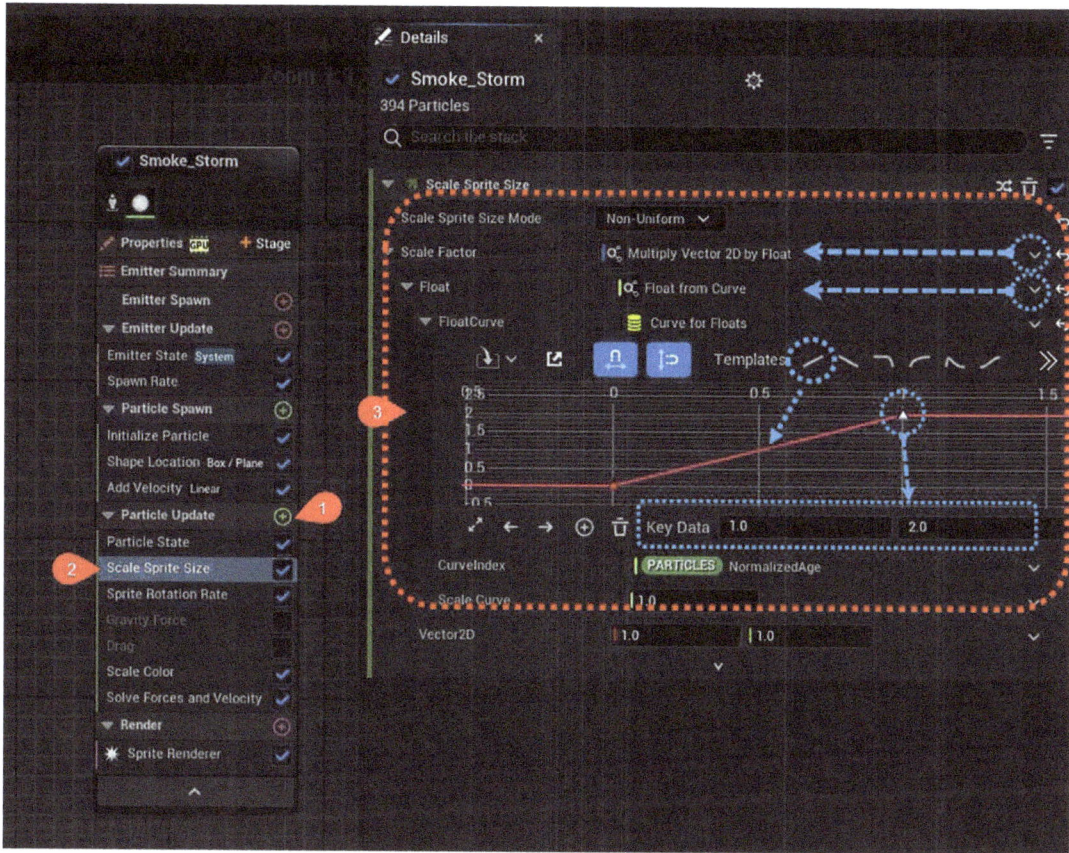

Figure 7-103. Scale Sprite Size for Smoke_Storm

Scale Color for Smoke_Storm

The "Scale Color" module in Niagara empowers creators to dynamically adjust particle colors over time, adding vibrant life and variation to effects. This tool is key for crafting visually engaging and evolving scenes. Figure 7-104 displays the "Scale Color" module within the "Smoke_Storm" emitter settings in a particle system editor to allow us to adjust the color properties of the particles over time.

- As shown in Step 1, we select the module to display the adjustment properties.

CHAPTER 7   HARNESSING THE POWER OF NIAGARA: PRACTICAL EXAMPLES IN UNREAL ENGINE 5

- Alpha Transparency Curve: As shown in Step 2, the highlighted curve in the "Float Curve" section represents the scaling of the particle's alpha transparency over its lifespan.

  - The alpha value determines how transparent the particle is; a lower alpha value makes the particle more transparent, while a higher alpha value makes it more opaque.

  - The curve starts at a lower value, increases to a peak, and then decreases again toward the end. This suggests that the smoke particles will become more visible as they "age" and then fade away before they disappear, simulating the natural dissipation of smoke.

The "Scale Mode" is set to "RGB and Alpha Separately," which allows independent control of the color and transparency scaling. This is crucial for creating a realistic smoke effect where particles need to change their visibility over time, providing a lifelike quality to the effect as the smoke appears to thicken and then thin out.

The "Scale RGB" settings are all set to 1.0, indicating that the RGB channels are currently not being scaled, and the particles will maintain their color throughout their life cycle unless adjusted further.

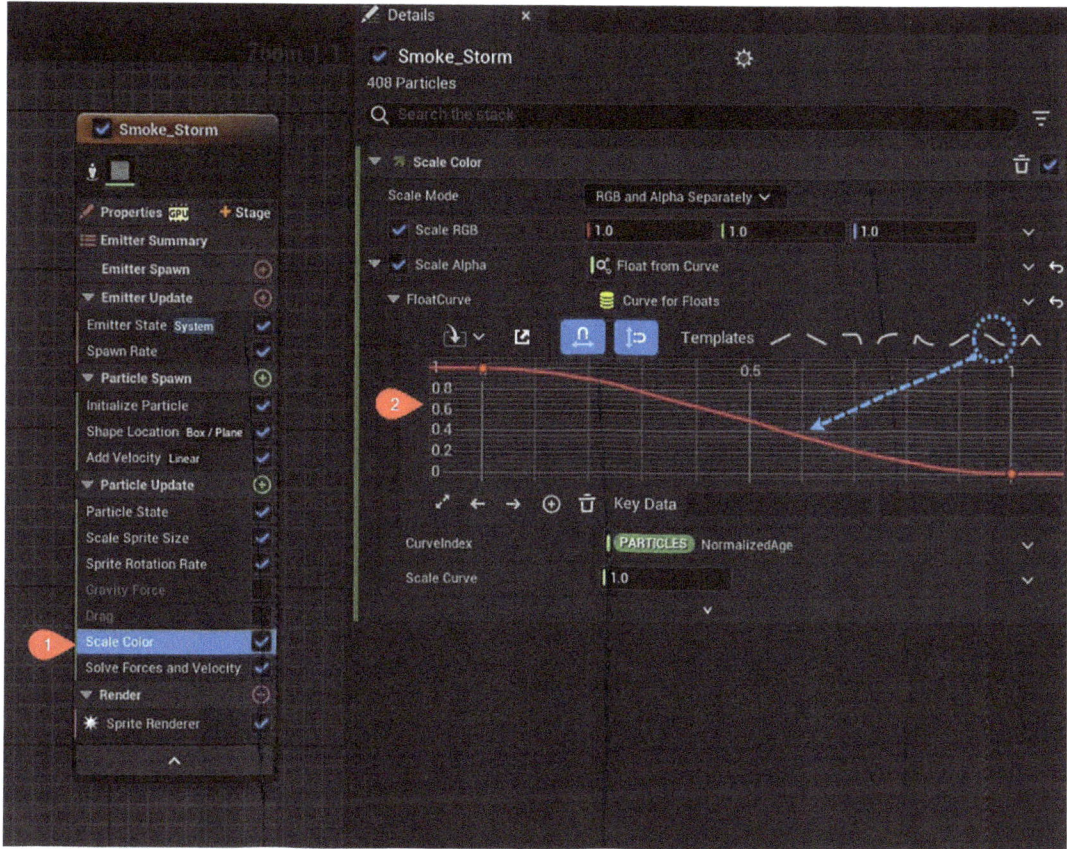

Figure 7-104. Scale Color for Smoke_Storm

Sprite Renderer for Smoke_Storm

The "Sprite Renderer" in Niagara is a core module for visualizing particles as 2D sprites within a 3D environment, enabling detailed control over their appearance and integration into scenes.

Figure 7-105 displays the same "Sprite Renderer" module settings as the previous emitter, "Smoke_From_Landing_Spaceship." The reader can refer back to the previous section if they wish to review the purpose of the Sprite Renderer settings. The final result, composed of "Dust_Rock" and "Smoke_Storm," can be viewed in Figure 7-106 as the "Dust_Storm" Niagara system.

CHAPTER 7 HARNESSING THE POWER OF NIAGARA: PRACTICAL EXAMPLES IN UNREAL ENGINE 5

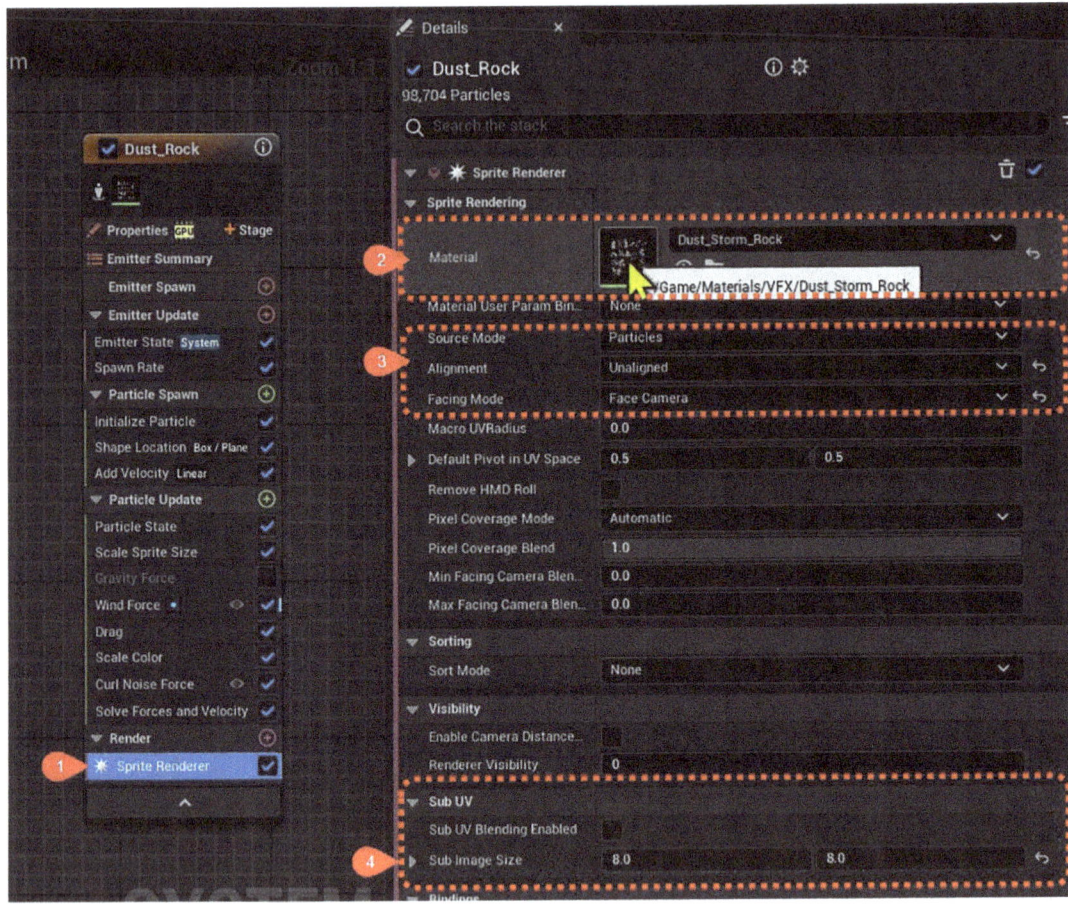

Figure 7-105. *Sprite Renderer for Smoke_Storm*

CHAPTER 7　HARNESSING THE POWER OF NIAGARA: PRACTICAL EXAMPLES IN UNREAL ENGINE 5

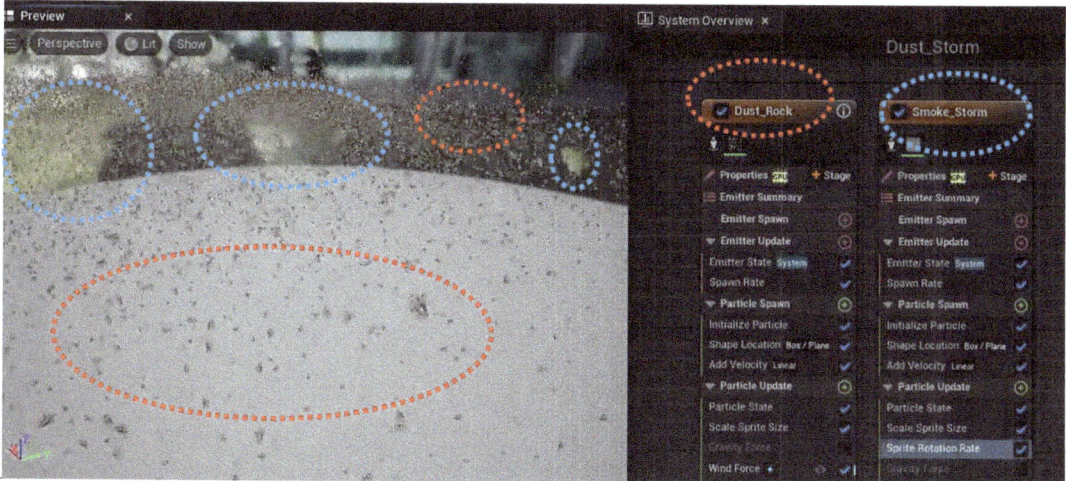

Figure 7-106. *End visual result for the Dust_Storm*

Placement of the Dust_Storm Niagara System

As shown in Figure 7-107, we are to place the Dust_Storm Niagara system in the level, as follows:

- Step 1: Use the Top perspective (if placement with mouse control is needed).

- We can place the Dust_Storm object in the scene using Select and Place Actor methods as shown in Figures 7-67 and 7-68, and we can follow Step 2 to place it at the following exact location:
  - Location X = −16,500, Y = −700, Z = −70,150

- Step 3 shows that we are also to scale up the particle system by setting the scale value of 10. The scale of this particle system will allow us to see the end result of each particle bigger in size.

We can now see the end visual result of both Dust_Rock and Smoke_Storm as shown in Figure 7-108.

405

CHAPTER 7 HARNESSING THE POWER OF NIAGARA: PRACTICAL EXAMPLES IN UNREAL ENGINE 5

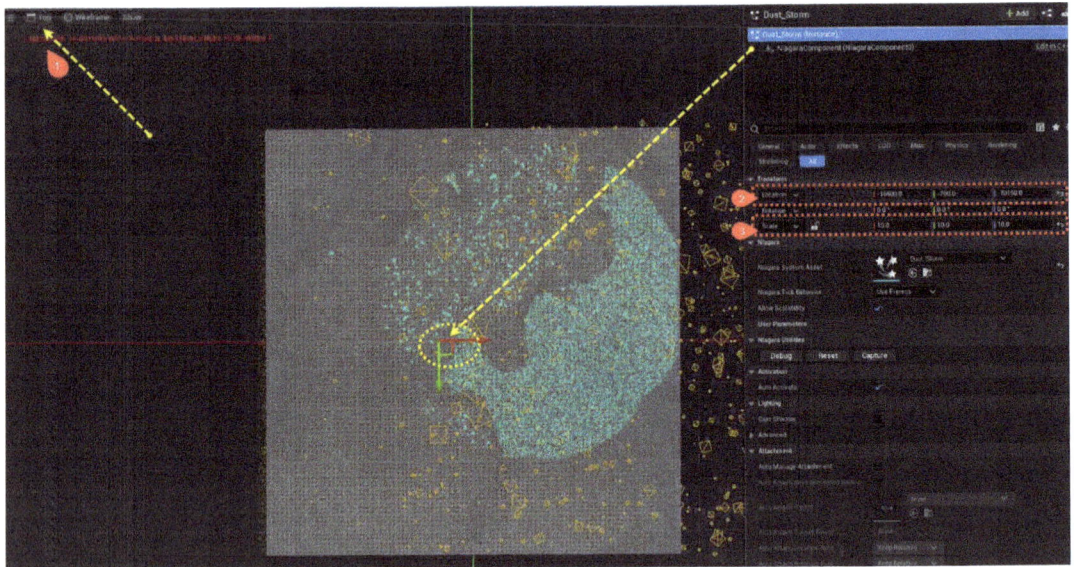

Figure 7-107. *Placement and scale for the Dust_Storm*

Figure 7-108. *Visual of the Dust_Storm*

Summary

This chapter explored the advanced capabilities of UE5's Niagara system, a state-of-the-art visual effects and particle simulation tool. It commenced with an introduction to Niagara's basic concepts and components, setting the foundation for more intricate explorations. Through practical examples, the readers were guided to understand the system's versatility, enabling them to craft complex particle systems that add realism to virtual environments.

The chapter was structured to serve as a foundational guide, focusing on essential features of Niagara, such as emitter deployment for particle system creation. It offered a deep dive into creating effects like smoke, spirals, and dust storms, providing insights to enhance the toolkit of game developers, visual effects artists, and enthusiasts.

Bringing Worlds to Life: Concluding the Design Phase

This chapter not only delves deep into the advanced capabilities of UE5's Niagara system but also marks the conclusion of this volume, which has been dedicated to the design phase of our project application. Throughout, we have explored the state-of-the-art visual effects and particle simulation tool, starting with the design of foundational concepts and moving toward the creation of complex particle systems that breathe life into virtual environments.

As we draw this volume to a close, we've journeyed through the expansive capabilities of UE5, delving deep into its design and visual effects toolkit. Each chapter has been a step in our exploration, starting with the foundational elements of UE in Chapter 1, where we were introduced to the environment and basic operations, setting the groundwork for our venture into game development.

Chapter 2 unfolded the complex yet fascinating process of crafting large, open landscapes, utilizing heightmaps and the engine's powerful tools to shape vast, immersive worlds. This chapter underscored the significance of creating realistic environments as the backdrop for our gameplay experiences. This allowed us to create more dynamic and expansive virtual spaces without sacrificing performance.

Chapter 3 explored UE's auto-blend materials, using height and slope data to merge textures for efficient and realistic terrain creation. It covered material setup, node usage for blending, and their application to landscapes, highlighting streamlined design and visual appeal.

CHAPTER 7   HARNESSING THE POWER OF NIAGARA: PRACTICAL EXAMPLES IN UNREAL ENGINE 5

Chapter 4 focused on enhancing UE5 scenes using Quixel Bridge and Megascans, outlining asset importation, texture replacement, and tiling improvements. It introduced the Procedural Content Generation plugin, detailing the use of Nanite and PCG nodes for realistic landscape elements like rocks and vegetation. This chapter aimed to streamline asset integration and elevate scene realism, demonstrating UE5's capabilities in environment creation.

Chapter 5 dove into the cutting-edge realm of Runtime Virtual Textures and material blending, demonstrating how to achieve seamless integration between 3D objects and terrains. This chapter revealed methods to elevate texture detail and visual fidelity, essential for crafting believable and engaging virtual worlds.

Chapter 6 explored Lumen in UE5, emphasizing its role in dynamic global illumination and reflections for creating realistic scenes. It guided through enabling Lumen, adjusting its settings for optimal performance, and using Post-Processing Volumes for precise lighting control. The chapter demonstrated Lumen's effects through examples, showcasing its ability to enhance visual realism and immersion in digital environments.

Chapter 7 delved deep into the practical application of UE5's Niagara system for creating advanced visual effects and particle simulations. It began with foundational knowledge of Niagara, progressing to intricate examples demonstrating its versatility. From crafting simple emissions to complex, interactive effects responsive to environmental factors, the chapter equipped the readers with the skills to fully utilize Niagara in the projects.

Throughout Chapters 2–7, we emphasized the system's capability to add realism and dynamism to virtual environments, enhancing the toolkit of developers and visual effects artists alike. With this volume, we've reached the conclusion of the design phase of our project, a comprehensive exploration of UE5's design capabilities.

But our journey doesn't end here. As we transition into the next volume of the series, we turn our focus to programming in C++, aiming to breathe life into our creations with interactive elements. The forthcoming chapters in the next volume will guide us through representing the player with interactive avatars from MetaHuman, implementing AI in adversaries like the Macrovirus, and empowering the player to combat these threats with magical antiviral balls.

This pivot from design to development marks a significant milestone in our project. The next volume promises a deep dive into the mechanics that make interactive experiences tick, from the nuances of C++ in UE5 to the implementation of game logic and player interactions. As we close this chapter of our exploration, we stand on the brink of turning our beautifully designed worlds into living, breathing ecosystems where story, action, and interaction converge to create immersive gameplay experiences.

Index

A

AutoBlend_Height_MAT material, visual improvement
    General_Template_MF, 115
    Material Function, 113, 114
    realistic surface textures, 113

Auto-blend height-based materials
    design
        bottom half layer, 86
        color gradient, 79
        first material result, 88, 90
        height position, individual pixel, 86
        lower elevational zones, 85
        material editor, 80–83
        middle elevational zones, 84
        upper half layer, 87, 88
        visualize result, 90, 92, 94
    final material node, 67
    import nodes, 68
        Constant node, 71
        Constant3Vector node, RGB color, 71, 72
        material parameter, 72–74
        MatLayerBlend_Standard, 75, 76
        named reroute declaration node, 69, 70
        Set Material Attributes node, 76–78
    open up material, 67

Auto-blend materials
    attached material, to landscape, 64, 65, 67
    height, 61
    setup, 62, 63
    slope, 61
        adding slope attribute, 95–98
        painting, 95
        visualize result, 98–100
    slope data, 407

B

Burst module, 295, 298

C

Central processing unit (CPU), 290

D, E

Digital signal processing (DSP), 5

Dust storms
    add another emitter
        add velocity module, 393, 395
        disable gravity/drag modules, 395, 397
        initialize particle, 390, 391
        layered effects, 386
        scale color, 401–403
        scale sprite size module, 399, 401
        shape location module, 391–393
        smoke storm, settings, 388, 389
        sprite renderer, 403–405
        sprite rotation rate, 397–399
        templates, 386, 387

INDEX

Dust storms (*cont.*)
    add velocity, 371, 372
    combo emitter, 362
    curl noise force module, 381, 382
    disable gravity force, 374, 375
    drag, 375, 377
    initialize particles, 367, 369
    placement, 405, 406
    rock and smoke storm, create, 361
    scale color, 379, 380
    scale sprite size, 373, 374
    setup, 364, 365, 367
    setup materials, emitter, 362–364
    shape location, 369–371
    sprite renderer, 383–386
    wind force module, 377–379

F

Fuel cell, 227, 228

G

Game development approach, 3
General_Template_MF
    AutoBlend_Height_MAT landscape, 125, 127–137
    improve texture tiling sampler, 120–123
    MF_MapAdjustments, 115, 116
    setup texture samples, properties, 116–120
    UV_Variation, 124, 125
GNU Image Manipulation Program (GIMP), 13, 27, 28
Graphics processing unit (GPU), 220, 290, 291, 388
"Ground_MF" function, 129, 130

H

Hardware Ray Tracing, 224
Heightmaps
    convert EXR to PNG, GIMP, 27–30
    create landscape, 31–38
    definition, 23
    2D grayscale image, 23
    ensure correct scale, 38–41
    external sources, 23
    preparing, 26
    setting up new level, 24, 25
    world partition
        conversion, 46, 48, 50
        definition, 42
        editor, 51, 53–55
        runtime settings, 56–59
        save level before conversion, 42–45
Hierarchical Level of Detail (HLOD), 42

I, J, K

Innovative lighting technology, 256
Integrated development environment (IDE), 10

L

Landed spaceship, smoke effects
    acceleration force module, 320–322
    add velocity module, 316–318
    color module, 322–324
    drag value, 319
    end result, 303
    gravity force, 318, 319
    import/setup spaceship, 303, 304, 306, 307

INDEX

initialize particles, 313, 314
scale sprite size, 325–327
setup material, 308–310
shape location module, 314, 315
SmokeFromLandingShip, 310–312, 330, 331
sprite renderer, 327–329

Lumen
features, 223
illumination and reflection system, 223
lighting effects
3D models, 227, 229
preparation tasks, 229, 230
light source, 224
post-processing volumes, 275
project settings, 223
ray tracing, 223
working, 224–226

M

Macrovirus, 227, 229, 242, 259–265, 408
Magical ball, spiral effects
add velocity, 342, 344
disable shape location, 341
drag, 348, 349
gravity force, 346, 347
import/setup, 332, 333
initialize particle, 334–338
rotate around point, 339, 340
scale color, 349–351
scale sprite size, 344–346
spiral in action, 353, 355–357, 359, 361
sprite renderer, 352, 353
Megascans, 101–102, 104, 105, 109, 115, 155, 178, 408

N

Nanite, 3, 101, 108–112, 137, 178, 223, 408
Niagara
dust storms, 361
features, 278
interactive effects, 408
Magical ball, spiral effects, 331
particle effect, creating, 278–282
particle emissions, 277
particle systems, 282

O

Open world landscape, *see* Heightmaps

P

Particle systems
burst effect, 295–298
burst frequency, adjusting, 298–300
customizing visual effects system, 300–302
definition, 282
emitter update, 293–295
local space option, enable, 285–289
Niagara system editor, 283
numbered labels, 282
rename emitter, 284, 285
simulation target, 289–293
Procedural content generated cave, *see* Procedural generation method
Procedural Content Generation (PCG)
large rock, procedural generate, 155–160
PCG graph, 154
procedurally generating gravel stones, 161–170

413

INDEX

Procedural Content Generation
(PCG) (*cont.*)
   procedurally generating
      vegetations, 170–175
   final result, 176, 178
   graph editor, 178
   import nodes
      bounds modifier, 151
      density filter, 150
      difference node, 151
      force regen button, 154
      input node, 145
      normal to density, 149
      point filter, 147
      projection node, 148, 149
      self-pruning, 150
      Static Mesh Spawner, 152, 153
      surface sampler, 146
      transform point, 147, 148
   setup volume, 138–145
   terrain, 137
Procedural generation method
   post-processing effects
      location and scale, 265, 267, 268
      Lumen effects, 272–275
      settings, 268–271
      tailored approach, 265
   preparation tasks
      actor tag, 237
      add tag, 232, 233
      import macrovirus, emissive
        effect, 259–265
      influence, 256–259
      location/scale, 236
      megascan, 230
      mud rock model, 238
      PCG volume, 238, 239
      planes, 233–235, 240–242

      separate planes, 231, 232
   set animate emissive materials
      3D models, 242
      importing model, 243–249,
        251–254
      place asset in level, 254–256

Q

Quixel Bridge, 408
   asset IDs, 104, 105
   3D assets/plants, 108–112
   definition, 101
   Megascans assets, 101
   permission, 104
   plugins, 101
   sign in, 103
   surface texture assets, 106, 107

R

Runtime Virtual Texture (RVT)
   color map, 182
   definition, 181
   GPU processing, 181
   material blends, 220
      asset's original materials, 212
      create alpha, 214–217
      3D assets, 206, 207
      declaration material nodes, 211
      duplicated material, 208–210
      final material/result, 217–220
      landscape's color RVT, 212, 213
   material editor, output
      node, 198
      result, 205, 206
      setup, 198–203, 205
   setup, assent in UE5, 185–187

INDEX

streaming methods, 181
volume
    inside landscape
        partitions, 195–197
    setup, 188-192, 194, 195
world height, 183, 184

S

"Screen Space" technique, 258
Signed Distance Fields (SDFs), 224
Simulation target, 289
Software Ray Tracing, 224
Spawn Count, 297
"Spawn Rate" module, 295

T

Temporal Accumulation (TA), 225
Temporal Super Resolution (TSR), 225

U

Unreal Engine 5 (UE5)
    benefits
        chaos unleased, 4, 5
        elevating control, 4
        expanding horizons, 2
        PCG, 3
        realistic visuals, 3
        sound designers, 5
        unleashing creativity, 4
        unleashing visual fidelity, 4
    definition, 2
    GIMP, 13
    installation/dependencies, 5–9
    prepare new project
        create empty C++, 15
        low-quality mode, run, 19, 20
        open setting, 18
        plugin browser, 16
        show engine content, 20, 21
        Visual Studio IDE, 14
        Visual Studio integration tool, 17
    Visual Studio Community 2022,
        installation, 10–13
UV_Variation Material Function, 124

V

Visual effects (VFX) system, 4, 308, 334, 363

W, X, Y

World Partition, 23, 42, 46–53, 56–60, 195, 223

Z

Z-Culling, 57, 59

GPSR Compliance

The European Union's (EU) General Product Safety Regulation (GPSR) is a set of rules that requires consumer products to be safe and our obligations to ensure this.

If you have any concerns about our products, you can contact us on

ProductSafety@springernature.com

In case Publisher is established outside the EU, the EU authorized representative is:

Springer Nature Customer Service Center GmbH
Europaplatz 3
69115 Heidelberg, Germany

www.ingramcontent.com/pod-product-compliance
Lightning Source LLC
LaVergne TN
LVHW080309260326
834688LV00038B/1032